AF759607

Springer Geography

The Springer Geography series seeks to publish a broad portfolio of scientific books, aiming at researchers, students, and everyone interested in geographical research. The series includes peer-reviewed monographs, edited volumes, textbooks, and conference proceedings. It covers the major topics in geography and geographical sciences including, but not limited to; Economic Geography, Landscape and Urban Planning, Urban Geography, Physical Geography and Environmental Geography.

Springer Geography – now indexed in Scopus

More information about this series at http://www.springer.com/series/10180

Timothy Tambassi
Editor

The Philosophy of GIS

Springer

Editor
Timothy Tambassi
University of Eastern Piedmont
Vercelli, Italy

ISSN 2194-315X ISSN 2194-3168 (electronic)
Springer Geography
ISBN 978-3-030-16828-5 ISBN 978-3-030-16829-2 (eBook)
https://doi.org/10.1007/978-3-030-16829-2

© Springer Nature Switzerland AG 2019
This work is subject to copyright. All rights are reserved by the Publisher, whether the whole or part of the material is concerned, specifically the rights of translation, reprinting, reuse of illustrations, recitation, broadcasting, reproduction on microfilms or in any other physical way, and transmission or information storage and retrieval, electronic adaptation, computer software, or by similar or dissimilar methodology now known or hereafter developed.
The use of general descriptive names, registered names, trademarks, service marks, etc. in this publication does not imply, even in the absence of a specific statement, that such names are exempt from the relevant protective laws and regulations and therefore free for general use.
The publisher, the authors and the editors are safe to assume that the advice and information in this book are believed to be true and accurate at the date of publication. Neither the publisher nor the authors or the editors give a warranty, expressed or implied, with respect to the material contained herein or for any errors or omissions that may have been made. The publisher remains neutral with regard to jurisdictional claims in published maps and institutional affiliations.

This Springer imprint is published by the registered company Springer Nature Switzerland AG
The registered company address is: Gewerbestrasse 11, 6330 Cham, Switzerland

Preface

The first time I approached, from a philosophical perspective, the topic of GIS, I could not help but note how the literature in the field was, at the same time, heterogeneous and scattered. Considering that I was dealing with a branch of knowledge on the border of geography, philosophy, and computer science, heterogeneity was not an element of surprise. Conversely, the scatterdness forced me to ask myself, at least, four different questions: (1) is there really something such as the "philosophy of GIS" or are there only sporadic reflections on the topic? (2) In the first case, what are its distinguishing features? (3) Should the philosophy of GIS be considered as a subpart of the philosophy of geography? (4) What is the difference, if any, between philosophy of GIS and philosophy of geo-ontologies?

As time goes by, I realized that the answer to the last question might seem trivial: the difference consists in the fact that GIS and geo-ontologies are two different things. However, this does not mean to deny that they can also share, at least in part, a common philosophical ground. But if the (analytical) ontology probably represents the main (maybe the exclusive) philosophical background of geo-ontologies, the same cannot be said for GIS that has also to do with other philosophical areas of investigation, as some of the articles of this book will show. Thus, the first thing I learned was to prevent the philosophy of GIS from collapsing into its ontological dimension.

The third question refers to a thorny issue that inflamed the debate over the nature of GIS: is GIS a (geographical) tool or a proper science? Choosing the first option probably means to consider the philosophy of GIS as a specific subpart of the philosophy of geography; conversely, the second option underlines a certain autonomy of the former to the latter. However, the two different opinions on the issue do not preclude to examine some peculiarities of GIS, avoiding the risk of pushing the philosophical questions arising from them into the background of a generalist reflection on geography.

In line with these remarks, the answer to the first two questions—generally shared by the contributors of this volume—was that there is really something such as a "Philosophy of GIS", with proper distinguishing features. Thus, the purpose for this book is collecting some articles, aimed at presenting the fundamental philosophical issues required by the reflection within and upon GIS. In particular, such a volume proposes an introduction to the philosophy of GIS from an analytical perspective, which looks at GIS with a specific focus on their fundamental assumptions, distinctions, and features. Accordingly, the first part of the book is devoted to explore some of the main philosophical questions arising from GIS and GIScience, which include, among others, investigations in ontology, epistemology, linguistics, and geometrical modeling. The second part concerns issues related to spatial and cartographical representations of the geographical world. The third part is focused on the ontology of geography, specifically in terms of geographical entities, objects, and boundaries. Finally, in the fourth part, the topic of GIS constitutes a starting point for exploring themes such as quantum geography and disorientation, and for defining professional profiles for geographers with competences in GIS environment.

Acknowledgements This book has been supported by a fellowship at the Research Institute of the University of Bucharest (ICUB). I wish to express my thanks to all the contributors for their cooperation during this book preparation. I am also extremely grateful to Robert Doe and Carmen Spelbos of Springer for all their assistance. My debts to other people are too numerous to list. However, I should like to record my particular gratitude to Raffaella Afferni, Ferruccio Andolfi, Margherita Azzari, Valeria Bizzari, Claudio Cerreti, Sorin Cheval, Carmen Chifiriuc, Matthew R. X. Dentith, Mihnea Dobre, Maurizio Lana, Giulia Lasagni, Diego Magro, Cristina Meini, Iulia Nitescu, Marcello Tanca, Italo Testa, Cristina Travanini, Achille Varzi, Paola Zamperlin and my sister, brother, and parents for their invaluable support. I dedicate this book to Giulia: wherever we are going to be.

Parma, Italy
February 2019

Timothy Tambassi

Contents

Part I Philosophical Issues in GIS and GIScence

1 **Unpacking the "I" in GIS: Information, Ontology, and the Geographic World** 3
Helen Couclelis

2 **Some Philosophical Issues Regarding Geometric Modeling for Geographic Information and Knowledge Systems** 25
Robert Laurini

3 **A Philosophical Perspective on Linguistic Paradigms in GIScience** 51
Qingyun Du

Part II Representing the Geographic World

4 **Space, Time and the Representation of Geographical Reality** 75
Antony Galton

5 **Mapping the Deep Blue Oceans** 99
Rasmus Grønfeldt Winther

6 **On the Distinction Between Classical and Nonclassical Geographies: Some Critical Remarks** 125
Timothy Tambassi

Part III Geo-ontological Perspectives

7 **Drawing Boundaries** 137
Barry Smith

8 **Geographic Objects and the Science of Geography** 159
Amie L. Thomasson

9 **What a Geographical Entity Could Be** 177
Timothy Tambassi

Part IV More Theory, Practice and Applications

10 **Is There a Quantum Geography?** 209
Thomas Bittner

11 **Disorientation and GIS-Informed Wilderness Search and Rescue** 241
Pablo Fernandez Velasco and Roberto Casati

12 **From Education to Job Opportunities. Defining Professional Profiles for Geographers with High Competences in GIS Environment** 253
Cristiano Pesaresi

Index 265

Part I
Philosophical Issues in GIS and GIScence

Chapter 1
Unpacking the "I" in GIS: Information, Ontology, and the Geographic World

Helen Couclelis

Abstract As a tool dealing with information rather than matter, GIS shares with other information technologies the conceptual challenges of its medium. For a number of years now, ontology development has helped harness the complexity of the notion of information and has emerged as an effective means for improving the fitness for use of information products. More recently, the broadening range of users and user needs has led to increasing calls for "lightweight" ontologies very different in structure, expressivity, and scope from the traditional foundational or domain-oriented ones. This paper outlines a conceptual model suitable for generating micro-ontologies of geographic information tailored to specific user needs and purposes, while avoiding the traps of relativism that ad hoc efforts might engender. The model focuses on the notion of *information* decomposed into three interrelated "views": that of *measurements* and *formal operations* on these, that of *semantics* that provide the meaning, and that of the *context* within which the information is interpreted and used. Together, these three aspects enable the construction of micro-ontologies, which correspond to user-motivated selections of measurements to fit particular, task-specific interpretations. The model supersedes the conceptual framework previously proposed by the author (Couclelis, Int J Geogr Inf Sci 24(12):1785–1809, 2010), which now becomes the *semantic* view. In its new role, the former framework allows informational threads to be traced through a nested sequence of layers of decreasing semantic richness, guided by user purpose. "Purpose" is here seen as both the interface between micro-ontologies and the social world that motivates user needs and perspectives, and as the primary principle in the selection and interpretation of Information most appropriate for the representational task at hand. Thus, the "I" in GIS also stands for the Individual whose need the tool serves.

Keywords GIS · Ontology · Semantics · Measurements · Context · User perspective

---

H. Couclelis (✉)
Department of Geography, University of California, Santa Barbara, CA 93103-4060, USA
e-mail: cook@geog.ucsb.edu

© Springer Nature Switzerland AG 2019

T. Tambassi (ed.), *The Philosophy of GIS*, Springer Geography,
https://doi.org/10.1007/978-3-030-16829-2_1

Introduction

Viewed as a tool, GIS transforms information rather than matter. It shares with traditional tools the purpose of serving some need of the user, but differs from these profoundly in the conceptual complexity and relative novelty of the stuff it manipulates. Little applied philosophy exists on the mysteries of matter, yet information—in this case, geographic information—has attracted the active attention of professional philosophers as well as engineers and others grappling with the possibilities and limitations of the tool. In this, GIS joins an array of information technologies that are turning to ontology as a means of figuring out practical theories applicable to different domains of interest.

What does an ontology really represent? The engineer and the metaphysician would give very different answers to this question. Both would claim that their work bears some relationship to the real world, yet at first sight, it might be hard to imagine finding much common ground between them. Literally meaning "the study of what exists" in the original Greek, Big-O-Ontology has been variously associated with questions about the world, the Creation, existence, identity, reality, the cosmos, universal flux, things both real and fictional, minds, ideas, consciousness, space and time, language, and Truth (Kavouras and Kokla 2011). But the engineer's questions are practical, originally focused on the technical problems of knowledge sharing, more specifically of interoperability among databases. The small-o ontologies resulting from that quest were meant to describe the finite artificial worlds defined by these databases. Yet before too long issues of cognition and language, of structure and meaning, of concepts and measurements, of physical and nonphysical entities, of space and time, of user needs and culture-specific interpretations, and of reality and philosophy also became part of the ontology discourse (Kavouras and Kokla 2008). Developed from an engineering perspective to meet engineering needs, foundational ontologies in particular have struggled with problems that are often part of mathematical philosophy if not of metaphysics. More recently the Semantic Web has brought additional dimensions of difficulty to the task because web knowledge has no primitives, no core, no fundamental categories, no fixed structure, and is heavily context-dependent. The new dilemma that ontology engineers face is that of serving highly disparate needs using data from highly disparate sources which are themselves part of highly disparate observation and interpretation contexts. Most challenging is the context dependency of web knowledge (indeed, of knowledge in general) since context is, by definition, what lies outside a representation or model. But even traditional kinds of data raise the issue of context as it relates to their interpretation and use if not also their origin. The problem appears intractable as it is the broader social, political, economic, cultural, institutional, and other societal factors that directly or indirectly constitute the context of each application, as well as more immediate factors of needs, resources, and practical constraints. In the tidy world of ontology engineering, the question then becomes how to treat the whole of society as context without running into the impossible task of having first to formalize it.

The notion of society as context is compatible with Gruber's (1993) widely quoted definition of ontology as a "formal, explicit specification of a *shared conceptualization*." Gruber clarifies: "That is, an ontology is a description ... of the concepts and relationships that can exist *for an agent or a community of agents*." Further: ontologies must "constrain the *possible interpretations* for the defined terms." (Emphasis added.) The implications are that (1), ontology is not directly about the real world but about some "shared conceptualization"; (2), the concepts and relationships defining an ontology are valid for an observer (agent) or "a community of agents"; (3), there may be several "possible interpretations for the defined terms." An ontology is thus by definition relative to specific agents facing specific tasks individually or as a community (e.g., a scientific or professional community), and its role is to pin down the fluid meanings of terms in ways that serve the needs of that individual or community at some specific time.

Recent advances in cyber-infrastructure development as well as inside the ontology field suggest that the traditional model of axiomatic foundational ontologies, as exemplified by DOLCE, SUMO, BFO, and other such major efforts (Kavouras and Kokla 2008), may need to be significantly amended if not replaced. More than a decade ago, protagonists of the DOLCE (Borgo and Masolo 2010) project and others were expressing reservations about the role of such heavy-weight ontologies in the age of the semantic web, and were proposing alternative, less centralized approaches (Gangemi and Mica 2003). The notion of multiple micro-ontologies, lightweight ontologies, or "microtheories," each developed for a particular use context, has been gaining traction (Adams and Janowicz 2011; Janowicz and Hitzler 2012). So is the idea of modular "local micro-ontologies" made up of interchangeable parts representing relatively uncontroversial pieces of knowledge, likely expressed in the form of engineering design patterns (Gangemi 2005; Gangemi and Presutti 2010). But such approaches raise questions of their own. One issue is concern about the relativism and knowledge fragmentation inherent in the proliferation of indefinite numbers of microtheories and design patterns lacking a common core. Another is the fact that the nagging philosophical questions raised by ontology engineering, some of which are mentioned above, are being pushed aside rather than confronted. The continuing debates over realism, conceptualism, instrumentalism, or constructivism in ontology development (Smith 2010; Scheider and Kuhn 2011), about the role of cognition and culture (Kuhn 2003), about whether nonphysical entities may or may not be primitives in an ontology (Gangemi and Mika 2003), and so on, attest to the persistent significance of the question of *representation* that underlies all forms of modeling, including ontology development.

What then if the representation in question is not directly *of* the real world, but of *information about* the real world? This paper presents a model of information based on a major modification and expansion of the conceptual framework for ontologies of geographic information developed by the author (Couclelis 2009a, 2010). The key modification consists of decomposing the notion of information into three distinct but interrelated "views": (1) that of *measurements* that derive from the empirical world, along with any formal *operations* applicable to these measurements; (2) that of *semantics* that may correspond to the measurements, and (3) that of the *context* of

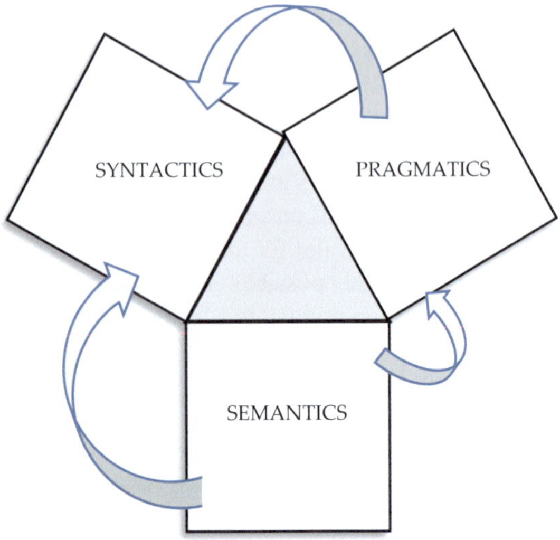

Fig. 1.1 The three views of the model and their connections. The central triangle represents the region where the micro-ontologies are generated

interpretation and use relating to the receiver of the information. Together these three aspects of information enable the construction of user-motivated micro-ontologies tailored to purpose-specific interpretations and selections of measurements (Fig. 1.1). To use a linguistics analogy, the raw measurements and their formalizations (view 1) correspond to the context-free and semantics-free *syntactic* aspects of information (something along the lines of Shannon's information), view (2) represents the *semantic* aspects, and the *pragmatics* view (3) helps define the specific context of each micro-ontology. In this model, the previously developed ontological framework (Couclelis 2010) is now the *semantics* view. The new model represents the necessary parts for the derivation of internally (and to some extent mutually: see section "The Syntactics and Pragmatics Views") consistent micro-ontologies in a universal-to-particular relationship. The role of the earlier framework as the semantics part of a "micro-ontology generation engine" is thus new to this paper. The relevant aspects of that earlier framework are outlined in section "Ontologies of Geographic Information: An Overview". In section "Purpose and Function in Micro-ontology Design", several hypothetical geographic examples are presented that tentatively illustrate how the model could be put to work. Section "Towards Implementing User-Centred Micro-ontologies" provides indications as to how it could be implemented, focusing on the model's connections with the literature and a few other potentially relevant aspects. The conclusion recapitulates the chapter's contributions and shortcomings and considers certain promising avenues for future research.

Ontologies of Geographic Information: An Overview

The framework presented in the paper with the above title (Couclelis 2010) is at the heart of the new model of information presented here. Note that the three parts of this new model are called "views", whereas the previously developed framework that now forms the semantic view is composed or "layers" or "levels"; also that the term "framework" is reserved for the earlier work. The new conceptual structure is called "model." This section summarizes three key principles of the framework that continue to be relevant in the new model. The main modifications that were made in the current project are indicated below and at the end of this section, and in the following.

Three Key Principles

Three aspects of the framework that are also a major part of the new model are presented here. These aspects are as follows: (a) The foregrounding of the perspective or motivation of the *user*; (b) the ability to distinguish a nested sequence of representational layers of varying degrees of *semantic richness*; and (c) the principle of *data filtering* through criteria resulting from the users' purpose-oriented semantic choices. While maintaining these key aspects of the framework, the new model introduces two major modifications. First, it treats the earlier unitary framework as one of three "views"—the *semantics* view—flanked by the views of *measurements* on the one hand, and of the *context* of information interpretation and use on the other. Second, it assigns special roles to the two layers of the semantics view that most directly ground the model: the "classification" layer on the one hand, which now serves as the bridge between the semantics and measurement views, and the "purpose" layer on the other, which now serves as the main connection between the semantics and context views (Fig. 1.1). To pursue the earlier linguistic analogy, the semantics, syntactics, and pragmatics of the representation interact with one another, as they do in language. The aspects of the framework relevant to the new model are as follows.

First, the framework is anchored at one end on the notion of the perspective or intention of the *user* of the information, since information is a binary concept that involves not only a source or sender but also an active receiver able to decode and use it. The user may be an individual, a community, or an institution, as determined by a particular requirement or kind of task, set in the context of specific societal interests and understandings, and subject to particular resources and constraints. For example, there may be many possible perspectives on the same transportation network as a real-world phenomenon, from that of the transportation engineer concerned about the flow of traffic to that of the biologist studying wildlife movement barriers and road casualties on segments of the network. The user perspective or purpose or more generally, intentionality (Searle 1983) as relating to the task at hand is molded by the context against which the relevance of a particular representation is defined, entailing a suitable interpretation, selection, and reification of available information (Adams and Janowicz 2011; Gangemi and Mika 2003; Scheider et al. 2010). Because user purposes as well as the phenomena of interest vary with time, the framework is inherently spatiotemporal, though only the basic static version is discussed here.

Second, the framework is a structure consisting of a number of nested layers representing cumulative degrees of "decoder capabilities"[1], which range from the full capabilities of a human observer to brute awareness. Just as we can build a hierarchy of descriptions based on decreasing spatiotemporal and attribute detail (*quantity* of available information), we can also envision a hierarchy based on decreasing semantic richness (*variety* of types of available information). For example, given a set of polygons geocoded with high accuracy, and the same set of polygons in rough outline along with the knowledge that it represents a college campus, the latter is considered to be semantically richer. Formally, the principle is somewhat analogous to the notion of "reduction" in rough set theory, iterated over seven levels (Pawlak and Skowron 2007).

Table 1.1 summarizes the basic version of the structure and indicates the names of the levels and of the key decoder capability that may be associated with each. In geographic information science, the most often studied transition is the one from L4 to L3—though it is more commonly treated in the opposite direction, from L3 to L4 (from data classifications to named objects)[2]—but most other transitions may be just as significant. For example, the move from L6 to L5 loses the ability to represent the notion of spatial *function* even though it is still possible to recognize and model relations among spatially disconnected or otherwise distinct objects. Think of the difference between—say—knowing that certain configurations of buildings and other installations constitute a water treatment plant, versus also understanding how that particular configuration relates to the actual functioning of the plant. The

Table 1.1 The expressivity of representations is gradually collapsed over levels 7–1 of the hierarchy as the semantic richness of the system contracts from top to bottom (principle of semantic contraction). Each level is associated with a characteristic property domain and also includes those at the levels below it (Adapted from Couclelis 2010)

Semantic resolution levels	Decoder capabilities	Forms of representation
7 Purpose	Intentionality	Objects
6 Function	Instrumentality	
5 Composite objects	Association	
4 Simple objects	Categorization	
3 Classes	Classification	Fields
2 Observables	Perception	
1 Existence in space-time	Awareness	

[1]This expression is used to avoid suggesting that the sequence described here has empirically established cognitive/psychological validity (although it may), or that human cognition is not relevant to ontology development (which of course it is).

[2]Fields and object are the two fundamental forms of representation in geographic information science. Much attention has been directed towards clarifying their logical relationships and formalizing their integration (Couclelis 1992; Galton 2001; Goodchild et al. 2007; Kjenstad 2006; Voudouris 2010). Note that the distinction between fields and objects qua representation forms is entirely a semantic issue. Raw measurements do not support one or the other interpretation.

third column marks the *reification* step that occurs between levels 4 and 3 and is usually considered the most critical.

The third important point results from the combination of the above two. The context of the representation problem, consisting of the fundamental duality of user perspective and phenomenon of interest, plus any pragmatics relating to data issues, time or resource constraints, and other situational factors, governs the selection of information at each level, based on the criterion of *relevance* to the task at hand. Thus, for the same phenomenon, different paths across the hierarchy will be traversed by different user selections. Similar paths will result in closely related representations, while very dissimilar paths will result in markedly different ones (Table 1.2). All paths end with the selection of the spatiotemporal framework (including geographic scale, granularity, extent, etc.) most appropriate for the task at hand. Note that the spatiotemporal framework must be discrete, since information is collected in discrete units at discrete locations, and the point analogues must be extended spatiotemporal "granules."

Table 1.2 Two different perspectives on a "road network" (Adapted from Couclelis 2010)

	A road map of region X	A map of roads in region X
7 Purpose	Facilitate vehicular travel planning and navigation	Identify and mitigate barriers to wildlife movements
6 Function	Represent possible routes from place A to place B	Represent the locations where wildlife corridors intersect with roads
5 Composite objects	A road network	A wildlife corridor network intersecting with a road network
4 Simple objects	Places, freeways, arterials, collectors, intersections, ramps, roundabouts, …	Roads, wildlife corridor segments, underpasses, culverts, high-conflict intersections, …
3 Classes	Fields of properties (corresponding to surface material, slope, network structure, …) aggregated in diverse geometrical patterns	Fields of properties (corresponding to incident frequency, barrier permeability, height, width …) aggregated in diverse geometrical patterns
2 Observables	Hard, rough, green, brown, wet, …	Open, blocked, green, hard, kill, dry, wet …
1 Space-time exist	"Task-relevant information exists here-now at such-and-such appropriate granularity"	"Task-relevant information exists here- now at such-and-such appropriate granularity"

Semantic Contraction and Geographic Information Constructs (GIC)

Unlike other stratified ontology models, the framework in Couclelis (2010) is not a taxonomy of aspects of reality (Frank 2003). Rather, it is a cumulative structure that may be systematically decomposed into simpler but still logically coherent substructures. Observed from the top down, the layers correspond to successively narrower domains of semantic content, as described in section "The Semantic Hierarchy". Each level includes its own characteristic domain of properties plus those of all the ones below it, so that internally consistent spatiotemporal representations may be obtained by merging information *from any level down*. One may imagine gradually draining the structure of semantic content by peeling away successive layers, until nothing is left but the idea of a spatiotemporal frame with information. This notion of step-wise semantic reduction of the structure is called "semantic contraction."

Semantic contraction thus results in truncated structures that may reach no higher than levels 3, 4 or 5. These are called here "geographic information constructs" ($\boldsymbol{GIC}_i$, i = 1–6) to allow differentiation from representations that span all seven levels. Table 1.3 indicates the relationship between a fully developed, 7-level representation, and a truncated one that reaches only to level 4 ($\boldsymbol{GIC}_4$). The table may be seen as an accounting sheet that registers the presence or absence of property domains, and of quantitative changes in property values within property domains. It is subdivided into seven primary columns corresponding to the seven levels of the hierarchy. Each of these columns is headed by one of the domains of properties $\{p_n\} \subset P$ that characterize the corresponding level over and above the properties at the next level down, and is further subdivided into as many columns as there are properties in that set, and then again, to whatever degree of detail is needed. The primary rows are

Table 1.3 The semantic information system. The full table represents a geographic information construct, $\boldsymbol{GIC}_7$, occupying topons g_1–g_m at chronon $\mathbf{x}_t$, complete through level 7. Columns $\{p_n\}$ may be removed sequentially from right to left to yield reduced but internally consistent representations of the original *GIC*. The white area represents a GIC reaching only to level 4 ($\boldsymbol{GIC}_4$). The *profile* of each topon *g* for a given chronon *x* is represented as a code (here: binary) extending across n ≤ 7 hierarchical levels. (Adapted from Couclelis 2010)

	x_t		$P = \{p_7\} + \{p_6\} + \ldots + \{p_1\}$				
		$\{p_1\}$	$\{p_2\}$	$\{p_3\}$	$\{p_4\}$	....	$\{p_7\}$
		p_1	$p_{21}, p_{22}, \ldots p_{2k}$	$p_{31}, p_{32}, \ldots p_{2j}$	$p_{21}, p_{22}, \ldots p_{2n}$	....	$p_{71}, p_{72}, \ldots p_{7i}$
	g_1	1	1 0 ... 1	1 0 ... 0	1 1 ... 1	....	1 1 ... 0
*GIC*i	g_2	1	1 1 ... 1	0 1 ... 1	1 1 0	....	1 1 ... 1
	...	....	....	....	....	....	....
	g_m	1	0 1 ... 0	1 1 0	1 1 0	....	1 1 0

labeled for the *GICs* of interest. Each primary row is subdivided into further rows, one for every granule of the space $g_1, g_2, g_3,\ldots g_i \in G$ that the corresponding GIC occupies at a specific (discrete) time granule.

As will be seen in the following, neither the spatial nor the temporal granules can be geometrically specified in advance independently of the highest semantic level at which a GIC is considered. To differentiate these extended atoms of space and time from ordinary spatiotemporal coordinates, they are given the names of *topons* and *chronons*, respectively. A topon (from the Greek τοπος, place) is thus the smallest 2- or 3-dimensional chunk of space over which properties of interest (from the perspective of a particular user purpose) may be measured. A chronon (from the Greek χρονος, time) is the smallest interval of time over which a change in some property of interest to the user may be observed. Spatial and temporal atoms are treated separately because the properties of time and space are not necessarily parallel at all levels. A cell C_{ij} of Table 1.3 thus records whether, for chronon x, topon g has been observed to have property p. The entry will be binary (yes/no, the property is or is not present) in the simplest case but generally it may be any value within the appropriate range. The table is read from right to left, in the direction of semantic contraction, beginning with the most semantically rich level 7, where all the property domains are available that are necessary for describing the relevant aspects, functions and purposes of any specific GIC. Reading across the rows we find ordered descriptions of what properties and values exist at each topon, and the corresponding vectors, or *profiles*, get shorter with every domain of properties being removed. These vectors represent the "paths" across the layers mentioned above.

Truncated GICs can be practically useful when the representational problem is fairly familiar and uncontroversial so that the explicit representation of the upper levels would be redundant. They are also theoretically significant because of the connection made in section "The Syntactics and Pragmatics Views" with the notion of geo-atom developed in Goodchild et al. (2007).

The Semantic Hierarchy

Very briefly, the characteristic properties of each level are as follows. Moving from level 7 downwards, the *semantics* of each level are built on its own level specific properties plus those of all the levels below it. The *purpose* of the representation serves as a filter providing for each level specific criteria that restrict what can (should, needs to) be expressed at that level. Terminating at level 1, the procedure picks the spatiotemporal frame (discrete-continuous, granularity, extent, etc.) most appropriate for the purpose-specific micro-ontology being generated.

- L7: *Purpose*. Purpose (more generally: intentionality, perspective, interest, and so on) is not a spatiotemporal concept but it is the interface between this spatiotemporal ontology framework on the one hand and the societal context motivating these purposes on the other. Purposes may be research-oriented, professional, institutional, governmental, etc.

- L6: *Function*. Function and purpose are tied together in a close means-ends relation. This relation is characteristic of the *design* perspective. Representations (models) must be designed such as to meet corresponding purposes (e.g., weather maps designed for scientific and TV presentation uses will be very different). Note that in addition to the purpose and function of the representation itself, many represented spatiotemporal phenomena such as cities, farms, transportation networks and harbors are *artifacts* with purposes and functions of their own, aspects of which may need to be explicitly represented for some uses. In this latter case, "function" also encompasses agency and action, both of which are teleological notions (doing something "in order to_").
- L5: *Composite objects*. Many objects in the world, both natural and artificial, are composed of spatially disconnected or otherwise heterogeneous parts. Recognizing that—say—a number of chairs and tables in specific spatial relations to each other are part of an office rather than a storage room, or that a large Christmas tree nursery is not a small pine forest, requires more "intelligence" than recognizing the constituent parts.
- L4: *Simple objects*. As the semantic contraction continues, descriptions have now reached the last level at which discrete objects may be recognized, categorized and named. How these objects are named (and understood) depends on the observer's perspective-specific criteria as these result from levels 7–5. As certain popular cartoons remind us, one person's playing field is another's building lot, and one person's fishing grounds is another's marine preserve. Different perspectives, and thus criteria, may also result in different object boundaries, e.g., the cartographer's valley may be different from the cyclist's or from of the Valley Hotel manager's.
- L3: *Classes*. This is the level of "pure" data. Here, objects have given way to corresponding classes of properties represented as (uninterpreted) fields of measurements. All data manipulations and analytic operations are in principle enabled here, as well as the aggregation of similar values into 0- to 3-dimensional geometries. Criteria specific to each user purpose, as derived from levels 7-4, restrict measurement specifics, manipulations, and operations to those suitable to the particular representation being generated. These same criteria also yield the *classifications* that are the semantic precursors of the simple objects of level 4.
- L2: *Observables*. This level serves to distinguish the necessary kinds of variables from the values these may take on (Goodchild et al. 2007; Probst 2007). It is the level of qualia, where spatial variation may be observed but not yet measured and communicated, let alone named. Levels 2 and 3 are as closely bound up with each other as levels 6 and 7 are at the top of the hierarchy.
- L1: *Spatiotemporal existence*. In and of themselves, the vanishing semantics of level 1 just barely suggest the existence of some spatiotemporal frame that may contain information relevant for some purpose. This extreme vagueness is resolved by the filtering procedure that started at level 7, which restricts possibilities to just the kind of spatiotemporal frame entailed by the original user perspective.

Thus the hierarchy is anchored at one end by the notion of a spatiotemporal framework to be specified by choices made at all the preceding levels, and at the other end, by the purpose of the representation, itself deriving from any number of societal influences and needs, which may be professional, institutional, socioeconomic, personal, educational, and so on.

In the new model, three modifications are made to the framework as described in Couclelis (2010) and summarized in this section. (a) As mentioned in the Introduction, the fundamental modification is the development of a model comprising three distinct views—syntactic, semantic, and pragmatic—of which the framework is now the *semantic* view. Also (b), there is a switch in the order of levels 2 and 3 in the semantic hierarchy, whereby the observables (required types of variables) are now picked up before the corresponding values. And (c), the two kinds of purpose-function pairs: of the representation itself, and of the entity represented (in the case of artificial entities, or more generally entities used to support some function), will be clearly distinguished in the following as type A and type B, respectively. Section "Purpose and Function in Micro-ontology Design" provides examples of these changes while focusing on the semantic view. The latter is discussed further in section "Towards Implementing User-Centred Micro-ontologies", along with the syntactic and pragmatic views.

Purpose and Function in Micro-ontology Design

Representing Natural, Artificial, and Abstract Entities

Representation is the basic underlying notion in all modeling, including the construction of ontologies seen as meta-models or templates for deriving particular kinds of models corresponding to specific requirements. The class of models of special interest in this research is that of the lightweight or micro-ontologies, seen as use-specific realizations of the conceptualization illustrated in Fig. 1.1. From an engineering perspective, all models are artifacts designed for particular purposes (clarifying, explaining, illustrating, supporting, facilitating, problem-solving, restricting meaning, etc.) and must function in ways that support these purposes. As such they share the general properties of designs as abstract or material tools developed for addressing the requirements of particular uses or users. According to Simon's seminal work on the design sciences (Simon 1969), artifacts constitute a separate ontological category from other things in the world because they would not have existed but for an agent's intentional action towards serving some purpose. Thus, models, ontologies, and micro-ontologies have purposes and functions irrespective of whether they represent natural entities such as mountains and rivers, or artificial entities such as road networks and college campuses, which of course have purposes and functions of their own. In the following, we will use the notation "A" for the purposes and functions of the representation itself, which are always present, and "B" for those of an artificial entity that may be the object of a given representation. In the latter case, the fact that the entity is artificial as opposed to natural may or may not be relevant depending on the purpose (A) of the representation. The example in Table 1.2 above (Sect. 2.1.1) along with two additional ones in this section will help clarify this point.

The examples below are selected to illustrate a number of different cases that may be distinguished, concerning: (1) artificial physical entities (section "Road network"), (2) artificial abstract entities (section "State"), and (3) natural entities (section "River"). In addition, it is shown that in all three cases very different representations (micro-ontologies) result depending on whether or not user needs call for the representation of the purpose and function of the artificial entity represented (examples "road network" and "state"), and whether or not user needs require that function be imputed to a natural entity (example: "river") that may be viewed, in this case, either as natural phenomenon or as transportation infrastructure.

"Road Network"

Table 1.2 illustrates a situation where a cartographer designs two different maps of the same region and phenomenon for two different purposes (type A): the first one for use in a road atlas, and the other for use by biologists studying wildlife road casualties. In the first case, the road network is approached as an artificial entity designed to facilitate movement between places, and these properties (type B) of the network are the ones of primary interest in the design of the map. In the second case, the network is treated as a collection of physical barriers and high-impact locations. The table indicates how these two different perspectives lead to very different selections of information across the hierarchy. Given appropriate data, possible queries include:

(a) Road map:

- Show me the two shortest routes from Goleta to North Fork
- Show me a route between Ventura and Mammoth with no grade above 7%
- Show me the average driving time between Corvallis, OR and San Francisco, CA on a summer weekend.

(b) Map of roads for use by wildlife biologist

- Show me the 10 road segments with the most wildlife fatalities in 2011
- Show me mountain lion casualties by year on the stretch from X to Y
- Show me where topography might allow wildlife passes under dangerous road segments.

Below I briefly outline two additional cases, the first concerning the concept of "State" (an abstract artificial entity), and the second is that of "river" (a natural entity).

"State"

What is a "state," or "country"? Do these two notions actually designate the same thing? The practical significance of these questions is highlighted in an article in *The Economist* (2010), which laments the difficulty of deciding how many states (or countries) there are in the world in the face of name ambiguities, disputed borders, rogue states, states with partial or no international recognition, stateless nations, states whose citizens use passports from another state, states made up of "just two nice buildings," and so on. As is the case with most complex concepts, the "best" definition of *state* or *country* depends to a large extent on who is asking: is it the United Nations (whose dilemma prompted the *Economist* article), is it a particular national government, is it a tourist agency, is it a cartographer, is it a political scientist or a historian, or is it an international immigrant seeking asylum in a foreign state?

The abstract concept of "state" has been discussed in the ontology literature and has been the topic of a debate as to whether, for example, a state is an organization or a legal person (Robinson 2010). The framework presented here suggests that a state can be both of these things or neither, and it may be many other things as well, depending on the purpose of the micro-theory being developed.

Let us take just two contrasting cases: that of a national state, and that of a stateless nation, both wishing to promote their interests in international negotiations. To be more specific, let us assume in addition that the national state is undergoing a severe economic recession, and that the stateless nation may be that of the Palestinians. In the latter case, the entity in question is not only abstract but also, at this point in time, hypothetical. In both cases, the main purposes and functions will be those of the representation (type A), and we will ignore in the first case the fact that states *qua* artificial entities also have purposes and functions of their own (type B), such as safeguarding the welfare of their populations and developing the institutional structures needed to do so. These aspects are however relevant for the second example, that of the stateless nation. With few exceptions, only spatial aspects are mentioned in these brief sketches.

a. National state:

- *Purpose*: highlight contributions of recent aid packages to regional progress, and point to remaining problems in order to help win further international economic and political support
- *Function*: emphasize factors that led to regional growth and/or decline, possibly using outputs from regional development models
- *Composite objects*: successful regions; backward regions; regions on the cusp of becoming successful; transportation networks; fast-developing communities
- *Simple objects*: infrastructure units, productive installations, productive land, tourist installations, rural and urban communities, border segments under immigration pressure, etc.
- *Observations*: required types of spatiotemporal information for describing the above objects, reflecting their relationships with growth, stagnation, and decline
- *Classifications*: classed values for the variables identified above

- *Spatiotemporal frame*: multi-scale (national, regional, and local as needed); temporal scale: 3–6 years (since time of last major international loan).

Given appropriate data, possible queries include:

- Show productive units in region X that benefited from earlier loan
- Show growth statistics for coastal communities that received tourism subsidies AND growth statistics for similar communities that did not receive such subsidies
- Show adjacent regions with contrasting ("high"-"low") levels of transportation infrastructure.

b. Stateless nation:

- *Purpose*: advance the cause of establishing a new sovereign state; argue for the feasibility of the desired territorial state; dramatize the consequences of failure
- *Function*: highlight and magnify the contradictions and conflicts resulting from the current spatial fragmentation of populations
- *Composite objects*: fragmented ethnic territories, adjacencies and overlaps with territories of other nations, disrupted transportation and supply corridors, regional urban distributions, regions of ethnic lands under foreign occupation, trans-border ethnic continuities, historic ethnic territorial distributions, etc.
- *Simple objects*: proposed national territory (–ies) covering all or most ethnic fragments, corresponding border(s); elements of the above composite objects; alternative proposed country borders
- *Observations*: types of variables required for representing and supporting alternative border proposals; evidence supporting the claim that proposals would lead to the eventual resolution of conflicts at current geographical hotspots;
- *Classifications*: values for the above variables
- *Spatiotemporal frame*: multiscale, emphasis on local. Temporal scale: short to medium term; ill-defined

Possible queries include:

- Show the locations with the most incidents resulting in casualties between 2009 and 2012
- Show the major routes where communications were disrupted by incidents since 2008, by duration and incident type
- Show the 20 most overpopulated ethnic areas and statistics on unemployment and building conditions
- Show alternative regions that enclose at least 90% of the ethnic areas OR at least 85% of the ethnic population.

Note that while no actual geocoded data exist for the focal point of this negotiation effort—the hoped-for national territory and its borders—hypothetical boundaries and any other relevant features may be indicated and geocoded on maps. While lacking physical reality, the potential new state is a (conceptual) artifact that implies (type B) purposes and functions, such as providing a permanent and safe home for people belonging to the stateless nation, and meeting minimal prerequisites of compactness, area, spatial organization, and infrastructure for proper functioning as a country.

"River"

Rivers are natural entities that, unlike artifacts such as roads and states, do not have their own intrinsic (type B) purposes and functions. Yet useful micro-ontologies representing rivers almost always reflect the purposes and functions of interest to their users. We may consider very briefly the contrasting cases of (a) a research hydrologist monitoring the discharge of a river over a multiyear period, and of (b) a company operating a fleet of tourist riverboats on that same river. The purpose of the research hydrologist's micro-ontology is to provide support in identifying relevant changes and trends in the river's quantity and quality of discharge, and their relationships with other physical factors such as rainfall and temperatures. The tourist boat agency, on the other hand, is interested in providing a positive client experience within given cost margins, and would want to know about water levels and river traffic at different times of the year, about distances between scenic stretches of the river, visual water quality, the quality and availability of shore facilities and landing areas, possibilities for shore excursions, the schedules of the competition, alternatives for very dry or very wet seasons, and so on. In other words—and avoiding the tedium of further analysis—these two perspectives on the same natural entity: the river as channel and water flow, and the river as transportation infrastructure, will likely have quite limited overlap. So will, of course, the corresponding micro-ontologies.

Towards Implementing User-Centered Micro-Ontologies

The model presented in this paper is conceptual and exploratory. Undoubtedly formalizing and implementing it will be challenging. However, some degree of optimism is justified considering certain marked affinities with research in ontology engineering and other areas in geographic information science. At this stage, such connections help provide an indirect evaluation of the work, by indicating how it may fit within the broader nexus of the literature.

From this angle, this section discusses the prospects of implementing the model. The previous section focused on semantics, aiming to demonstrate the effect of different user perspectives on representations of the same kind of entity. However, semantics is only one of three views of the new model, next to those of measurements and formal operations (syntactics view) and contexts (pragmatics view). All three views must eventually be implemented in a consistent manner for the model to be of use. Here, I indicate certain encouraging connections with the literature and then address a few broader aspects of the model.

The Semantics, Syntactics, and Pragmatics Views

The Semantics View

Possibly closest in spirit to the work presented here, though deriving from a very different perspective, is the paper by Gangemi and Mika (2003) on Descriptions and Situations (D&S). Having discussed the need for lightweight ontologies as the preferred way to harness the power of the semantic web, these authors propose the D&S ontology as a plug-in to DOLCE (Borgo and Masolo 2010). The D&S ontology makes a clear distinction between "Situations" S, which are uninterpreted data configurations, and "Descriptions" D, which provide the agents' conceptualizations or interpretations of the Situations "*based on a nonphysical context*" (p. 690, original emphasis). The Situations themselves are derived from "States of Affairs" (SoA), defined as non-empty sets of assertions that are constituted by statements about the world. The authors indicate that notions cognate to SoA are "flux, unstructured world, or data" (p. 694). The D&S strategy achieves the goal of providing the needed flexibility for supporting multiple conceptualizations of the same situation without lapsing into relativism, as it is built on top of the DOLCE foundation ontology.

The model proposed here has several aspects in common with the D&S ontology. Here too, the agents' alternative interpretations or conceptualizations are clearly distinguished from the empirical data being interpreted. Indeed, the entire semantic view in the model presented here corresponds to the Descriptions (conceptualizations, interpretations) aspect of D&S in Gangemi and Mika (2003), with the semantic hierarchy representing discrete degrees of sophistication ("intelligence") in these conceptualizations. Moreover, in both cases, these interpretations are "based on a nonphysical context" in the form of the agents' intentionality. However, the present model also recognizes low levels of semantic content in the uninterpreted data (Levels 1–3), distinguishing, between, on the one hand, "States of Affairs," which are about "flux" and "unstructured world" (Level 1), and on the other hand, Situations, which are about "Gestalt" and "Setting" (Level 3) and "configuration or structure" (Level 2). Also, in both Gangemi and Mika (2003) and in the present model, the key generative mechanism is reification (Adams and Janowicz 2011; Scheider et al. 2010).

Note that what is proposed here is very different from traditional AI work on goal-oriented knowledge representation, which tends to be concerned with the knowledge necessary for solving a specific problem or for carrying out a specific practical task. The limitations of these ad hoc structures were part of the motivation for the development of coherent foundational ontologies to represent general knowledge (Guarino 1995). Here, the task-specific, purpose-oriented representations are systematically derived from a unified framework that is closer to foundational ontologies in structure, if not in philosophy and content. The central role played in the model by the semantic view and the semantic contraction procedure (section "Semantic Contraction and Geographic Information Constructs (GIC)") should contribute to the internal semantic consistency of micro-ontologies developed on its basis.

The Syntactics and Pragmatics Views

In the model, the measurements provided by the empirical world are gathered in the *syntactics* view. It is called "syntactic" because all formal manipulations that may be performed on quantities are possible in this view, regardless of any interpretations. At least one approach from the geographic information science literature appears very suitable for formalizing the syntactics view of the model presented here. This is the "General theory of geographic representation" proposed by Goodchild et al. (2007), and in particular the notion of "geo-atom." A geo-atom is "a tuple <x, Z, z(x)> where x defines a point in space-time, Z identifies a property, and z(x) defines the particular value of that property at that point."(p. 243). This is analogous to the definition of the *topon* **g** in the present study (see section "Semantic Contraction and Geographic Information Constructs (GIC)" and Table 1.3). Indeed, the topon may be written as a tuple $\langle g, P_n, \{p_n\}(g)\rangle$ where *g* is a granule of 1, 2, or 3d space at time (*chronon*) x_t, P_n is the cumulative set of properties from level n down to level 1, and $\{p_n\}(g)$ is a vector of values for these properties at that topon and chronon. Note that unlike in Goodchild et al. (2007), time is here treated separately from space because of qualitative differences in temporal behaviour from level to level in the dynamic version of the model.

Further connections exist between this framework and the work in Bouquet et al. (2003) on *context* in knowledge representation and reasoning. These authors distinguish two kinds of theories of context that they call *divide-and-conquer* and *compose-and-conquer*. The former implicitly assume some kind of a "global theory of the world" that is subdivided into collections of contexts. The latter assume only local theories, each of which represents a view point of the world. Neighboring local theories are partially compatible but there is no expectation of global consistency. The *pragmatics* view of the model presented here is closely related to just this kind of local theories ("microtheories") and user-oriented viewpoints of the world. While the parallel may not be perfect, the complete, 3-part model appears to synthesize the two contrasting perspectives in Bouquet et al. (2003). It serves as a "global theory" from which specific applications may be derived, while the microtheories generated against the specific context provided by the "pragmatics" view, and realized with information and operations from the syntactic view, may be seen as the local theories. Similar ones will appear in clusters, and others will be far apart, forming a network floating against an unstructured, multidimensional space of professional and broader societal purposes and understandings (Fig. 1.2). It is evident that the relative positions of the microtheories in that space are largely a function of user perspective, which determines what constitutes relevant similarity by selecting the properties of interest in the context of each specific application.

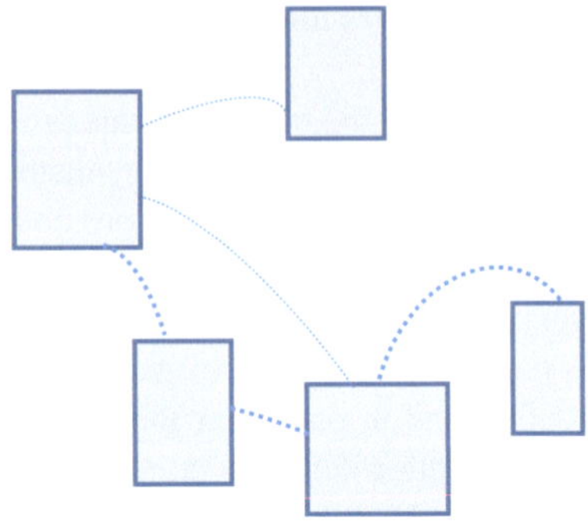

Fig. 1.2 Micro-ontologies as local theories of context (adapted from Bouquet et al. 2003). Distances among micro-ontologies correspond to degrees of similarity, and are a function of user interests

Broadening the Perspective

Micro-ontologies as Designs

A further characteristic of the model is the integration of at least five different semantic "zones," from the stage "of an animal at the vanishing point of intelligence" (Peirce 1878/1998), (L1), to the point where full-fledged mathematical spatial analysis may be conducted (L2 and L3), to declarative knowledge of spatiotemporal entities (L4 and L5), to design (L6 and L7), up to the often unacknowledged influences of societal factors as the source of user motivations (L7). Practical implications for the development of micro-ontologies may derive from the connections established in the model between (scientific) analysis and design, since these two complementary approaches to the world are often contrasted as opposites that are difficult to reconcile (Couclelis 2009b).

Viewing micro-ontologies as engineered (designed) artifacts implies that these must always express their intended users' purposes and functions. By integrating (a) the design perspective, characterized by goals, objectives, and a synthetic, problem-solving stance, with (b) the analytic perspective of science focused on observation, analysis, and representation, the model presented in this paper suggests the broad outlines of a procedure for generating task-appropriate micro-ontologies. The steps below, which correspond to the layers of the semantic core of the model, are along the lines of the design process in general.

- Clarify the purpose of the representation relative to the entity of interest
- Decide on a function or functions for the representation (type A)

  If the entity of interest is artificial, determine which aspects of its own purpose and function (type B) should be included in the representation

- Using function as a selection criterion, identify which configuration of (spatial) parts and relations should be represented
- List and name the individual parts and relations of the above configuration

- Determine what information (types of variables) is needed for representing the above in ways appropriate for the chosen function
- Obtain data of appropriate quality, quantity, format, and resolution

  If certain necessary data are not available, revise the design

- Represent in the appropriate geometry
- In practice of course the design process is not linear, and there may be several feedbacks among layers.

Looking Forward

Beyond the reassuring connections with other work, the model has a number of desirable properties. First, it makes little distinction between physical and abstract, natural and artificial, and even between actual and unrealized or imaginary entities to the extent that these are part of purposefully developed representations. This helps avoid several philosophical problems bound to arise in ontologies that purport to directly model the real world. Second, in its role as "micro-ontology generating engine," the model provides one type of answer to the concern that a proliferation of micro-ontologies lacking a common core would lead to knowledge fragmentation. Further implications being investigated concern the notions of time and uncertainty, since these manifest themselves differently on each layer of the semantics view.

Next to these welcome properties there are some major questions. Unlike the D&S theory in Gangemi and Mika (2003), the model does not have the backing of some traditional foundational ontology, and indeed it may not need to. It is not clear at this point what might be required instead for implementation purposes. Partial solutions come to mind. Domain ontologies come close to the "formal, explicit specification of a shared conceptualization" required by Gruber's (1993) definition, though it is clear that "shared conceptualization" can be quite a relative notion even within well-defined scientific or professional communities (Adams and Janowicz 2011). Also, most future users of the semantic web will likely not belong to well-defined professional communities. Similarly, an expanding library of ontology design patterns (Gangemi 2005, Gangemi and Presutti 2010) seems achievable and desirable for addressing the very large number of phenomena whose representation should be more or less uncontroversial. These could be matched to the levels of the framework and expanded or adjusted as needed while retaining internal consistency. Or perhaps the semantic web itself, as a repository of mostly common-sense knowledge, could take the place of a foundational ontology. Recent work on non-axiomatic logic (Wang 2013) might help make such a prospect realistic. Interestingly, the same "compose-and-conquer" imagery (Bouquet et al. 2003) mentioned above for its parallels with the model discussed here, also appears to apply to the semantic web: locally consistent, globally not.

The model's complement could also turn out to be some combination of other theoretical work in geographic information science, such as the data-oriented general theory of geographic information in Goodchild et al. (2007), as briefly discussed above, or the linguistics- and cognition-oriented perspectives in Kuhn (2003), (2009), Scheider (2011), and Scheider and Kuhn (2011). The latter two lines of research are especially promising since they correspond to two of the three layers where the model presented here is most clearly grounded in the empirical world: the layer of data (L2), and the layer of objects that can be identified and named (L4). Some other work, possibly from logic or philosophy, may be found to support the third such empirical layer, that of intentionality or purpose (L7).

Conclusion

This paper presented a new conceptual model that may be suitable for the generation of lightweight ontologies for GIS and other fields, capable of expressing a broad range of different user needs. That model supersedes the framework in Couclelis (2010), which, from the perspective of the current work, was developed exclusively around the semantics of geographic information, and addressed neither the question of the context of ontology use, nor that of measurements and related operations. In its new role as semantic view, the older framework mediates between the measurements provided by the empirical world on the one hand, and the possible context-dependent representations and interpretations of that world, on the other. From within the latter, different agents' perspectives act as filters to select, interpret, manipulate, and reify the information that is relevant to the task at hand, thus generating micro-ontologies or lightweight ontologies.

A basic premise of this research is that ontologies are not about representing the real world, but instead, about enabling the construction of useful and internally consistent representations of the *information about* that world: they are, in effect, meta-theories for the derivation of empirical models in the broad sense, or micro-ontologies in the case of this research. The strategy adopted in this study is thus to focus primarily not on real-world entities and data, but on the realm of purposes and interpretations relating to entities, and the representational and informational requirements resulting from these. Indeed, while the empirical world may be one, models are built through the lenses of a myriad different perspectives. The users' intentionality is treated as the main generator of microtheories tailored to specific professional, scientific, institutional, and other kinds of interests. A procedure here called "semantic contraction" generates seven nesting, semantically distinct levels of representation, starting with user purpose and ending with a vague notion of a space-time, to be clarified in the context of each specific application as a function of each user's representational needs. The approach has several properties, such as: the avoidance of a number of philosophical problems that affect attempts to represent the real world directly; the ability to handle physical, abstract, artificial, natural, and hypothetical entities as needed; an integrated approach to analysis and design; and evidence of connections with formal theories developed in related areas.

This being work in progress, a host of issues remain, among which the questions relating to implementation are the most pressing. The indications of multiple connections with other work in related fields are encouraging. Even more promising is the evidence of convergent thinking among researchers departing from quite different premises, as in Gangemi and Mika (2003), or in Kuhn (2003), (p. 407), where we read: "…how can we reasonably decide on the contents [of information sources], let alone the representations, before specifying their use? … What conceptualisations occur in an application? … How can a common ontology be constructed for them?"

The democratization of the web that accompanied recent major developments in cyber-infrastructure has often been understood in terms of technical user-friendliness: easy to find, easy to download, easy to upload, easy to change, easy to carry around, easy to probe, easy to update, easy to annotate, easy to share. Less attention has been paid to the fact that the momentous broadening (and deepening) of the user base for online resources also calls for greater *conceptual* user-friendliness, that is, for better adaptability of informational resources to the vast array of new needs and perspectives that these resources could potentially serve. The recent growing interest in lightweight ontologies, especially when seen as tools for mining the semantic web from any desired angle, is a major step towards greater web content usability. It is also a major research challenge, calling for new approaches from the ground up, ranging from the very practical to the philosophical. This paper contributes to the discussion towards that goal.

References

Adams BJ, Janowicz K (2011) Constructing geo-ontologies by reification of observation data. ACM SIGSPATIAL GIS. ACM Press, Chicago, pp 309–318

Borgo S, Masolo C (2010) Ontological foundations of DOLCE. In: Poli R, Healy M, Kameas A (eds) Theory and applications of ontology: computer applications. Springer, Dordrecht, pp 279–295

Bouquet P, Ghidini C, Giunchiglia F, Blanzieri E (2003) Theories and uses of context in knowledge representation and reasoning. J Pragmat 35(3):455–484

Couclelis H (1992) People manipulate objects (but cultivate fields): beyond the raster-vector debate in GIS. In: Frank AU, Campari I, Formentini U (eds) Theories and methods of spatio-temporal reasoning in geographic space. Springer, Berlin, pp 65–77

Couclelis H (2009a) Ontology, epistemology, teleology: triangulating geographic information science. In: Navratil G (ed) Research trends in geographic information science. Springer, Berlin, pp 3–16

Couclelis H (2009b) The abduction of geographic information science: transporting spatial reasoning to the realm of purpose and design. In Stewart Hornsby K, Claramunt C, Denis M, Ligozat G (eds), Spatial information theory, 9th international conference COSIT'09, pp 342–356. Springer, Berlin

Couclelis H (2010) Ontologies of geographic information. Int J Geogr Inf Sci 24(12):1785–1809

Frank AU (2003) Ontology for spatio-temporal databases. In: Koubarakis M, Sellis T (eds) Spatio-Temporal databases: the CHOROCHRONOS approach. Springer, Berlin, pp 9–77

Galton A (2001) A formal theory of objects and fields. In: Montello DR (ed) Proceedings, international conference COSIT 2001. Springer, Berlin, pp 458–473

Gangemi A (2005) Ontology design patterns for semantic web content. In: Gil Y, Motta E, Benjamins VR, Musen MA (eds) The semantic web. Springer, Berlin, pp 262–276

Gangemi A, Mika P (2003) Understanding the semantic web through descriptions and situations. In Meersman R, Tari Z, Schmidt D (eds), On the move to meaningful internet systems 2003: CoopIS, DOA, and ODBASE. OTM 2003, pp 689–706. Springer, Berlin
Gangemi A, Presutti V (2010) Towards a pattern science for the semantic web. Semant Web J 0:1–7
Goodchild MF, Yuan M, Cova T (2007) Towards a general theory of geographic representation in GIS. Int J Geogr Inf Sci 21(3):239–260
Gruber TR (1993) A translational approach to portable ontology specifications. Knowl Acquis 5(2):199–220
Guarino N (1995) Formal ontology, conceptual analysis and knowledge representation. Int J Hum Comput Stud 43(5–6):625–640
Janowicz K, Hitzler P (2012) The digital earth as knowledge engine. Editorial. Semant Web 1(1):1–10
Kavouras M, Kokla M (2008) Theories of geographic concepts: ontological approaches to semantic integration. CRC Press, Boca Raton
Kavouras M, Kokla M (2011) Geographic ontologies and society. In: Nyerges TL, Couclelis H, McMaster R (eds) The SAGE handbook of GIS and society. SAGE, London, pp 46–66
Kjenstad K (2006) On the integration of object-based models and field-based model in GIS. Int J Geogr Inf Sci 20(5):491–509
Kuhn W (2003) Semantic reference systems. Int J Geogr Inf Sci 17(5):405–409
Kuhn W (2009) Semantic engineering. In: Navratil G (ed) Research trends in geographic information science. Springer, Berlin, pp 63–76
Pawlak Z, Skowron A (2007) Rudiments of rough sets. Inf Sci 177:3–17
Peirce CS (1878/1998). The order of nature. Pop Sci (1878). Reprinted in Cohen MR (ed.) Chance, love, and logic: philosophical essays, Charles Sander Peirce. University of Nebraska Press, Lincoln
Probst F (2007) Semantic reference systems for observations and measurements. PhD thesis, University of Münster
Robinson EH (2010) An ontological analysis of states: organizations versus legal persons. Appl Ontol 5(2):109–125
Scheider S (2011) Grounding geographic information in perceptual operations. PhD thesis, University of Münster
Scheider S, Kuhn W (2011) Finite relativistic geometry grounded in perceptual operations. In: Egenhofer MJ, Giudice NA, Moratz R, Worboys MF (eds) Proceedings of 11th international conference COSIT'11. Springer, Berlin, pp 371–390
Scheider S, Probst F, Janowicz K (2010) Constructing bodies and their qualities from observations. In AM. Galton and R. Mizoguchi (eds) *Formal Ontology in Information Systems* (pp 131–144): Proceedings of the Sixth International Conference, FOIS 2010
Searle JR (1983) Intentionality: an essay in the philosophy of mind. Cambridge University Press, Cambridge
Simon HA (1969) The sciences of the artificial. The MIT Press, Cambridge
Smith B (2010) Ontological realism: a methodology for coordinated evolution of scientific ontologies. Appl Ontol 5(3–4):139–188
The Economist (2010) In quite a state: defining what makes a country. Print edition, 62–63 Apr 2010
Voudouris V (2010) Towards a unifying formalisation of geospatial representation: the object-field model with uncertainty and semantics. Int J Geogr Inf Sci 24(12):1811–1828
Wang Pei (2013) Non-axiomatic logic: a model of intelligent reasoning. World Scientific, Singapore

Chapter 2
Some Philosophical Issues Regarding Geometric Modeling for Geographic Information and Knowledge Systems

Robert Laurini

Abstract It is common to state the importance of geometry in geographic information systems. But with the advent of the knowledge society, it is important to revisit some philosophical aspects that were traditionally the backbone of GIS. Indeed, the necessity to build robust systems for automatic geographic reasoning implies that several issues must be reexamined, especially due to the existence of new types of sensors which continuously measure some phenomena under interest: two sensors which will measure any phenomenon will give values a little bit different for various reasons. And we have to integrate those aspects. Now, with the appearance of new systems based on geographic knowledge, mathematic modeling of reality is again in the critical path of research. In this paper, we will examine rapidly the philosophical background of the common modeling used in GIS and try to propose new directions especially in the vision of requirements for geographic knowledge systems.

Keywords GIS philosophy · Computational geometry · Query processing · Geographic knowledge

Introduction

From a conceptual point of view, the use of mathematics in geographic information systems is fundamental for several reasons. Etymologically speaking, the word geometry means, in Ancient Greek γεωμετρία, the measurement of the Earth, or the measurement of terrains. It could be traced to Babylonians and Egyptians for surveying, two millennia BC. Remember that in geography (γεωγραφία), the Greek work γραφία means both drawing and writing about the Earth.

In other words, geometry can be considered as a basis and an ancestor of Geographic Information Systems. But, now with the use of information technologies

R. Laurini (✉)
KSI, Overland Park, KS, USA
e-mail: roberto.laurini@gmail.com

University of Lyon, Lyon, France

© Springer Nature Switzerland AG 2019

T. Tambassi (ed.), *The Philosophy of GIS*, Springer Geography,
https://doi.org/10.1007/978-3-030-16829-2_2

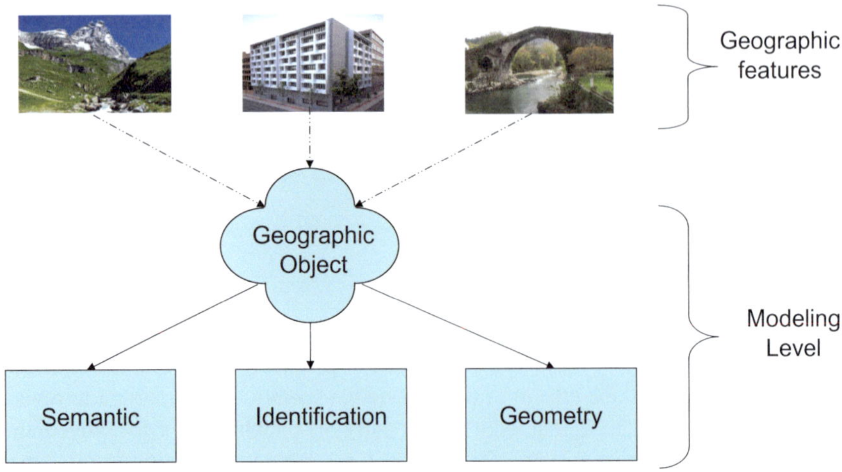

Fig. 2.1 The various points of view for modeling geographic features

and especially of knowledge engineering, it is important to revisit the mathematical backgrounds of geoprocessing.

In a previous book (Laurini 2017a), I have shown that any geographic feature can be modeled according to three points of views (Fig. 2.1):

- A **semantic point of view**, that is, the nature of the feature (road, river, mountain, etc.) which can be categorized or even subcategorized; the role of geographic ontologies is to offer adequate and relevant categorizations; do not forget that those categorizations have different cultural backgrounds leading to various categorizations in various languages;
- An **identification point of view**, that is, the name of the feature (Germany, Lady Liberty, Eiffel Tower, California, etc.); again, some linguistic aspects can be considered since features can have different names in different languages; for instance, the city of "Venice", Italy, is also known as "Venezia", "Venise", and "Venedig", respectively, in Italian, French, and German;
- A **geometric point of view** corresponding to the shape and location of the feature.

In this chapter, only the philosophical issues of the geometric point of view will be developed, more precisely, will be examined the mathematical backgrounds for modeling the characteristics of geographic features and their relationships. Then, the importance of spatial analysis will be developed leading to automatic geographic reasoning.

About Geometric Modeling of Geographic Features

It is common to state that geometric objects can be modeled at 0D (point), 1D (lines), 2D (areas) and 3D (volumes), and we can add the temporal dimension when necessary. But, is anybody able to show 0D and even 1D objects on the Earth? As far as I know, we can only mention the North and the South poles as points. Regarding lines, the so-called linear objects as roads and rivers have some width, leading to consider them as areas. However, concerning geodesy, equator, parallels, and meridians are theoretical lines which have no visible reality: they could be considered as geographic objects, but not as geographic features.

Remember that before the expression "GIS", the specialists were speaking about "Computer-Aided Cartography." The idea was to easily generate new maps by changing scales and colors. But the data were essentially coming from digitizing existing planar maps. Then, some other acquisition devices were created and used. But the original background is still on planar maps, planar plotters, and planar screens. In other words, the world looks planar. After, it becomes apparent that the major interest was not only to generate maps, but to store geographic information in the so-called GIS.

As for practical problems, for instance, at urban level, it is acceptable to consider a flat world. But what is the limit?

Earth Rotundity

The Earth is not a sphere; indeed, due to centrifugal force, it must be considered as an ellipsoid whose larger radius is located at the Equator. Now it is modeled as a geoid.

Consider now a projection plane tangent to the Earth at point C, and a point P in the Earth. There are several methods to project a spherical point onto a plane. Figure 2.2 shows two examples of projections, T as the intersection of an Earth radius and the projection plane, and S as an orthogonal projection. It is easy to see that the position of T is given, respectively, by $Rtg(\theta)$ and of S by $Rsin(\theta)$ in which R is the radius of the Earth. Remember that the real distance of P to C is $R\theta$. By doing so, we commit an error, respectively $\varepsilon_T = |R\theta - Rtg(\theta)|$ and or $\varepsilon_S = |R\theta - Rsin(\theta)|$. Using the first terms of Taylor series, namely, $tg(\theta) = \theta - \theta^3/3 + 2\theta^5/15 + \cdots$ and $sin(\theta) = \theta - \theta^3/6 + \theta^5/120 + \cdots$ we can compute an approximation of the errors; or conversely, by accepting any error level, we can compute the size of a place in which the planar assumption is acceptable. So, $\varepsilon_T = R\theta^3/3$ and $\varepsilon_S = R\theta^3/6$ and consequently $\theta = \sqrt[3]{3\varepsilon_T/R}$ or $\theta = \sqrt[3]{6\varepsilon_S/R}$.

Now, based on those models, let's consider a squared place and let's, for instance, tolerate an error of 1 cm. Respectively, we obtain 20 and 25 km wide places; and for 1 mm, we get 9.4 km and 11.8 km.

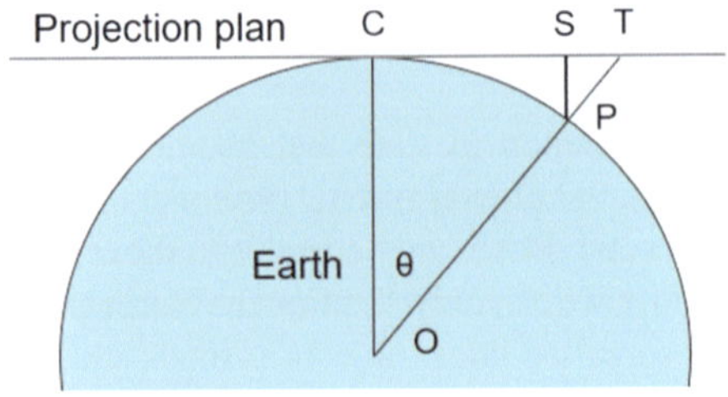

Fig. 2.2 Example of some projections

As a conclusion, we can state that if we consider a town or city with such a size, and accept those error limits, we can consider a flat world for this city or region. But, outside, Earth rotundity must be taken into account. Of course, based on other models of projections, the results can be a little bit different.

Regarding GIS, if the jurisdiction of the owner is small enough, the planar assumption is valid.

A second aspect is the origin of the measurement unit. Historically speaking, length measurement seems to be linked to the human body. A traditional tale tells the story of Henry I (1100–1135) who decreed that the yard should be "the distance from the tip of the King's nose to the end of his outstretched thumb".[1] So, this unit varied in time. To avoid such problems, during the eighteenth century, some people argue to have a more rationalistic definition , for instance, based on the Earth. During the French Revolution, in 1791, the French National Assembly decided in favor of a standard that would be one ten-millionth part of a quarter of the earth's circumference: the well-known meter unit was thus created. But now, some more precise definition is standardized, based on physics. Finally, from a philosophical point of view, the origin of length measurement was based on a geographic reasoning.

Characteristics of Geographic Features

According to Prolegomenon #1 (3D +T objects): "*All existing objects are tridimensional and can have temporal evolution; lower dimensions (0D, 1D and 2D) are only used for modeling (in databases) and visualization (in cartography)*" (Laurini 2017a). Indeed, rivers can change their bed, mountains can have earth slides, continents move (continental drift), roads can be enlarged, buildings can be demolished, and so on. As a consequence, their mathematical model must integrate also a temporal dimension. But in this chapter, this important problem will not be addressed.

[1] http://www.npl.co.uk/educate-explore/factsheets/history-of-length-measurement/.

Euclidean and Spherical Geometry

Remember that what we usually call 2D geometry or rather Euclidean geometry is not a true 2D geometry since the eyes are located in the third dimension. Suppose your eye is really located in the 2D plane, everything will be modeled by segments, whether it is a line or an area. As a consequence, the objects are no more distinguishable. As previously shown, Euclidean planar geometry can be applied in smaller zone, whereas spherical geometry in larger places, perhaps with the help of projection.

Remember that, by definition, cadastral data only use x and y coordinates, never the elevation. In other, it means that the cadastral surface of a sloping terrain is not the soil surface, but its projection unto a plane, say the horizontal surface. For instance, a cliff has a horizontal surface close to zero, whereas the vertical surface can be important. But, now, some countries intend to build 3D cadasters.

About Points, Lines, and Areas in Our World

At the beginning of geoprocessing, several models of polygons were in competition; among them, some were based on a set of points and other on a set of segments. In this regard, the role of SORSA (Segment-Oriented Referencing System Association, then Spatially Oriented Referencing System Association) was important in the 70s by promoting the segment-oriented approach essentially because it was a more efficient model to deal with consistent tessellations. Alas, the object orientation mode in the 80s imposes a definition of geographic object as delimited by a set of points, and it was the basis of the OGC geographic model (See Fig. 2.3). But this approach leads to difficulties to tessellations and overall to secure for their consistency.

As previously told, in the nature, points and lines do not exist. In a previous work, I have introduced the concept of ribbons (Laurini 2014) which can be defined as a line with a width. Remember that in mathematics, lines have no width, as ribbons can be seemed as a good starting point for modeling roads, rivers, etc. But rivers and roads can have various widths. So, several types of ribbons must be defined, rectangular ribbons, curved ribbons, and loose ribbons as given Fig. 2.4 with an application in urban context in Fig. 2.5.

Crisp Boundaries and Indeterminate Boundaries

Smith and Varzi (2000) distinguish two categories of geographic objects, those whose boundaries are natural, such as island, continents, rivers, and those which were defined by humans, for instance, property line, national boundaries, etc. Using Latin expressions, they call the former, *fiat*, and the latter *bona fide* objects.

But the situation is a little more complex. For instance, some states in the USA have some *fiat* boundaries along rivers and *bona fide* boundaries sometimes defined by parallels or meridians.

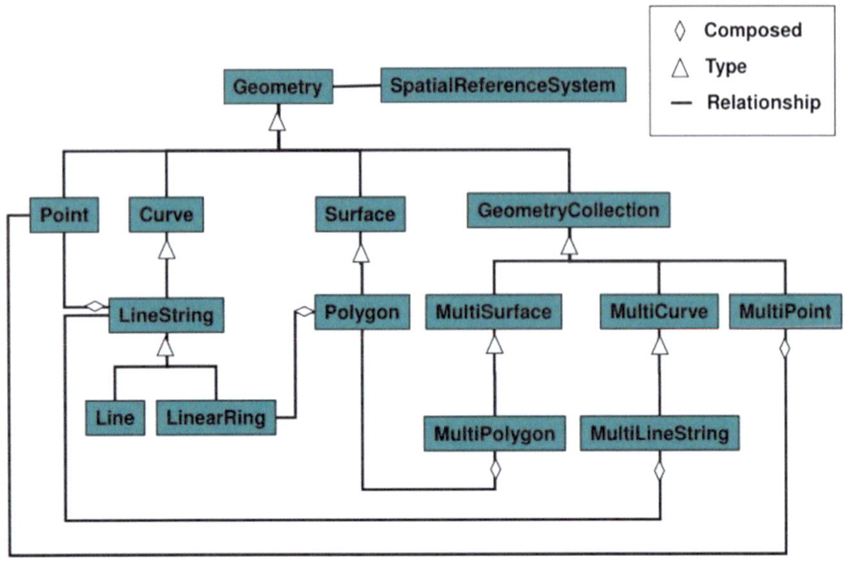

Fig. 2.3 OGC spatial model

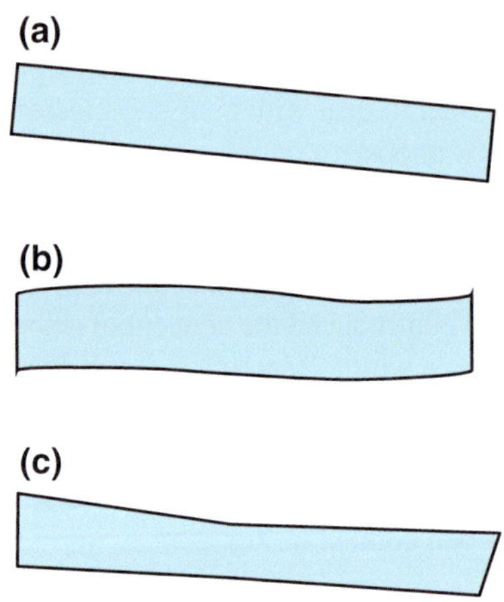

Fig. 2.4 Various types of ribbons. **a** rectangular ribbon; **b** curved ribbon; **c** loose ribbon

Now, consider mountains. Where do they begin, where do they end? Similarly, for deserts, mangroves, etc. Here, fuzzy set theory (Zadeh 1965; Pantazis and Donnay 1998) can help model those objects with indeterminate boundaries: in the very core of the mountains, points can be said to belong 100%, farther 80%, or even 20%. As a consequence, some features can be modeled by means of fuzzy sets (Fig. 2.6).

But the manipulation of fuzzy sets in geography is not so easy. For those objects, fuzzy set theory can be used in which some membership grades can be defined (Fig. 2.7) (Zadeh 1965). An interesting model (Cohn and Gotts 1996) is the "egg-yolk" model with two parts, the core (the yellow part) and the extension, the white

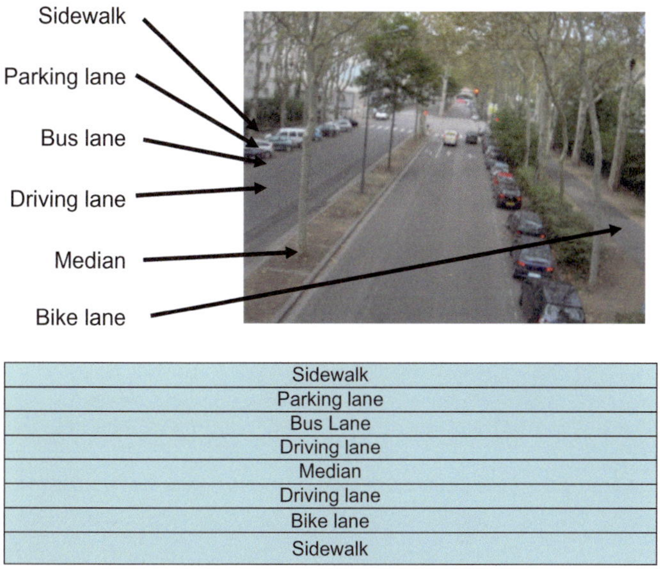

Fig. 2.5 Identification of ribbons in an urban context

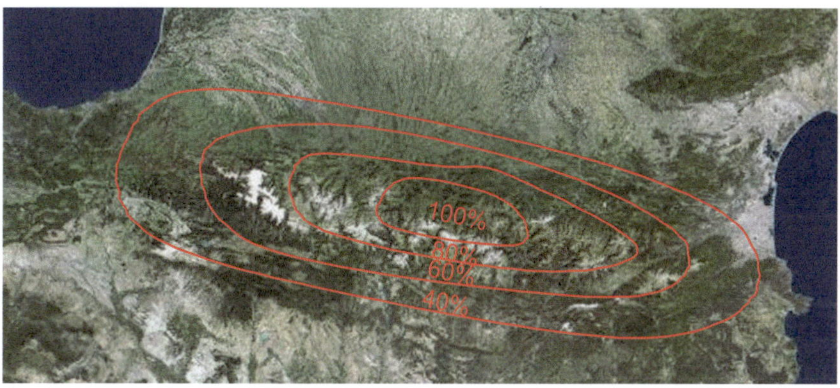

Fig. 2.6 Example of mountains modeled by fuzzy sets

part of the egg. For instance, for a river, the "yolk" represents the minor bed, whereas the "egg" modeled its major bed. Another example is given in Fig. 2.8 in which the mangrove and the jungle are modeled with the egg-yolk representation.

Again, the egg-yolk model can be used to model ribbons: so, a fuzzy ribbon can have a wider ribbon including a narrower ribbon. This model can be applied to modeling rivers each of them with their minor bed and major bed as exemplified in Fig. 2.9.

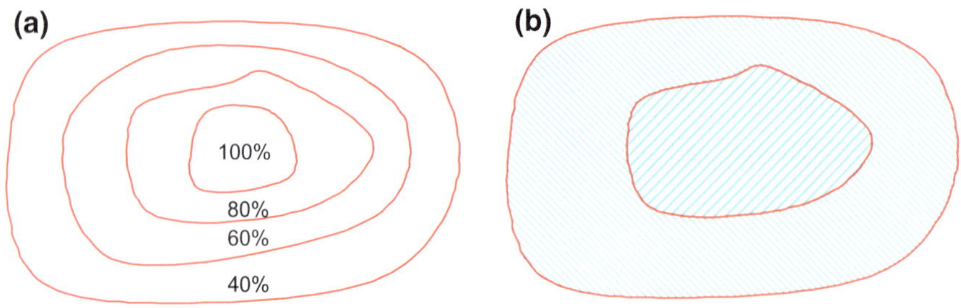

Fig. 2.7 Fuzzy geographic object. **a** Different membership grades. **b** The egg-yolk representation

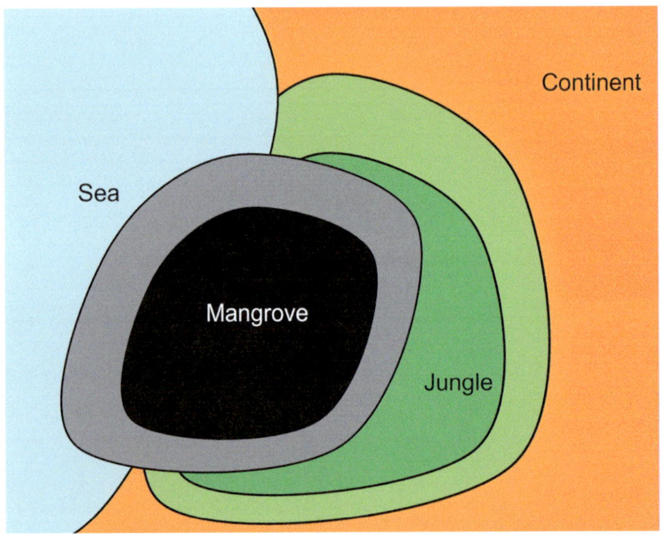

Fig. 2.8 Fuzzy geographic features

Fig. 2.9 River modeled with the egg-yolk representation, emphasizing the minor bed and the major bed

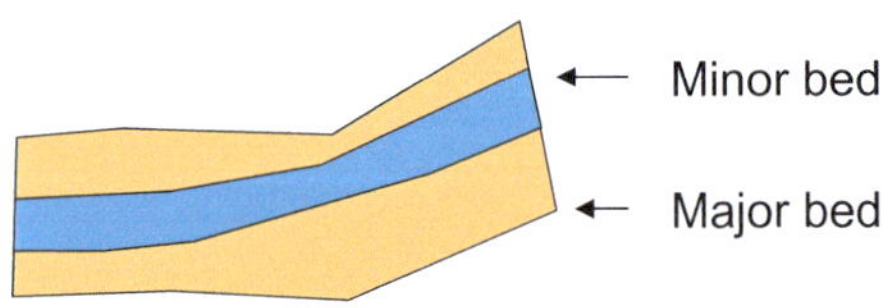

Multiple Representations

Now, continue for example to consider a river (Fig. 2.10). Various mathematical models can be assigned depending on the context. In cartography, this is generally a line; for navigation, an area and a volume. And considering tributaries, a so-called hydrographic network, or more exactly a hierarchical network can be defined. But if we add canals, this network is a little bit more complex. Moreover, during floods, the river bed is enlarged leading the existence of minor bed and major beds. As a

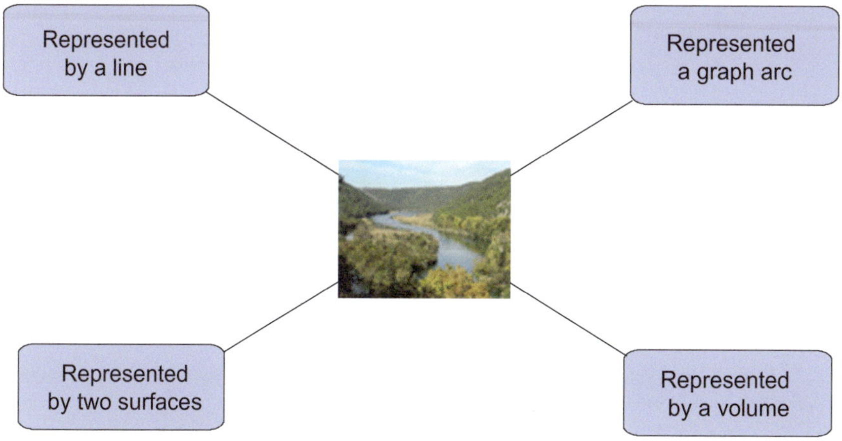

Fig. 2.10 Multiple representations of a river

consequence, multiple mathematical representations of the same geographic feature can be offered.

Now, consider an island, and ask several people to measure this island. Some, a little bit lazy, will only give the coordinates of a 100-point polygon, whereas other can give 1000 or more points. Even if two people are giving both 100 points, those points could be different. Finally, we get different measures for the same feature, leading to different computer geographic objects.

Suppose that we have now two different databases in which we have in both the representation of the same feature, namely, O_1 and O_2. Obviously, the stored data will be different, but representing the same object. If the stored measures are exactly the same, we can easily write $O_1 = O_2$. But, if only one data is a little bit different, this equality does not hold anymore! Whereas they represent the same feature! As a consequence, equality is a too strong concept for comparing geographic objects, and we need to weaken it.

Between two objects, A and B, a homology relation is a relation that is reflexive and symmetric which defines a sort of similarity between two things. Let us denote ₪ this relation, so that one can write A ₪ B. Therefore, both (A ₪ B) and (B ₪ A) hold. Remark that an equivalence relation ($\equiv$) is a homology relation, which is also transitive. Figure 2.11a illustrates two homologous polygons and Fig. 2.11b two homologous lines.

As an example, consider two persons (Mark and Fran) taking the coordinates of the same island as exemplified in Fig. 2.12. When we overlay those two polygons, we can see that they are not exactly the same, thus a relation of homology must hold between those two polygons.

Another aspect must be taken into consideration, namely linked to specifications. Back to the previous example, consider that one is measuring at low tide and the other at high tide: if the island has no beaches, no additional problem, but when

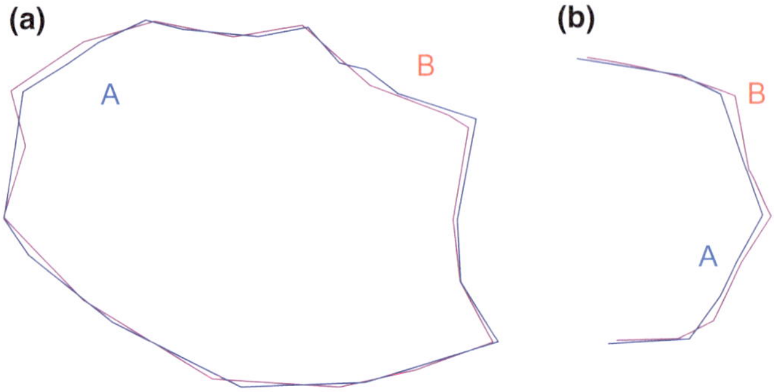

Fig. 2.11 Two homologous geometric descriptions of the same geographic objects. **a** Polygon homology. **b** Line homology

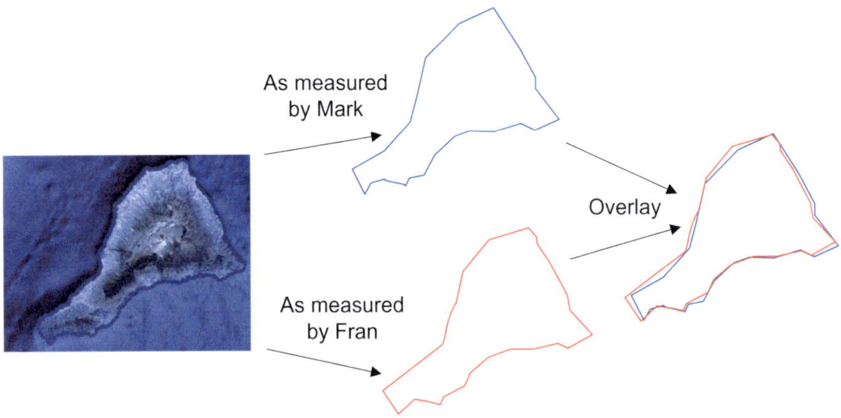

Fig. 2.12 Two descriptions of the same island, one by Mark and one by Fran and their overlay

there are several beaches, the size of the foreshore will change the measured data. Similarly, consider a house (Fig. 2.13); in the cadaster, measures are taken in the soil (soil coordinates), whereas in aerial photo, there are roof coordinates for which the difference can reach easily one foot.

Remember fractal geometry. In his seminal book, Mandelbrot (1967) shows that, depending on the length of a yardstick, the perimeter of an island varies: the shorter the yardstick, the longer the perimeter. And finally, the perimeter tends toward infinity whereas the area converges to a finite value.

As a consequence, depending on the application, the necessary number of points of a polygon vary. Suppose that for cartography, the limit is decided 0.1 mm, and we have a 10 m yardstick for measuring an island. The threshold scale s_0 will be 0.1 mm/10 m = 10^{-5} for using all acquired points. If the yardstick is shorter, we can remove points because they become useless for this application.

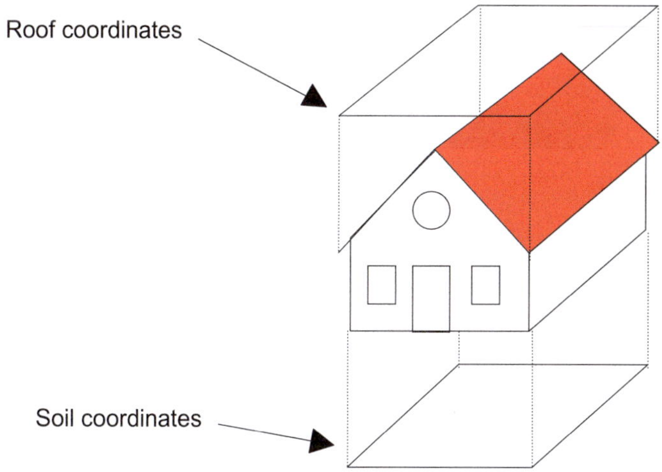

Fig. 2.13 For a house, soil, and roof coordinates

Connected and Non-connected Objects

A connected object is defined by the following: if we consider any couple of points belonging to this object, there is a path inside the object. Otherwise, it is said non-connected.

Naively speaking, a geometric object is defined with different sub-objects and holes. Consider Italy: it has several islands (Sicily, Sardinia, etc.) and two holes (San Marino and Vatican), so this country is said non-connected from a geometric point of view.

Back to the discourse about *fiat* and *bona fide* objects, we can state the following: fiat objects are always connected whereas bona fide can be non-connected, such as countries having islands or containing various pieces. Another example is the USA, with Puerto Rico and Alaska and other smaller territories throughout the world. In contrast, an archipelago is a set of isolated islands with a single name.

About Polygons and Tessellations

There are several ways to encode a polygon, and all of them have consequences regarding tessellations. Remind that each polygon may be defined either as a set of points connected by segments of lines or by a set of segments. Let call them, respectively, point-oriented polygons and segment-oriented polygons.

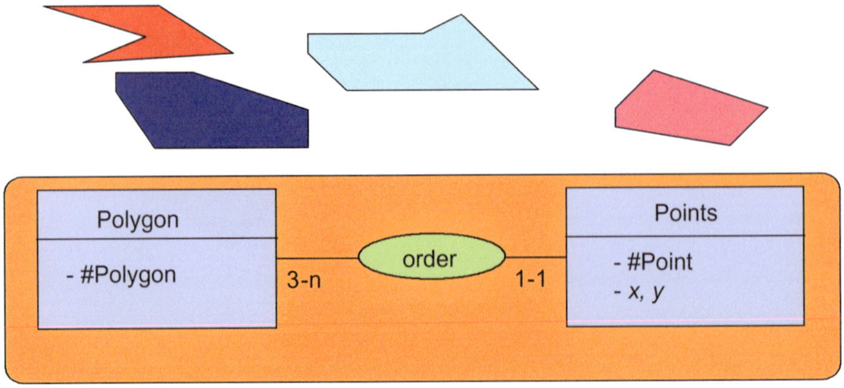

Fig. 2.14 Description of isolated polygons by a set of points by means of the entity–relationship model

Point-Oriented Polygons

A very simple way to encode polygons is to give an ordered set points, for instance, either in the clockwise order, or in the trigonometric order. As this representation is excellent for a non-connected polygon, there are difficulties for connected polygons. For example, consider again Italy; in addition to the main body of this country, we need additional polygons for islands such as Sicily, Sardinia, etc., and for the two holes, i.e., San Marino and Vatican City.

This representation is very common in GIS, but the great disadvantage is the difficulty to tackle consistent tessellations especially due to errors forming the so-called sliver polygons. An example is given in Fig. 2.14 using the entity–relationship approach (See Laurini and Thompson 1992) for more details. So, for checking the consistency of tessellations consisting of point-oriented polygons is a very complex task.

The more common way to store the coordinates is to consider a new abstract data type generally named "geometry". Anyhow, in the well-known standard OGC model,[2] this representation was chosen.

Segment-Oriented Polygons

In this representation, a polygon is seen as an unordered set of segments, each segment being limited by two points. Among advantages, we can see that there are no problems regarding islands and holes. But one of the disadvantages is that the reconstruction of a polygon is a little more complex task. See Fig. 2.15. Another advantage is that the tessellations are consistent.

[2] http://www.opengeospatial.org/.

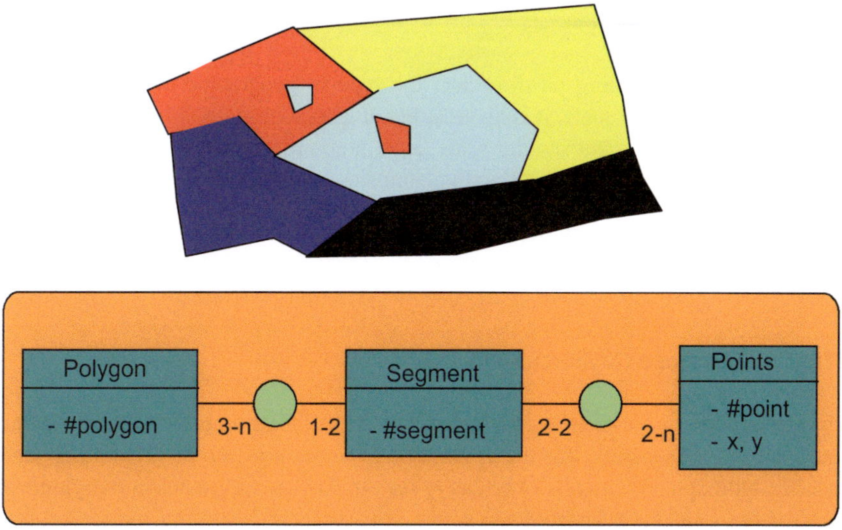

Fig. 2.15 Description of a tessellation of non-connected polygons with the segment orientation

Conclusions About Polygons

The OGC standards regarding geographic object modeling, based on point representations, have pushed to solved efficiently many practical problems. But by facing new applications, some models must be revisited. In the past, several other models have been proposed, either based on segments or allowing to manage consistent tessellations more easily.

Another aspect is the well-known Euler–Poincaré formula stating that $V + F = E + \mathrm{S}$ in which

- V is the number of vertices (previously called points),
- F is the number of faces (previously called polygons),
- E is the number of edges (previously called segment), and
- S is the number of disconnected tessellation sub-objects (holes and islands).

This formula can be used as an integrity constraint to state whether the tessellation is correct. Alas, this condition is necessary, but not sufficient because an extra vertex can balance an extra edge. Anyhow, such a formula must be extended to secure tessellations, and overall to take errors into account. In a cadaster, this formula can be written as $P+V=CB+E$ in which CB and P stand respectively for the number of city-blocks and parcels. While this formula is easy to check in the segment-oriented representation, it is very complex for the object-oriented one.

As a conclusion, all the discarded models must be examined again to test whether they can be more powerful for solving new salient problems.

Geometric-Type Mutation

In cartography, depending on the scale, the type of geographic objects can vary. Consider a city, at one scale, it is an area, but at a smaller scale, it becomes a point, and again at a smaller and smaller scale, as it is no more mentioned, it disappears. This can be formalized by the following chains:

- Area ==> Point ==> null object,
- Ribbon ==> Line ==> null object.

What About Shape Grammars

Shape grammars were initially conceptualized by the Italian architect Palladio (albeit not using this expression) for the systematic generation of rooms in buildings and façades, although repetitions can be easily seen in Ancient Egyptians or Babylonians constructions and cultural artifacts.

More generally, shape grammars allow the defining of iterative geometric objects (Stiny 1978, 1980), whereas fractal geometry defines recursive objects. Indeed, repetitions are commonly found in man-made environment. Look, for instance, the models of cities (Fig. 2.16) as defined initially by Hippodamos of Milet, successively refined by L'Enfant for Washington D.C. or Benoît for La Plata in Argentina. Other models do exist (Laurini 2017a, b) for designing populated communities, schools, barracks, hospitals, campuses, etc. (See Halatsch et al. 2008 for an example in Master Planning or Schirmer and Kawagishi 2011).

Beyond the Vector/Raster Debate

Two classes of representations exist, vector or raster. As the vector representation is based on points, lines, etc., the raster representation is based on grids and generally on squared cells organized into rectangular grids: this is the case for satellite images and photographs. The main theoretical problem is that with collections of squared and rectangles, it is impossible to continuously cover the geoid when squared are big. But if you accept the tolerance as explained in section "Euclidean and Spherical Geometry", this becomes possible.

Another important aspect of using the raster model is for the representation of phenomena generally modeled by field orientation (terrains, meteorology, etc.). In other terms, we can state that for any point, a function $f(x, y, z)$ can determine a value, for instance, in meteorology. This aspect can be extended from scalar fields to vector fields.

Remember that a 3D continuous field is governed by Laplace equation:

$$\frac{\partial^2 f}{\partial x^2} + \frac{\partial^2 f}{\partial y^2} + \frac{\partial^2 f}{\partial z^2} = 0$$

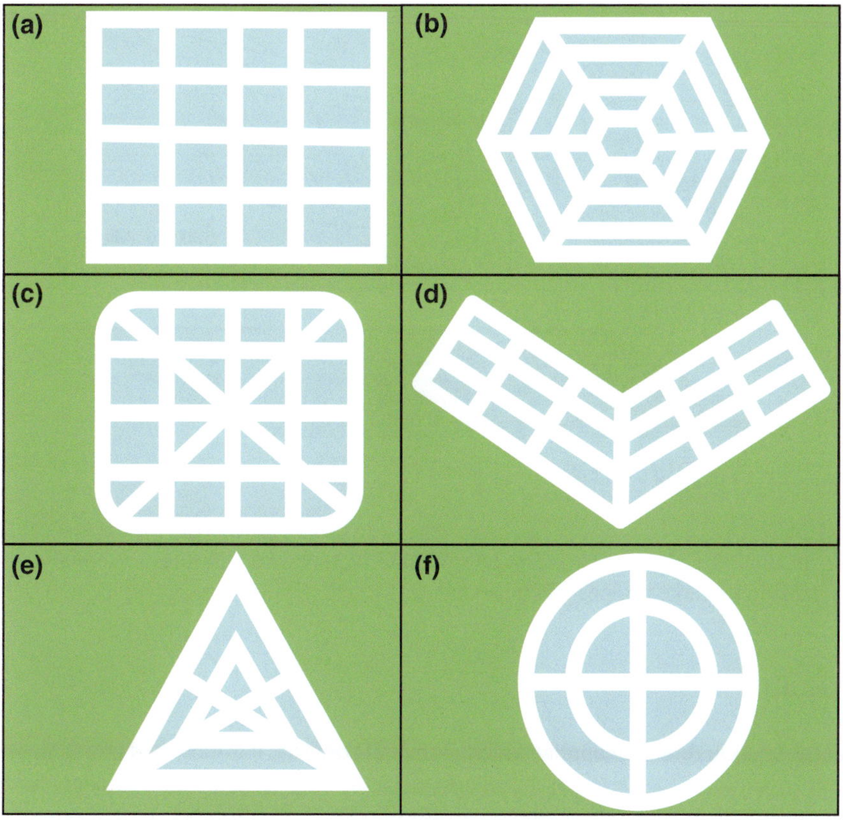

Fig. 2.16 Illustration of spatial patterns in cities. **a** Plan designed by Hippodamos of Milet. **b** Palmanova. **c** La Plata. **d** Brasilia. **e** Erice in Sicily. **f** Beijing's ring roads

Couclelis (1992), by stating "is the world ultimately made up of discrete, indivisible elementary particles, or is it a continuum with different properties at different locations?" showed that field orientation as can be a nice way to model environmental phenomena. Then Kemp (1996) explicated more precisely the variables, and Laurini et al. (2001) proposed a complete model for applications in meteorology. See Fig. 2.17.

Eventually, three GIS models are in competition, the vector model for applications such as cadaster, traffic, etc.; the raster model especially for remote sensing; and finally, the field-oriented approach for modeling environmental phenomena. Anyhow, two tracks of research may be followed:

1. a theoretical path which must tend toward a unique and integrative mathematical model; for the moment being and as far as we know, no clue is given to explore this solution;

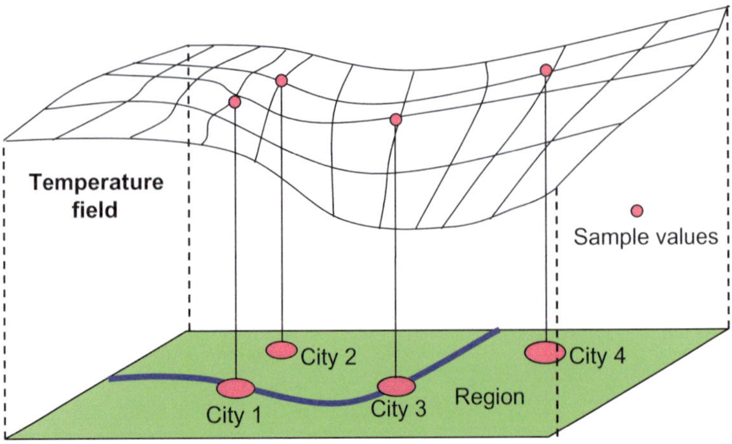

Fig. 2.17 Example of a field-oriented model for temperature

2. pragmatic solutions based on interoperability software products, essentially by trying to make bridges when needed.

And 3D

In the continuation of mapping which prefers 2D objects, it is more and more important to study 3D aspects. As an intermediary to sore terrains, 2,5D models have been created in which the altitude z is considered as an attribute. As it is possible to store easily typical terrains and mountains with this solution, it is impossible to model caves and also some cliffs, because to one couple (x, y), there may correspond several z's.

Different methods exist to acquire terrain models. From conventional theodolites, one can measure altitudes of points, for instance, organized along a squared grid as illustrated in Fig. 2.18a, whereas Fig. 2.18b depicts the case of terrain altitudes acquired through a distance laser. In the first case, we deal with point heights and in the second case with "pixel" heights.

Afterward, the resolution must be taken into account. If the resolution is less than 10 cm, the two models appear equivalent; but when the resolution is larger, for instance, 100 m, the situation is totally different leading to "scalp" mountain summits.

Many authors have faced full 3D models. See for instance (Van Oosterom et al. 2008). The objective is not only to offer 3D models of the Earth but also to handle practical 3D models for cities and inside buildings. The present challenge is to offer seamless models for outdoor and indoor applications. For the description of building, the BIM standard (Building Information Modeling)[3] is presently mainstream, but

[3]https://www.nationalbimstandard.org/.

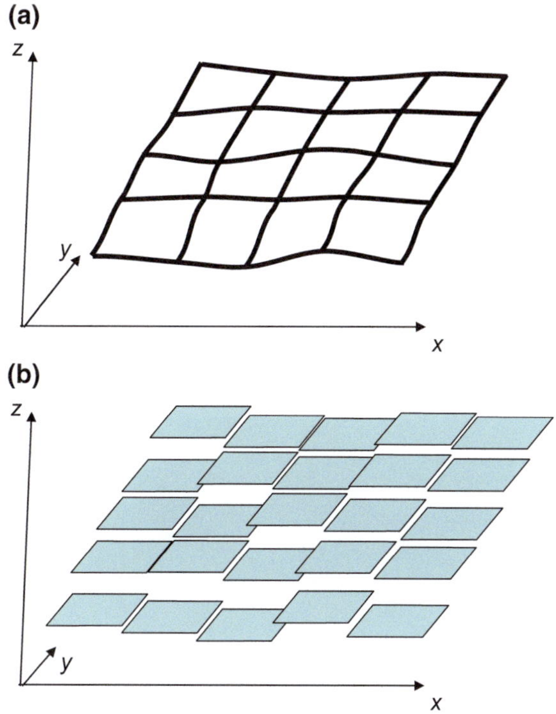

Fig. 2.18 Examples of terrain models. **a** Based on points located along a grid. **b** Based on squares (pixels) whose altitudes come from laser beam

sometimes modified by additional national options. One of the issues is the link with Computer-Aided Design (CAD); as the links with CAD in architecture is obvious, it is not the case with mechanics. When designing a new plant with a lot of robots, the possible connections between architectural and mechanical CAD software products must be envisioned, within a sort of interoperability mechanism.

So, the big question is "what is the limit from a geometric point of view?" Presently, the limit is something like 0.1 mm, but the interoperability of building modeling with the objects located inside buildings can lead to other solutions.

After having reexamined geographic objects and their mathematical modeling, let's revisit topology.

Topological Relations

Topology is the study of the relative positioning of two geometric objects. The word topology comes from the Greek τόπος, place, and λόγος, study. From a mathematical point of view, there are several domains of usage of this term sometimes with different meanings. In this chapter, I don't want to annoy the reader with too complicated notions, but only to show some practical difficulties.

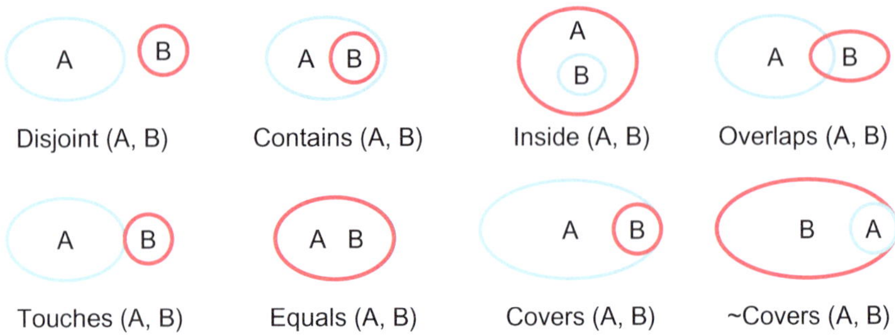

Fig. 2.19 Egenhofer topological relations at 2D (Egenhofer 1994)

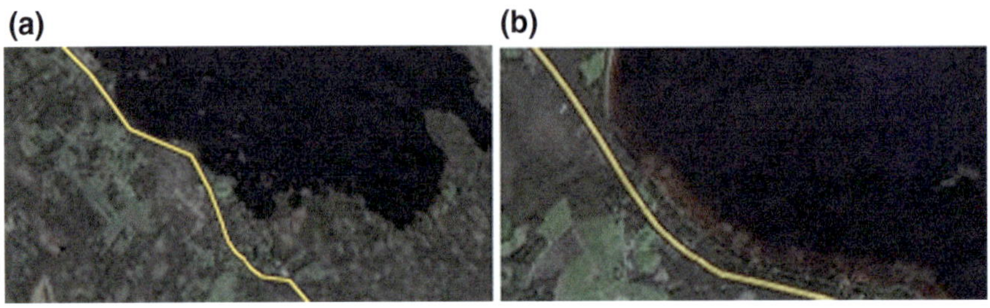

Fig. 2.20 According to scale, the road *Touches* or not the sea

Usually, in GIS, the Egenhofer (Egenhofer and Franzosa 1991; Egenhofer 1994) model is used as exemplified in Fig. 2.19 for defining the relationships between two objects *A* and *B*. Sometimes, the so-called RCC is also in use (Randell et al. 1992), but their characteristics are similar.

But, due to both the difficulties of measuring and the scaling consequence, we need to revisit this model.

Scaling Effects

Consider a road going along the sea; so, implying a *Touches* relation between the road and the sea. But if we carefully consider those features, sometimes there are small beaches between the road and the sea (Fig. 2.20). From a cartographic point of view, the type of relation will vary: indeed, at a scale of 1:1000, the relation is *Disjoint*, whereas at 1:100,000, there is a *Touches*, since the beach is discarded. More generally, the concept of granularity of interest will enlarge the concept of scale.

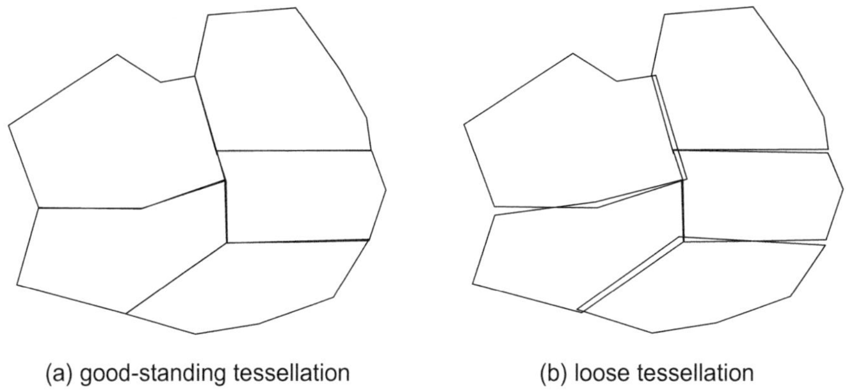

Fig. 2.21 Examples of irregular tessellations. **a** A mathematical good-looking tessellation (valid). **b** A practical tessellation (loose tessellation) with sliver polygons in which errors are voluntarily exaggerated

Measuring Effects

Consider two neighboring countries. Officially, there is a *Touches* relation between them. But since the boundary coordinates were acquired differently, they are different; as a consequence, small overlaps or holes may exist.

Back on Tessellations

Let's continue the reflection considering tessellations, i.e., composed of many polygons. Considering that each of them has errors, all the boundaries between them do not coincide. In Fig. 2.21, two cases are illustrated, the first case (Fig. 2.21a) of a good-looking or consistent tessellation, and in (Fig. 2.21b) the case of a loose tessellations in which boundaries do not coincide. Thus, specific algorithms must be run to correct this tessellation.

Indeed, by applying the OGC standards for polygons, this problem is common.

Another problem in tessellation comes from very small polygons. For instance, in a small European map, it is common not to consider small countries some as Andorra.

In reality, the situation is a little bit more complex, because different cases can occur. Figure 2.22 depicts those cases.

1. The first one is a very small polygon inside a bigger (let's call it Vatican style). In this case, this polygon can be discarded.

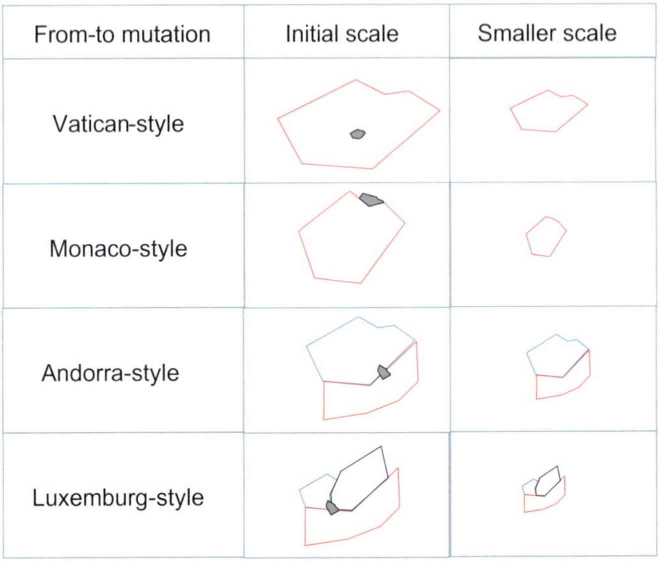

Fig. 2.22 Different cases of polygon disappearance due to scaling in tessellations

2. The second is when the small polygon is located at the boundary of the tessellation (Monaco-style); in this case, it could be either discarded or absorbed by the neighbor.
3. The third one is when the smaller polygon has only two neighbors (Andorra-style). In this case, for instance, each neighbor absorbs 50% of the smaller polygon.
4. The latter case is when the smaller polygon has three neighbors (Luxembourg style). Likewise, it can be split and absorbed.

To conclude this paragraph, let's say that novel algorithms must be designed.

Toward New Topological Relations

Taking these characteristics into account, some new kinds of relations must be defined. Starting from Egenhofer relations, the relation "*Equals*" is already transformed into a homology. When objects are very far, the "*Disjoint*" relation holds on; but when the boundaries are very close or overlapping a little bit, the problem is more getting complex. Figure 2.23 illustrates those new relations.

From a formal point of view, we have the following statements, provided that some thresholds are provided:

1. DISJOINT: If *A* far from *B*, no problem; but if A is very close to *B*, the relation can become TOUCHES.

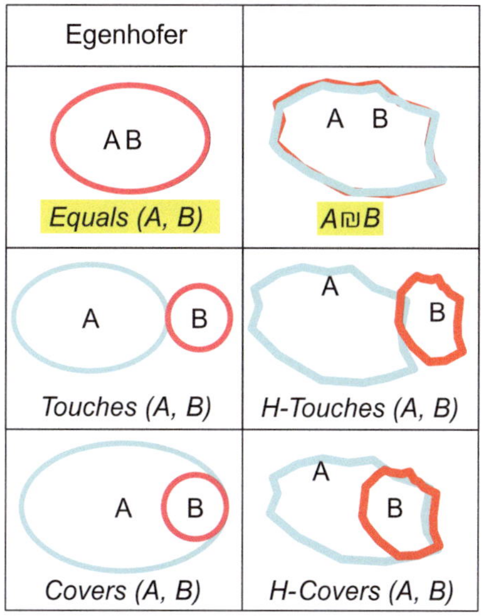

Fig. 2.23 New types of topological relations

2. TOUCHES: taking measuring difficulties into account, this relation practically never holds.
3. OVERLAP: if this is a very small overlap, the new TOUCHES relation can hold.
4. COVERS: taking measuring difficulties into account, this relation practically never holds.
5. CONTAINS: if the distance to the boundary from the smallest object to the biggest object is very small, maybe a COVERS can hold.
6. INSIDE: similar as CONTAINS, but exchanging *A* and *B*.
7. COVERBY: similar as COVERS, but exchanging *A* and *B*.
8. EQUALS: see homology.

Defining exactly those new relations is outside the goal of this chapter; nevertheless, we can say that one or several thresholds must be defined. The big difficulties stay in their values. Are those values unique for any kind of geographic objects, or several must be given? For instance, when comparing parcels and countries the areas of which are very different, a threshold given as a percentage could be of interest; perhaps 3%, less or more. The question is delicate and implies more investigations: they are known as sliver polygons in tessellation.

Now, let examine the problem of encoding geospatial knowledge.

Mathematical Requirements for Automatic Geospatial Reasonings

Two very different aspects must be considered, first, the actual way of encoding geospatial knowledge and second, the requirements for automatic reasoning.

Encoding Geospatial Knowledge

In conventional logic, assertions (or statements) are usually written based on predicates (Boolean expressions). Let us defined A_i and B_j some predicates (Boolean conditions), a rule is expressed as

$$A_1 \wedge A_2 \wedge A_3 \wedge A_n \Rightarrow B_1 \vee B_2 \vee B_3 \vee B_m$$

that is a conjunction of predicates implies a disjunction of other predicates. As the conjunctions are clear from a practical point of view, the disjunctions are not very clear. Does it mean that all B_i must be true, or only some of them? So, a clarification of the semantics is needed.

In Business Intelligence, those assertions are encoded as rules. According to Ross (2011), two types of rules exist, IF-THEN-fact and IF-THEN-action in which the first one corresponds to the creation of a new fact, for instance, the value of an attribute or even the existence of a predicate, and the second for an action to run, maybe by a computer, a human or any kind of machine. But in our case, those statements are too limited. Hence, in Laurini (2017a, b), concerning geoprocessing, new other types of rules can be distinguished:

- IF-THEN-Zone, for the creation of a zone from scratch, for instance, the administrative creation of a recreational park;
- Metarules such as "IF some conditions hold, THEN apply *RuleC*";
- among the latter a special case is located rules such as "IF in the place A, THEN apply *RuleB*", meaning that when we are in the place A, the *RuleB* holds;
- colocation rules the meaning of which is "if something here, then another thing nearby";
- bilocation rules such as "IF something holds in place P, then something else in place Q"; in other domains, this rule is similar to the well-known butterfly effect.

Regarding colocation rules, since in a lot of towns, a church is located at the vicinity of the town hall (say within 500 m), the encoding can be as follows:

$$\begin{gathered} \forall T \in GO, \exists C \in GO, Type(T) = \text{“Town hall”}, Type\,(C) = \text{“Church”} : \\ Distance\,(Centroid(Geom(T)), Centroid\,(Geom(C)) < 500 \\ \Rightarrow Colocation(T, C) \end{gathered}$$

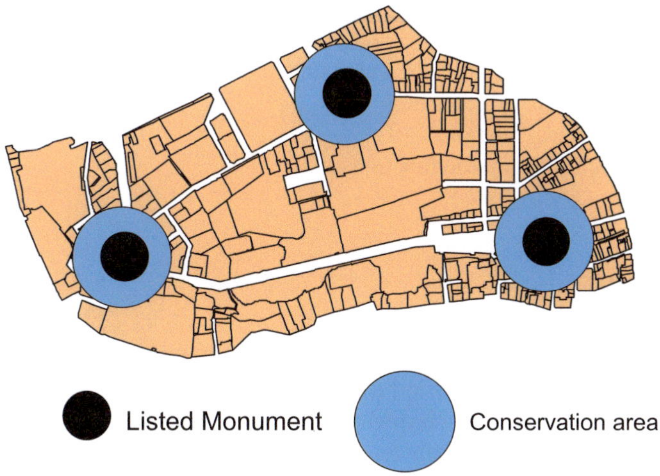

Fig. 2.24 At the vicinity of listed monuments, it is prohibited to construct a new building within the conservation area

In which

- *GO* corresponds to the set of Geographic Objects,
- *Type*, to a type as defined in an ontology,
- *Geom*, a function for the geometry of an object, and *Centroid* for defining its centroid,
- *Distance*, an operator to compute the distance between two points, and
- *Colocation*, a relation of colocation between two objects.

As example in urban planning, let us consider the case of somebody having a project to construct a new building within the conservation area of a listed monument. Practically, in all countries, such new construction is prohibited (Fig. 2.24).

To deny the approbation of this building, the rule can be encoded as follows (distance equals 100 m):

$$\boxed{\begin{array}{c} \forall Terr \in EARTH, \forall B \in PROJECT, \forall M \in GO, \\ Type\,(B) = \text{“Building”}, \\ Type\,(M) = \text{“Listed\_Monument”} : \\ Inside\,(Geom(B), Terr) \\ \wedge Inside(Geom(M), Terr) \\ \wedge Inside(Geom(B), Union(Buffer(Geom(M), 100))) \\ \Rightarrow \\ State\,(B) = \text{“LM\_Denied”} \end{array}}$$

In which

- *Terr* represents the territory onto which this rule applies,
- *PROJECT*, the set of projects,
- *Inside*, a topological relation,
- *Union* and *Buffer*, geometric functions.

Requirements

With the increasing use of artificial intelligence and knowledge engineering in a great variety of domains, it could be interesting to re-examine how conventional mathematical background must be revisited to allow automatic reasoning. In fact, as it was thought that this problem was answered decades ago, mathematical issues must be considered again to be at the edge in the critical path of research in geoprocessing. Among those problems in automatic reasoning, the more salient to be solved are as follows:

- Encoding of geospatial rules and mechanisms to deduce new knowledge chunks or to suggest new actions to be made (Laurini et al. 2016);
- Considering big data in smart cities, create an efficient framework for deep learning, i.e., starting from examples and observations to derive mechanisms for better solutions.

So, a research program must be set to exhibit novel solutions in the following domains:

1. find a unified representation covering all aspects of geographic features (2D, 3D, time, multi-representations, etc.), robust enough to take rid of measurement uncertainties;
2. based on this new model, design powerful algorithms for all conventional geographic queries (point-in-a-polygon, topological queries, graph queries (minimum path, etc.), spatial analysis, what-if models);
3. present and experiment models for encoding geographic rules, able to overpass uncertainties, to allow deduction;
4. innovate in geovisualization; and
5. allow deep learning.

Conclusions

Historically speaking, the future will be on smart cities and territorial intelligence. Facing this evolution, the goal of this chapter was to revisit geometric modeling for geographic applications from a philosophical point of view. It has been shown that, as geometry was born to measure land (i.e., for geoprocessing before the word was existing), practical problems usually issued from data acquisition imply to revisit the main concepts of geometry applied to geoprocessing, and geographic reasoning.

Geographic reasoning demands very robust theories to get rid of difficulties derived from the various devices of data acquisition. In the chapter, it has been shown that comparing geometric entities imply the weakening of conventional mathematical operators (for instance from "=" to "₪"), of topological relations between geographic objects.

Four main remarks must be presented concerning GIS geometry.

The first remark is the necessity to revisit the well-known quadruplet (point, line, area, volume) by integrating the concept of ribbons which is more appropriate to model the so-called GIS linear objects.

The second remark is the necessity to create more robust mechanisms to compare geographic objects based on the multiplicity of representations and measurements.

The third remark concerns topology. Conventional topology (f.i. Egenhofer topology) must be revisited to properly integrate differently measured objects.

To define a library of adequate functions and relations is to be easily handled for automatic reasoning.

To conclude this chapter, let me make some final comments:

1. In the context of smart cities, and especially of geometric reasoning, it is important to revisit geometric modeling to provide more robustness (to get rid of measuring uncertainties) and more independence from geometric representations.
2. The increasing role of sensors in urban context must imply some new real time approaches in spatial analysis.
3. From big data, we need to design new technologies to integrate reasoning from deep learning in geoprocessing applications, not only to understand but also to explore new scenarios of development.

References

Cohn A, Gotts N (1996) The "Egg-Yolk" representation of regions with indeterminate boundaries. In: Burrough P, Frank A (eds) Geographic objects with indeterminate boundaries. Taylor & Francis, London-Bristol (PA)

Couclelis H (1992). People manipulate objects (but cultivate fields): beyond the raster-vector debate in GIS. In: Proceedings of the conference: theories and methods of spatio-temporal reasoning in geographic space, international conference GIS—from space to territory: theories and methods of spatio-temporal reasoning, Pisa, Italy, 21–23 September 1992, pp 65–77

Egenhofer M, Franzosa RD (1991) Point-set topological spatial relations. Int J GIS 5(2):161–174

Egenhofer M (1994) Deriving the combination of binary topological relations. J Vis Lang Comput (JVLC) 5(2):133–149

Halatsch J, Kunze A, Schmitt G (2008) Using shape grammars for master planning. In: Gero JS, Goel AK (eds) Design computing and cognition DCC'08. Proceedings of the third international conference on design computing and cognition. Springer, Dordrecht. https://doi.org/10.1007/978-14020-8728-8_34

Kemp K (1996) Fields as a framework for integrating GIS and environmental process models. Part 2: Specifying field variables. Trans GIS 1(3):235–246

Laurini R (2014) A conceptual framework for geographic knowledge engineering. J Vis Lang Comput 25:2–19. https://doi.org/10.1016/j.jvlc.2013.10.004i
Laurini R (2017a) Geographic knowledge infrastructure: applications for territorial intelligence and smart cities. Elsevier and ISTE, London
Laurini R (2017b) Nature of geographic knowledge bases. In: Fazi S, Mahmoudi K (eds) Handbook of research on geographic information systems applications and advancements. Hershey (PA), IGI-Global
Laurini R, Gordillo S, Mostaccio C (2001) Architecture of a new field-oriented database system for meteorology. In: XXVI general conference of the European geo-physical society, Nice, France, 25–30 March 2001, EGS Newsletter, Number 78
Laurini R, Servigne S, Favetta F (2016) An introduction to geographic rule semantics. In: Proceedings of the 22nd international conference on distributed multimedia systems (DMS 2016), Salerno, Italy, 25–26 Nov 2016. Published by Knowledge Systems Institute
Laurini R, Thompson D (1992) Fundamentals of spatial information systems. Academic Press, London
Mandelbrot B (1967) How long is the coast of Britain? Statistical self-similarity and fractional dimension. Science 156(3775):636–638. https://doi.org/10.1126/science.156.3775.636
Pantazis D, Donnay JP (1998) Objets géographiques à limites indéterminées. Modélisation et Intégration dans un modèle conceptuel de données. Revue Internationale de Géomatique 7(2):159–186
Randell DA, Cui Z, Cohn AG (1992) A spatial logic based on regions and connection. In: Nebel B, Swartout W, Rich C (eds) Proceedings of 3rd international conference on knowledge representation and reasoning. Morgan Kaufmann, Los Altos (CA)
Ross RG (2011) More on the If-Then format for expressing business rules: questions and answers. Bus Rules J 12(4). http://www.BRCommun2002ity.com/a2011/b588.html
Schirmer P, Kawagishi N (2011) Using shape grammars as a rule-based approach in urban planning—a report on practice. In: 29th conference on education in computer aided architectural design in Europe (eCAADe 2011), Ljubljana, Slovenia, Sept 21–24
Smith B, Varzi AC (2000) Fiat and bona fide boundaries. Philos Phenom Res 60:401–420
Stiny G (1980) Introduction to shape and shape grammars. Environ Plan 7(3):343–351
Stiny G, Mitchell WJ (1978) The palladian grammar. Environ Plann B 5:5–18
Van Oosterom P, Zlatanova S, Penninga F, Fendel EM (eds) (2008) Advances in 3D geoinformation systems. Lecture notes in geoinformation and cartography. Springer, Berlin-Heidelberg
Zadeh LA (1965) Fuzzy sets. Inf Control 8(3):338–353

Chapter 3
A Philosophical Perspective on Linguistic Paradigms in GIScience

Qingyun Du

Abstract To date, the prevailing carto-linguistics studies have taken either macrocosmic or social perspectives. In this paper, the research into the linguistic paradigms is concerned with methodology; it regards the representation of spatial information as an analogy to language and establishes phonetic, semantic and syntactic theories of spatial information. There are at least two reasons for this analogy: one is the wide acceptance of carto-semiotics and carto-linguistics, and the other is that we still need a unified paradigm for the disciplines of digital mapping and GIScience. A more methodologically oriented linguistic paradigm could potentially fill this gap. In this paper, taking the viewpoints of ontology, methodology and a qualitative approach, we construct a conceptual model of the internal linguistic structure of spatial information based on phonetics, semantics and syntax. We believe that this methodological approach to the carto-linguistic paradigm will enhance its implementation rather than weaken its influence on digital cartography and GIScience because it lends these fields a perspective that integrates the 'morphology, meaning and structure' aspects of spatial information.

Keywords Philosophical perspective · Linguistic paradigm · Carto-linguistics

Introduction

Similar to other computational systems, such as database systems, GIS lacks a sound paradigm and a conceptual model. In contrast to the deep research interests in computational implementations, ontological concerns with spatial information are lacking, which in turn results in a mismatch between the formal model and human spatial cognition and inhibits progress in spatial data handling (Chrisman et al. 1989; Openshaw 1990). In contrast, contemporary linguistics has achieved great successes after flourishing for more than a hundred years. Human beings are approaching an era of

Q. Du (✉)
Laboratory of GIS, Ministry of Education, School of Resource and Environmental Science, Wuhan University, Wuhan, China
e-mail: qydu@whu.edu.cn

© Springer Nature Switzerland AG 2019

T. Tambassi (ed.), *The Philosophy of GIS*, Springer Geography,
https://doi.org/10.1007/978-3-030-16829-2_3

ubiquitous machine translation, and natural language understanding lies at the heart of Artificial Intelligence. Contemporary linguistics has implications beyond being a stand-alone discipline: it is becoming a universal methodology that is applied widely in both the natural and social sciences, including such fields as cognitive science, sociology, philosophy, psychology, computer science, and many others. As a major channel of human communication, the progress with linguistics and computers will certainly provide models for automatically addressing geographic information.

A Philosophical Perspective in Theoretical Cartography and GIScience

Cartography has a long history as a science that addresses the representation, communication and exploration of spatial knowledge. Cartography is primarily concerned with map making and map use, and it has evolved considerably since the introduction of computational technologies. GIS first became prevalent in the 1960s and progressed further, into GIScience, around the 1990s. The characteristics of information integration mean that GIS goes far beyond mapmaking or even pure spatial knowledge. However, cartography is still important because spatial knowledge plays an exclusive role in GIS. The two main components of cartography are maps and people, while in GIS, various types of map data (e.g. digital maps, images, attributes, multimedia, etc.), machine components (computers, networks, software) and people (GIS designers, GIS users, society at large) are the key players. In GIS, machines function to replace aspects such as paper-based recording and to augment human spatial thinking ability. Machines, along with computation technology, prevailed in these aspects and have undergone rapid development for decades; hence, other aspects of cartography have been somehow overlooked and lagged behind.

Traditionally, theoretical cartography has been closely related to different schools of philosophy. In fact, it is difficult to even list all the different philosophical approaches to theoretical cartography. Epistemology is concerned with human cognitive and aesthetic aspects as they relate to map symbols, colours, patterns and layouts. Semiotics is concerned with conceptual model of maps and their surroundings. Metaphysics was also adopted to build a theory of cartography (e.g. meta-cartography), and ontology has also played a role. To date, semiotics seems to prevail in the theoretical cartographic community because it has the ability to both explain and assign value to the methodology. In contrast, ontology is becoming popular in GIScience via information science and knowledge engineering. Philosophical perspectives always dominate in cartographic and GIScience research.

Maps and cartography have a history nearly as ancient as that of humans, while GIS has existed for only the past few decades, and GIScience has an even shorter history. GIScience has learned much from previous efforts in theoretical cartography. To some extent, maps are to GIS as theoretical cartography is to GIScience. The analogy is both meaningful and significant. Cartography addresses not only maps but also map users; it considers the human mind and cognition. GIScience addresses spatial data associated with geographic meaning and knowledge and involves the ability to process and interpret them. This comparison extends to aspects such as the social space of a map vs. the cyberspace of a geographic information service.

Advances in the Philosophical Aspects of GIScience

Ontology, epistemology and linguistics are three main branches that have developed in philosophy. These aspects represent different approaches for understanding the universe and humanity: ontology is concerned with existence, epistemology with knowledge, and linguistics with existence via knowledge.

Ontology

Ontology is the earliest philosophical branch concerned with being and existence; the general rules and constructs in the universe; the science of what is; the types and structures of objects; and their properties, events, processes and relations in every area of reality. Ontology generally inspires the descriptions of reality, even the representation of reality in our minds (i.e. our conceptualization of the world).

In GIScience, ontology can help in building better models of geographic space, including upper level ontologies, domain ontologies and task ontologies. Ontology is becoming a knowledge framework in information science.

Apart from a universal ontology, ontologies often change from time to time, place to place, domain to domain and task to task. An ontology can be seen as a common communal conceptualization. An ontology cannot be built in an intuitive or empirical way. Cartography can help build better geographic ontologies because a map itself is a conceptual system that represents human cognition of the geographic space at different historical points, cultures, locations and domains and for different purposes. This map spectrum is the sole geographic ontological base represented by map language. To mine geographic ontology from different maps could be a feasible way to build a more intelligent GIS.

Epistemology

Epistemology is related to our cognitive ability and our knowledge system. A mental map is the best result of the epistemological process. Epistemology, in turn, reflects the difficulty of understanding existence when leaving our cognition aside. We look inward to ourselves to determine what we can understand, how we can represent and organize the reality in our mental world. Epistemology is strongly connected with aesthetic and psychological processes.

In GIScience, epistemology is closely related to algorithmic and analytical models. Cartography has a long history of research on human perception and cognition regarding map symbols, colours and spatial patterns, and it is strongly connected with aesthetic and psychological processes. Map creation and map reading are two inverse cognitive processes that encode and decode spatial knowledge, respectively, mediated by cartographic representations. The interaction between a human and map, incorporated based on the geographic circumstances in which the map was made and used, is a perfect model for communicating knowledge.

Linguistics (Semiotics)

In information science, linguistic paradigms play a main role in connecting ontology and epistemology because ontology stems from conceptualization by epistemological processes and requires language to formalize it—i.e. language represents observations of reality (ontology) through the filter of our minds (epistemology). Maps, which are a type of spatial language or semiotic system, have dual functions; they both record our knowledge about the world and provide knowledge about the world. A map as a cognitive model, as a communication channel, as a spatial index tool and as a spatial analytical tool reflects its linguistic function perfectly. Syntactics, semantics and pragmatics are three aspects of the language system that provide different methodological approaches by which we can perform in-depth investigations of the structure of a language system.

A map archive is a knowledge base that provides not only descriptive knowledge but also declarative and procedural knowledge.

Thus, considering GIS as a linguistic system is a more useful concept than considering it as a database or set of geographic objects. Using a linguistic perspective, GIS will become more structural, meaningful and useful. Numerous linguistic approaches developed over the more than 100-year history of modern linguistics can be applied to geographic data; these include phonetics, phonology, morphology, syntax, semantics, psycho-linguistics, sociolinguistics, historical linguistics, mathematical linguistics, applied linguistics, visual linguistics and many others.

Various Approaches to Linguistic Paradigms in GIScience

The linguistic paradigm is not new in geographic information-related disciplines. The most important is map language theory, which stemmed from Bertin's retinal variables (Bertin 1967) and is one of the three most popular cartographic theories. The map language paradigm regards maps as analogues of either natural language or semiotic systems through which the association of content and expression or referent or that of communication and interpreter can be investigated (Blaut 1954; Decay 1970; Schlichtmann 1979, 1985, 1994, 1999; Head 1999). The main concern of this approach regards conceptual models of maps and their digital forms (Schlichtmann 1999).

Another linguistic paradigm approach stems from the viewpoint of cognition, which reaches back to Russell's idea that the structure of language corresponds somehow to the structure of the objective world. The typical example of this approach was advanced by initiative 2 of NCGIA—a linguistic aspect of spatial relations (Mark 1988, 1991; Mark et al. 1989; Frank and Mark 1991; Mark and Frank 1992; Egenhofer and Shariff 1998).

A third approach that is still popular in disciplines such as pattern recognition and digital image processing and their descendants in cartography and GIS (Youngmann 1978; Taketa 1979; Nyerges 1991; Du 1997, 1998) takes a more methodological viewpoint.

Although linguistic paradigms are widely accepted in geographic information-related circles, a complete and genuine linguistics-oriented investigation is absent, especially from a microcosmic viewpoint. Thus, the research into the linguistic paradigm in this chapter is primarily concerned with methodology at the microcosmic level, which considers the representation of geographic information as an analogy to natural language to construct a phonetic, semantic and syntactic theory of geographic information as a two-dimensional graphic language.

In contemporary linguistics, different schools have their own terminology. However, they do have something in common. The following is the essential acceptance of the language system.

- A language system is composed of three structures, i.e. phonetics, syntax and semantics. Among these three independent structures, phonetics and semantics are the two 'poles' of the symbol unit of language. All language components consist of these two poles.
- However, the two poles of phonetics and semantics cannot form language components by themselves; they are only the elements of language components. Only when combined by these two elements can language components such as lexical and syntactic components be formed.
- Lexics and syntax are not two poles; they are two levels of language symbol.

These linguistic principles form the basis for constructing a linguistic system, which is embedded with widely utilized linguistic terminology such as phonetics, semantics, lexics and syntax, and in which each element and component possesses a more complicated internal structure.

Spatial Phonetics and Phonology

In Linguistics, phonetics addresses the physical characteristics of language—the sound in verbal language and the 'stroke' of a character in written language. In a graphic representation of geographic information, the physical characteristics are carried by graphic ink strokes rather than sound; thus, research into the physical characteristics of graphics is analogous to a phonetic analysis of a spatial information language.

Physical Characteristics of Geographic Information

In theoretical cartography, concern with the physical characteristics of a map began with Bertin's retinal variable theory. Many map semiotic and linguistic scholars regard retinal variables as the minimally distinct features (DF) of map symbols.

Dimensionality is another physical characteristic of spatial information. Points, lines, area and volume are four geometrical components with different dimensions. Among these, point s are indivisible, while the others are extended components that can be further subdivided.

As the representation of geographic reality, the appearance of a geographic object is the full presentation of its physical characteristics. In fact, the appearance is dominated by three factors: geographic reality, cognitive restrictions and geometric rules.

Phoneme and Phoneme Combinations in Geographic Information

Phonetic analysis of language is the analysis of language's physical characteristics through the application of linguistic rules. Although geographic reality has a continuous distribution over space, humans generally tend to discretize the essentially continuous phenomena.

Phonetic syncopation based on retinal variables such as distinctive features is the most important procedure in phonetic analysis. Different syncopation methods stem from different disciplines and have different purposes, among which are geometric and psychological syncopation in pattern recognition, mathematic morphology and cartographic syncopation.

A phoneme is the minimal (and meaningless) phonetic element that functions as a signal that allows the meaning of higher language units to be distinguished. In geographic information, retinal variables and dimensionality have the potential to act as DFs that comprise phoneme-based syncopation. After analysing eight retinal variables, we find that they are still not atomic enough for phonetic analysis, among these, only colour, brightness, size and orientation are truly atomic features. Regarding dimensionality, points and line segments have two different dimensions that cannot be further divided (i.e. dividing a line segment results in two line segments). Thus, we can see that point or line segments with variable colour, brightness, size and orientation form the phonemes of the graphic language system. All other phonetic aspects of the linguistic component are combinations of these phonemes.

The power of graphic symbol expression comes from the various combinations of the phonemes, as well as allophones of a given phoneme. Combining these phonemes results in minimal meaningful units (morphemes) or in still-meaningless units (syllables). In the context of geographic information, creating phoneme combinations has traditionally been the domain of map symbol design.

Suprasegmental Phonemes in Geographic Information

Just as the characteristics of phonetics underly the entire speech stream in natural language, the graphic characteristics of spatial information are reflected in the entire distribution pattern. After a morpheme is constructed that can carry certain semantics, general phonetic analyses at the levels of morphemes, words, phrases or sentences is the task of suprasegmental phonology. Here, geographic reality and its conceptual system take the position of geometrical rules as the main functional factors. For example, the curve of a line or a minimal simple polygon of an area could be suprasegmental phonemes.

Spatial Semantics

Semantics addresses the meaning of language, and in geographic information, the semantics lie in the association of geographic information and geographic reality. Two main semantic theories exist, one of which is concerned with the semantics within the language system (i.e. how words and sentences are mutually connected) and the other with semantics outside of the language system (i.e. how words and sentences are connected with their referent objects and processes).

Component Analysis of Semantics in Geographic Information

A component analysis of spatial information involves decomposing the 'meaning' into 'semantic features', in other words, the analysis of the internal semantic features that reflect the objective essence by an empirical understanding of geographic features and phenomena. The inner semantic features are independent of a concrete language context; they are associated with the ontology of reality that the words express.

Thus far, in linguistics, an efficient theory for semantic feature extraction is lacking: empiricism and introspection are two main approaches. Regarding geographic information, the corresponding geographic ontology is much confined in comparison with the reality corresponding to natural language; thus ontological investigation is an efficient approach.

Borgo et al. (1996) and Guarino (1997) proposed the concept of 'Ontological Strata' (Fig. 3.1) for the construction of large-scale ontologies. 'Objects have an intentional criterion of identity, in the sense that they are more than mere sums of parts. Within objects, further distinctions can be made according to the identity criteria ascribed to them.' For geographic information, static biological strata have less influence, and mereological and physical strata not only have effects on internal semantic features but also function over the entire semantic structure. We can extract the following categories of semantic features:

- Matter (mereological): water, soil, clay, stone, sand, vegetable, artificial material, etc.
- Appearance (morphological): flowing, standing, naturally curved, regular appearance, dimensionality, etc.
- Size (morphological): large, medium, small, etc.
- Function (functional): transportation, obstruction, inhabitancy, cumulating, tourism, breeding, etc.
- Social class (social): political, economic, cultural, etc.

As an example, we can decompose the following into semantic geographic information features.

- River—[water] + [flowing] + [naturally curved] + [transportation] + [linear];
- Lake—[water] + [standing] + [tourism] + [breeding] + [area];
- Highway—[artificial material] + [transportation] + [constraint curved] + [linear] + [economic];
- Fence—[artificial material] + [obstruction] + [regular] + [linear];
- Building—[artificial material] + [inhabitancy] + [regular appearance] + [area] + [obstruction] + [political and economic meaning];

Of course, using only the identifying criteria of everyday world objects for geographic entities and phenomena has limitations. For geographic concepts such as a

Fig. 3.1 Ontological strata (Guarino 1997)

Static *(a situation)*
Mereological *(an amount of matter)*
Physical
Topological *(a piece of matter)*
Morphological *(a cubic block)*
Functional *(an artefact)*
Biological *(a human body)*
Intentional *(a person or a robot)*
Social *(a company)*

bay, a cape, or political boundaries (Smith and Mark 1998), we still extract their semantic features from a syntactic framework.

Structural Analysis of Semantics in Geographic Information

Structural semantics adopts implicational-lexical relations. Its main argument is that some words are implicitly associated with others in a language system; therefore, the concept is an intrinsic language issue. The meaning of a word is dependent on its position in the lexical system.

According to this argument, semantic relations fall into four categories, i.e. synonymy, hyponymy, meronymy and antonymy (Buitelaar 2001).

Synonymy

Synonymy refers to words with the same or similar meanings. When representing geographic information, we use a strictly artificial language system. The represented objects are first classified, and then, the lexical system is prescribed upon the represented objects; thus, strict synonymy does not exist in such systems. However, if we look more closely at the potential lexical system of geographic information (both analogue and digital), similar semantics with multiple representations are quite common. It includes the following:

- Multiple expressions of the same entity: variations of shape, colour and weight result in multiple expressions of the same entity.
- Multiple expression of the same spatial relation: for example, the different expressions for a highway intersection (Fig. 3.2).

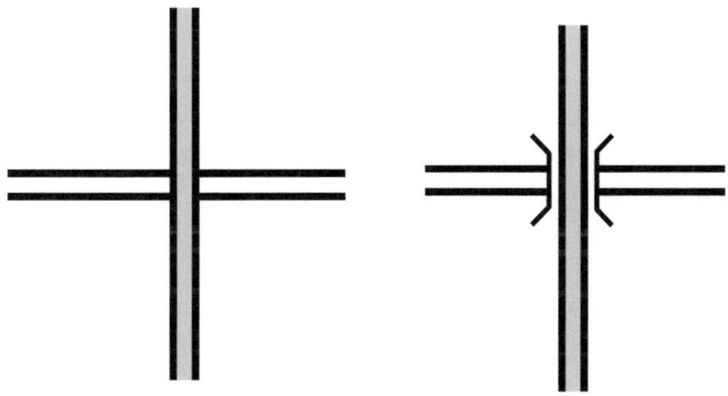

Fig. 3.2 Synonymy of the same spatial relation

Hyponymy

Hyponymy also refers to a similarity relationship but to the similarity of classes. It involves the inclusion of classes. For example, the class 'vehicles' include both motor vehicles and non-motor vehicles.

It can be said that the deep foundation for word hyponymy is the hierarchical structure of concepts. Geographic ontology itself is a hierarchical structure (Smith and Mark 1998); thus, hyponymy has a reason to be an important relation in the lexical relations of geographic information. In terms of expression of geographic concepts, phonetics, lexics and syntax have the effect of constraint. However, the actual number of categories for language units is much smaller than the number of geographic concepts; consequently, they do not have a one-to-one relation. The following situations are possible:

- A geographic concept has no related word; for example, fire does not appear on most map symbol systems;
- One geographic concept has one related word, e.g. chimney, cave, etc.;
- One geographic concept has multiple related words, e.g. rivers can be denoted by single lines, double lines, etc.;
- One geographic concept is a combination of multiple words; e.g. slope and valley;
- Multiple geographical concepts are included in one word; for example, stadium includes the concepts raceway, stand, exit, etc.
- Multiple geographical concepts correspond to one word; e.g. pond and lake have only one corresponding word.

These differences are the essence of representation: an infinite number of concepts can be expressed by a finite number of symbols.

Meronymy

Meronymy refers to the part–whole relation of objects or object classes as represented by words. In the linguistic structure of geographic information, meronymy is especially important for any geographic feature and it is essentially compound, consisting of many parts. Smith and Mark (1998) regard mereology as one of three basic tools for ontological investigations of geographic types, and mereology is actually the main concern of formal ontology in knowledge engineering (Simons 1987; Guarino 1995).

We can find meronymy in a geographic lexic system as follows:

- Some words are composed of many parts by themselves. For instance, a block is actually composed of buildings, streets, squares, grass and so on, and a reservoir is composed of water bodies, boundaries, dams, etc.
- Some words can be defined only in context. For instance, an exit must be defined as a part of a construction, a water boundary must be defined as a part of hydrological features, and so on.

While hyponymy has epistemic attributes, meronymy has strong ontological attributes. Meronymy is more dominated by the internal physical rules of reality.

Antonymy

Antonymy refers to the contrary relations of objects. There are different categories of antonymy, such as gradable antonyms, binary antonyms and relational antonyms. Antonymy is also reflected in geographic information as follows:

- Morphology-driven antonymy—curve/straight, regular/irregular
- Class-driven antonymy—Block/street, sea/land, mountain/plain, valley/ridge, urban/countryside
- Attribute-driven antonymy—Highway/street, long-standing river/seasonal river
- Relation-driven antonymy—above/below, overlaps/overlapped, contains/contained.

Spatial Syntax

Syntax is an important concept from linguistic systems viewpoints such as 'phonetics-lexics-syntax', 'phonetics-syntax-semantics' or 'syntax-semantics-pragmatics', where syntax lies at the heart of the system. The function of syntax is to integrate language units with different physical features and conceptual meanings into higher language units that conform to rules and convey certain meanings.

Similar to phonetic and semantic structures, syntax also has an internal structure. In syntax (grammar), lexis and syntax are the two different units used to build words and sentences, respectively.

Elementary Spatial Relations in Geographic Information

The earth's surface is an infinitely complex system. The linguistic paradigm regards geographic information as a hierarchical structure constructed by multiple levels of language units associated with various combination relations. As in natural language, revealing the internal structure of geographic information must start from the most elementary combination relations.

The types of spatial relations that exist between spatial entities is always a hot topic in GIScience (Egenhofer 1989). The following binary relations that reveal some combination relations from various angles.

Topological Relation

Topological relations refer to those relations that remain unchanged under topological transformations such as shifting, rotating and scaling, which were commonly used in early node-arc-polygon representations in digital maps (ESRI 1995).

The point-set based topological relation (Egenhofer and Franzosa 1991) greatly advanced scientific recognition of topological relations. The 4- and 9-intersection models mathematically, logically, and completely enumerate these topological relations. However, after it was discovered that they are somewhat less restricted in real-world situations, more metrical factors based on the idea 'topology matters, metric refines' were introduced to refine these relations (Egenhofer and Mark 1995; Shariff et al. 1998).

Topological relations are the most stable and yet most important relations in geographic features; fortunately, that they are easy to cognize and store in computation systems. However, more combination relationships of different types must be investigated based on the following reasons.

- First, topological relations are loose relations whose requirements are easy to meet. To further refine the spatial relations, we must seek other types of relationships. In language, some words can be defined within their full context.
- Second, disjoint relations occupy the bulk of spatial relations in the real world. Thus, we also need other types of relations to describe these disjoint entities.
- Third, as highly formalized descriptions, topological relations need semantic and ontological constraints in pragmatic contexts.

Metric Relation

Although topological relations can be considered as having been thoroughly investigated, we have little knowledge about what are roughly called metric relations.

Metric and topological are not opposing concepts; they are two aspects of the same spatially distributed phenomena. Different forms of topological and metric relations hold between any two spatial entities, and both can be computed from their positions. Topological relations provide the qualitative aspect, while metric relations provide other aspects of quantity. The only way to describe metric relation is to approximate graphics as accurately as possible in analogue representations.

Direction relations and distance relations can be defined between any two points, lines or areas. Because metric relations are difficult to process qualitatively way (unlike topological relations), metric relations are neglected in most geographic information research. Some more complicated definitions come from the qualitative definitions of the direction and distance relations and are intended to simulate human qualitative reasoning (Peuquet and Zhan 1987; Papadias and Sellis 1994).

Combinational Qualitative Definition of Metric Relation

In a spatial distribution, we instinctively feel that some spatial relations are stronger than others: it is easy to form new features from some features; for others, it is more difficult; and for some, it is impossible. Thus, spatial relations have a combinatory function from a linguistic viewpoint:

(1) Spatial relations can act as verbs that combine words into spatial propositions. Consider the two propositions in Fig. 3.3: 'The bridge crosses the river' and 'The village is to the east of the river'. Under no circumstances can these be combined into a new proposition.
(2) Language units with the same or similar meaning (synonymy or hyponymy) can form new sentences and can also easily be replaced by new higher language units.
(3) Combination can occur only between those language units with certain spatial relations.
(4) When 2 and 3 meet, the phonetics begin to take effect. The physical feature of the language units constrained the combination. Then, phonetics attribute is so-called Morphology, or Gestalt attributes (Guarino 1997).
(5) Combinations are also constrained by surrounding structures; for example, combining city blocks is constrained by the structure of the street network.

Here, we find that topology plays a less important role than does the building pattern, and metric relations are more obvious here. To create further definitions of the spatial combinations of geographic entities, we need to further refine our spatial information theories, starting with a qualitative definition of metric relation.

Some scholars have noted that topological relationships in geographic domains are not genuinely topological in a topological sense (Smith and Varzi 1997). In fact, the most common topological examples involve shifting, rotating and scaling. Map projections are also included, but most of the projections are below second-order transformations.

Fig. 3.3 Spatial sentence and phrase

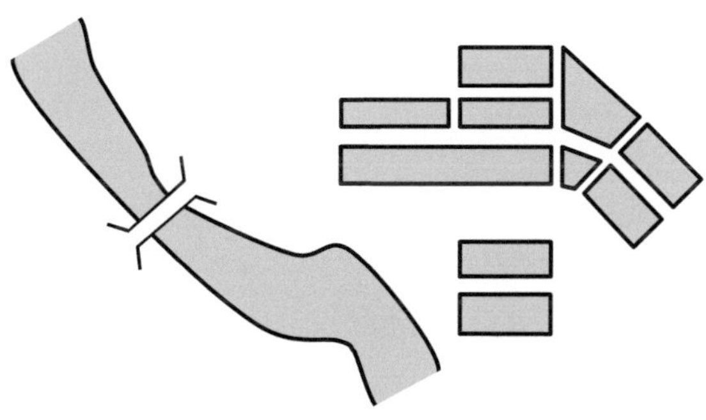

If we suppose the following:

- First-order transformations are limited to shifting, rotating and scaling, and that
- map projections are below second-order, and their spatial scale is relatively small, then, the two values.

Definition 1 $R_l = \begin{cases} d_{34}/d_{12}, \text{ while } d_{12} \geq d_{34} \\ d_{12}/d_{34}, \text{ while } d_{12} < d_{34}; \end{cases}$

where d_{12} and d_{34} are distances between point 1 and point 2 and between point 3 and point 4, respectively, and

Definition 2 $D_a = |a_{12} - a_{34}|$

where a_{12} and a_{34} are the azimuth values between point 1 and point 2 and between point 3 and point 4, respectively, will remain basically unchanged. We can define an intermediate geometry, called combinational qualitative geometry (CQG), based on these two invariants, the length ratio and the difference in azimuth. From these two atomic properties, more qualitative aspects of the metric relation can be deduced.

Distance Combinational Relation

The distance syntagmatic relation is based on the invariant $\boldsymbol{R_l}$. Suppose we have two line segments o_1 and o_2. We can distinguish the following two relations:

Definition 3 if $R_l = 1$, o_1 and o_2 are said to have an *equal-length relation*, notated as o_1 ***El*** o_2.

Definition 4 if $R_l = 1/2$, o_1 and o_2 are said to have a *double-length relation*, notated as o_1 ***Dl*** o_2.

The equal-length relation is a very common combinational relation in geographic space. Its ontological basis is that most spatial features—especially artificial features—have a statistically invariant magnitude range, such as the width of a street or the length of a river branch. In spatial associations, the occurrences of equal-length relations will increase dramatically.

Direction Combinational Relation

The direction combinational relation is based on $\boldsymbol{D_a}$. We can distinguish the following two relations:

Definition 5 $\boldsymbol{D_a} = 0°$. In this case, o_1 and o_2 are said to *have a parallel relation*, notated as o_1 ***Pa*** o_2.

Definition 6 $\boldsymbol{D_a} = 90°$. In this case, o_1 and o_2 are said to have a *perpendicularity relation*, notated as o_1 ***Pp*** o_2.

Parallel and perpendicularity relations are also very common relationships in geographic space. Their ontological basis is that the direction relation belongs to a circular measurement level, which has a much smaller variance range than does length. Second, the human ability to discern tiny angular differences is limited. Third, inherent physical rules result in more parallel and perpendicular distributive patterns in geographic space. Fourth, gestalt rules also influence artificial constructive features.

Composition of Combinational Qualitative Relations (CQRs)

The composition of CQRs with other spatial information results in a dramatic increase in the number of spatial relations. Generally, suppose we have two objects o_1 and o_2 that meet relations ***R*** and ***S***, respectively. Their sum is denoted as ***R*** + ***S***, thus

$$o_1(\boldsymbol{R}+\boldsymbol{S})o_2 \equiv o_1\boldsymbol{R}o_2 \vee o_1\boldsymbol{S}o_2 \text{ (where } \vee \text{ means "or")}$$

in atomic relations will result in a 2^n sum of relations. Among these, the products of ***R*** and ***S*** are defined as follows:

$$o_1(\boldsymbol{R}\circ\boldsymbol{S})o_2 \equiv o_1\boldsymbol{R}o_2 \wedge o_1\boldsymbol{S}o_2 \text{ where } \vee \text{ is "and"}$$

Simple Composition of CQRs with Topological Relation $\mathbf{M_l}$.

Noting those topological relations in which two linear features with intersected boundaries but null-intersected interior are present as $\mathbf{M_l}$, we can define a simple composition as shown in Fig. 3.4,

Pa ∘ $\mathbf{M_l}$-Based Multiple Composition.

With ***Pa*** ∘ $\mathbf{M_l}$-based multiple composition, we can define colinearity, sequence and direct neighbourhood relations.

Definition 7 For a set of line segments $S = \{o_1, o_2, \ldots, o_n \mid n \geq 3\}$ if at least one $o_j \in S$ results in an o_i (***Pa*** ∘ $\mathbf{M_l}$)o_j for each $o_i \in S$. Here, S is called a line colinearity set. Any number of elements above two have a *line colinearity relation*, notated as ***Cl***.

Definition 8 For a set of points $P = \{p_1, p_2, \ldots, p_n \mid n \geq 3\}$, if the set $S = \{l_{i,j} \mid 1 \leq i < n, 1 \leq j < n\}$ composed of the connection of any two elements of set P is a

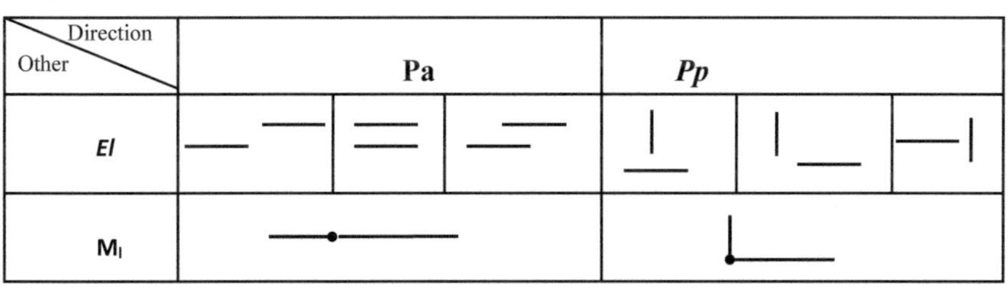

Fig. 3.4 Composition of CQRs with topological relation

line colinearity set, then P is called a point colinearity set. Any number of elements above two have a *point colinearity relation*, denoted as *Cp*.

Definition 9 For any sequence $s_o = \{o_j \mid o_j \in S\}$ that is a subset of a colinearity set S, if $o_j(Pa \circ M_l)o_{j+1}$ and $Xo_j \lozenge Xo_{j+1}$ (where $\lozenge$means '>' or '<') hold for each element o_j and its successive element o_{j+1}, s_o is called a *sequence relation* based on set S and notated as S*p*. In contrast, $Nd = \{(o_{j,}\ o_{j+1})\}$ is a *direct neighbourhood relation*.

CQRs based on metric relations can be regarded as the atomic spatial relations of extended geographic features. They are immediately applicable to artificial and natural features such as block and pipe systems but may vary with when applied to most natural features with real geographic configurations. CQRs are to further lose their constraints.

Linguistic Anamorphosis of Spatial Relations

By defining CQRs, the deficiencies resulting from the too-loose constraints of topological relation and the too-strong constraints of metric relation can be overcome to some extent. CQRs can act as a combinatory mechanism between different levels of language units, acting in concert with topological and metric relationships to form a complete language structure.

Thus far, most GISs have depended heavily on Euclidean geometry. However, much research has revealed that the structure of geographic information is not merely a geometric matter; instead, it is an ontological, epistemic and (natural) linguistic matter. Spatial relations provide us with syntactic rules that can establish linguistic structures that convey real and meaningful information only when integrated with phonetics and semantics.

Phonetic Anamorphosis of Spatial Relations

Compared with natural language systems, one of the particularities of geographic information as a linguistic system is that spatial relations as syntax is strongly related to the physical properties of their associated objects as phonetics, where scale and distance both play important roles.

The representation of spatial relations begins at the phonetic stage. As shown in Fig. 3.5, most spatial relations are presented by the usage of phonetic units.

At the suprasegment stage, based on curves and minimal simple polygons, we can define more spatial relationships that are the phonetic anamorphoses of both topographic relations and CQRs, such as the sequences of the same level of curves, containing different levels of curves and contacting multiple levels of minimal simple polygons as in a double-line river system. In particular, CQRs lose their phonetic constraints to accommodate natural distribution situations, where curve-based parallel, perpendicularity relations and equal-length relations can be defined.

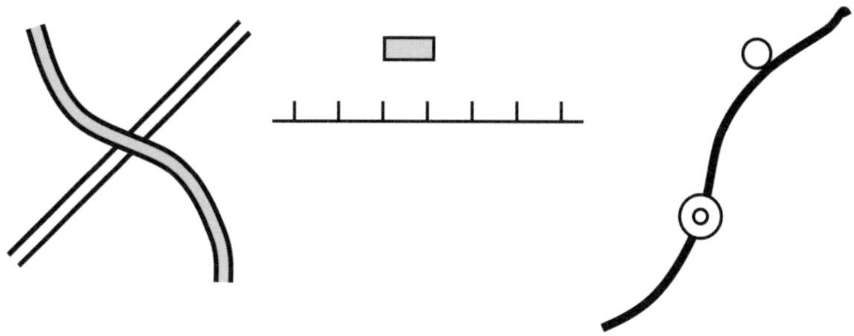

Fig. 3.5 Spatial relation by phonetics

Syntactic Anamorphosis of Spatial Relations

In syntax, a spatial relation has two functional aspects. Based on the context in which it occurs, spatial relations can be grouped into two linguistic units, one is verbs and the other is constrained between language units. As a sentence, geographic information conveys 'what is where', and involves the following sentence types:

- The building is there
- The building is beside the river
- The building is to the north of that building.

In a sentence of type 1, the words are implicitly connected with a reference system that reflects a spatial situation. A sentence of type 2 gives the spatial relation between two objects with different semantics; here, the relative positions of the objects are more important than their absolute positions. A sentence of type 3 gives the spatial relation between two objects with the same semantics. Although they are different points, the objects in this sentence are liable to merge into a new language unit when some neighbourhood condition is satisfied. In this case, the spatial relation tends to be constrained between the language units rather than acting as a verb.

Semantic Anamorphosis of Spatial Relations

From a geometric viewpoint, spatial information has various geometric components. One combination of the components can define only one spatial relation. As in the 9-intersection model, CQRs spatial relations stem more from the perspective of this type of geometric component analysis.

However, the semantics of spatial information also have an ontological basis. Using linguistic approaches, we can investigate ontological semantics based on the participants in relations as follows:

- Dimensionality property: A certain spatial relation requires dimensionality for its participants. As an example, a containing relation requires an area object that acts as the container;

Table 3.1 Semantic anamorphosis of the same spatial relation in different contexts

Spatial relation	Participant	Semantic anamorphosis
Line/Area overlapping	Road, park	Go through
	Bridge, river	Cross over
	Dam, river	Block the water of
	Ferry course, river	Transit the surface of
	Waterfall, river	Fall as part of
Area/Point containing	Reef, sea	Submerged in
	Elevation point, road	On the surface of
	Buoy, river	Floating on
	Cave, mountain	In

- Active/passive: Some spatial relations imply an active/passive relation. For example, lines may be more active than areas in line/area relations: for example, roads pass through parks.
- Vertical relation: Although two-dimensional spatial relations are defined on a plane, the participants do not always exist on a plane. The following situations are possible:
  - On one plane: e.g. a road and an adjacent farm field;
  - Above: e.g. a bridge and a river
  - Below: e.g. a tunnel and a mountain
  - Uncertain: e.g. highways at a crossroads.
- Compatibility: Some spatial relations require the participants to be compatible; otherwise, they are impossible in reality. For example, conceptual objects such as political boundaries can share locations with rivers, while highways cannot.
- Spatial constraint property: Some spatial relations imply spatial constraints and correlations, such as parallel relations.
- Cause property: In spatial relations such as colinearity, equal-length implies artificial construction. Some patterns imply certain natural causes.

Understanding the semantic features of geographic information is a prerequisite for humans to better understand geographic space. These features are linguistic knowledge that stands apart from concrete geographic configurations. Tablc 3.1 provides an example of how certain spatial relationships obtain their semantic anamorphosis in an ontological context.

Conclusion

In this paper, even merely from a microcosmic and technological approach, it is clear that the linguistic paradigm of geographic information has at least four aspects of significance for the development of GIScience, i.e. its paradigm potential, its ontological concern, its methodological guidance and its qualitative approach.

As a paradigm, linguistic research will promote geographic information into an information category with the most cognition and communication functions; thus, it is an excellent theoretical platform for professionals in the fields of cartography, GIS, cognitive science, linguistics, computer science and Artificial Intelligence, and many other domains, and can accommodate research on both macroscopic and microcosmic levels, and from both functional and formal aspects. As a form of methodological guidance, linguistic research will inherit linguistic approaches in their entirety that have been developing for more than a hundred years; these include addressing geographic information in an integrated way, from phonetic, semantic and syntactic aspects—and even further—from pragmatic aspects. The comparative, structural and formal methodology that prevails in linguistics helps to reveal the internal structure of geographic information and can further guide the generation and understanding of geographic information in a computing environment. The philosophical perspective and qualitative approaches have been particularly concerned with linguistics for a long time, and they are increasingly concerned with knowledge engineering and GIS studies in the context of integrating knowledge into computing systems to gain robust computational and reasoning abilities.

In the framework of the linguistic paradigm, we have conducted some technical studies on map symbol recognition (Du 1998) and the linguistic analysis of a multimedia electronic atlas (Du 2001). Further research will include investigations of how physical features and ontological knowledge can be integrated into spatial relations to enhance the automatic understanding of geographic information systems and to evaluate how they can benefit spatial data mining and knowledge discovery. Moreover, the possibility of an investigation of map semiotic systems that include spatial, temporal and cultural coverage with the aim of revealing the evolution of human spatial cognition is also in our sights.

Part of this chapter was read at the International Cartographic Association during the 21st International Cartographic Conference (Duban, August 2003).

References

Bertin J (1967) Sémiologie Graphique. Les diagrammes – les réseaux – les cartes., Paris and Den Haag: Gauthier – Villars and Mouton

Blaut JK (1954) The language of map. Prof Geogrs 6(1):9–11

Borgo S, Guarino N, Masolo C (1996) Stratified ontologies: the case of physical objects. In: Proceedings of ECAI-96 workshop on ontological engineering, Budapest, pp 1–15

Buitelaar P (2001) Lexical semantics in information systems. http://dfki.de/~paulb/lexsem/2001/lexsem.html
Chrisman NR, Cowen DJ, Fischer PF, Goodchild MF, Mark DM (1989) Geographic information systems. In: Gale GL, Willmott CJ (eds) Geography in America. Merrill, Columbus, OH, pp 776–796
Decay MF (1970) Linguistic aspects of maps and geographic information. Ont Geogr 5:71–80
Du QY (1997) Linguistic characteristics and automatic understanding of cartographic information. In: Proceeding of the 18th international cartographic conference, Stockholm, pp 357–364
Du QY (1998) Map language based cartographic understanding. In: Proceeding of '98 ACSM annual conference and exhibition, Baltimore, pp 253–257
Du QY (2001) From micro to macro: an exploratory into the structure of multimedia electronic atlas. In: Proceeding of the 20th international cartographic conference, Beijing
Egenhofer M (1989) A formal definition of binary topological relationships. In: Litwin W, Schek H (eds) Third international conference on foundations of data organization and algorithms (FODO). Lecture notes in computer science, Paris, France, vol 367. Springer, New York, pp 457–472
Egenhofer M, Franzosa R (1991) Point-set topological spatial relations. Int J Geogr Inf Syst 5:161–174
Egenhofer M, Mark DM (1995) Naive geography. In: Frank AU, Kuhn W (eds) Spatial information theory: a theoretical basis for GIS. Proceedings of the second international conference. Springer, Berlin, Heidelberg, pp 1–15
Egenhofer M, Shariff R (1998) Metric details for natural-language spatial relations. ACM Trans Inf Syst 16(4):295–321
ESRI (1995) ARC/INFO: the world's GIS, ESRI White Paper Series
Guarino N (1995) Formal ontology, conceptual analysis and knowledge representation. Int J Hum Comput Stud 43(5/6):625–640
Guarino N (1997) Some organization principle for a unified top-level ontology. In: Working notes of AAAI spring symposium on ontological engineering. AAAI Press, Stanford, CA
Head CG (1999) Warp and woof: carto-semiotics and carto-linguistics in the English-language literature. In: Map semiotics around the world. ICA Publication, pp 15–64
Mark DM (1988) Cognitive and linguistic aspects of geographic space: report on a workshop. NCGIA technical paper 88-3
Mark DM et al (1989) Languages of spatial relations: initiative two specialist meeting report. National Center for Geographic Information and Analysis, Santa Barbara, CA, Report 89-2
Mark DM (1991) Spatial representation: a cognitive view. In: Maguire D et al (eds) Geographic information system: principles and applications. Longman Scientific & Technical, New York, NY
Mark DM, Frank AU (1992) NCGIA initiative 2 "languages of spatial relations" closing report
Mark DM, Smith B, Tversky B (1999) Ontology and geographic objects: an empirical study of cognitive categorization. In: Freksa C, Mark DM (eds) Spatial information theory: a theoretical basis for GIS. Lecture notes in computer sciences. Springer, Berlin, pp 283–298
Nyerges TL (1991) Geographic information abstractions: conceptual clarity for geographic modeling. Environ Plan A 23:1483–1499
Openshaw S (1990) A spatial analysis research strategy for the regional research laboratory initiative. Regional Research Laboratory Initiative Paper 3, Department of Town and Regional Planning, University of Shefield
Papadias D, Sellis T (1994) Qualitative representation of spatial knowledge in two-dimensional space. Very Large Data Bases J (VLDB) Spec Iss Spat Databases 3:479–516
Peuquet D, Zhan C-X (1987) An algorithm to determine the directional relationship between arbitrarily-shaped polygons in a plane. Pattern Recogn 20:65–74
Schlichtmann H (1979) Codes in map communication. Can Cartogr 16(1):81–97
Schlichtmann H (1985) Characteristic traits of the semiotic system 'map symbolism'. Cartogr J 22(1):29–30

Schlichtmann H (1994) Map symbolism revisited—units, order, and contents. In: Pravda J et al (eds) Cartographic thinking and map semiotics. Geogr Slovaca 5:47–62
Schlichtmann H (1999) An inventory of research in map semiotics. In: Map semiotics around the world. ICA Publication, pp 1–14
Shariff AR, Egenhofer M, Mark D (1998) Natural-language spatial relations between linear and areal objects: the topology and metric of English-language terms. Int J Geogr Inf Sci 12(3):215–246
Simons PM (1987) Part: an essay in ontology. Clarendon Press, Oxford
Smith B (1998) Basic tools of formal ontology. In: Guarino N (ed) Formal ontology in information systems. IOS Press, Amsterdam
Smith B, Mark DM (1998) In: Poiker TK, Chrisman N (eds) Ontology and geographic kinds. International Geographical Union, Columbia, pp 308–320
Smith B, Varzi AC (1997) Fiat and bona fide boundaries: towards an ontology of spatially extended objects. In: Hirtle SC, Frank AU (eds) Spatial information theory: a theoretical basis for GIS, vol 1329. Lecture notes in computer science. Springer, Berlin, pp 103–119)
Taketa RA (1979) Structure and meaning in map generalization. Unpublished PhD dissertation, University of Washington, Seattle, WA, USA
Youngmann CE (1978) A linguistic approach to map description. In: First international advanced study symposium on topological data structure for GIS, vol 7. Harvard University

Part II
Representing the Geographic World

Chapter 4
Space, Time and the Representation of Geographical Reality

Antony Galton

Abstract Geographical information science is interesting from a philosophical point of view because the distinctions that its practitioners find themselves compelled to make have important resonances with distinctions that have been proposed in other contexts. An example is the dichotomy between object-based and field-based presentations of geographical data. This paper explores the relationships amongst a set of closely aligned distinctions which have appeared in the literature on both spatial and temporal reasoning in philosophy, cognitive science, geographical science, linguistics and other fields. Any systematic account of such distinctions must inform the construction of a workable ontology for spatio-temporal sciences such as geography.

Keywords Geographical information · Raster versus vector · Field versus object · Continuous versus discrete · Absolute versus relational · Mass versus count · Spatio-temporal representations

Author's Retrospect, 18 Years on

This paper was first published in the journal Topoi *in 2001, and is reprinted here without alteration. Looking back on the paper after 18 years, I find that although my thinking on many of these issues has developed since then, I remain broadly in agreement with what I then wrote. In this prefatory note, I draw attention to a number of points where, if I were to rewrite the paper today, I might do things differently.*

In relation to the raster/vector distinction, discussed in section "The Raster/Vector Distinction in GIS", *I failed to mention the origination of these terms in the context of computer graphics, in which raster graphics generates an image by specifying the colour to be displayed at each pixel of the array, whereas vector graphics generates an image by specifying its components as geometrical figures anchored on points with specified coordinates. This highlights the fact that in its early days GIS struggled to emancipate itself from the idea that its primary concern was with the display of geographical information in cartographic form.*

A. Galton (✉)
Department of Computer Science, University of Exeter, Exeter, UK
e-mail: apgalton@ex.ac.uk

© Springer Nature Switzerland AG 2019

T. Tambassi (ed.), *The Philosophy of GIS*, Springer Geography,
https://doi.org/10.1007/978-3-030-16829-2_4

The discussion of continuous and discrete representations in section "Continuous Versus Discrete Representations" *is undoubtedly oversimplified, presenting the two terms as a straight dichotomy and thereby missing any consideration of more nuanced classifications such as the early nominal/ordinal/interval/ratio sequence of* Stevens (1946), *and, in a specifically geographical context, the series of ten levels introduced by* Chrisman (1998).

Again, the discussion of absolute and relational views of space was oversimple. The positions specified in a field-based model can be relative *positions, for example, specified by coordinates on the earth's surface, and there is no need to posit any kind of 'absolute' space to accommodate this. This is mentioned, albeit rather obliquely, in the penultimate paragraph of* "Absolute vs relational views of space", *but should perhaps have been highlighted more explicitly.*

Finally, it is worth mentioning a couple of areas where I have developed further some ideas that were only embryonic in this paper. Towards the end of "Mass terms vs count terms, matter vs objects", *I briefly discussed the idea of a texture defined by the pattern and distribution of colours (or other attributes) over a surface. This idea, with an extension to the temporal domain, is discussed in more detail in* Galton (2018). *Likewise, in* Galton (2011), *I have more to say on the topic of spatio-temporal analogies discussed in* "Spatio-temporal analogies" *—in particular, the extent and limitations of such analogies.*

I have in front of me on my desk a road atlas of Europe, opened at the page showing the area of southern Europe spanned by the cities of Lyon, Marseille and Torino. What a wealth of information the map provides! At the bottom of the page is a uniform pale blue expanse representing the Mediterranean sea, but above this, where the land begins, all is a riot of words, symbols and patches of colour. The mountainous areas of Savoie and Haute-Provence are picked out by irregular patches of light grey shading giving a suitable impression of uneven topography. Certain individual mountains are indicated by means of little triangles annotated with their height in metres. There are many wooded areas indicated by patches of pale green, and rivers represented by winding blue lines. As it is a road atlas, these natural features merely serve as a background to the enormous number of man-made features that are depicted. There are cities, towns and villages in abundance: most of them are shown as circles of various sizes, but the larger cities are represented as expanses of yellow indicating, at least approximately, the true shape and extent of the built-up area. And there is, of course, an intricate web of roads, from the motorways shown as bold yellow lines bordered in red, to a succession of lesser roads in red, yellow, or white, their thicknesses varying to indicate their relative importance. In addition to all this, of course, there are conventional markings such as administrative boundaries, grid lines, and a great many names.

Everything I see on the map can be described as geographical information. It is obvious that such information comes in many different forms. Representing a town by placing a small circle at a specific location on the map is quite different from showing the extent of woodland by colouring areas of the map green. How should the different kinds of geographical information be classified? Can we divide them into a small number of basic kinds, or are we faced with a plethora of uniquely different sorts of information that resists any attempt at systematisation? If we settle for a few basic kinds, how are these kinds related, and how should we decide which kind to use in any particular case?

Questions such as these lie at the heart of the discipline of Geographical Information Science (GIS), which has emerged in recent decades as a result of the increasing use of computers for storing and processing geographical information. GIS is concerned with both the theory and the practice of the digital representation and manipulation of geographical information. Its practical products are Geographical Information Systems (also called GIS!), which are computer applications providing the ability to handle geographic data in ways deemed desirable by those for whom the systems are designed. These end users can include not just geographers but also map-makers, town planners, environmentalists, travel agencies, biologists, geologists, mining companies and many other groups of people with interests in what can be found—or constructed—on the Earth's surface.

From a philosophical point of view, what is especially interesting about GIS is that the distinctions that its practitioners find themselves compelled to make have important resonances with distinctions that have been explored in other contexts. In this paper, I propose to explore the relationships amongst a number of these distinctions which have appeared in the literature on both spatial and temporal reasoning (in philosophy, cognitive science, geographical science, linguistics, among others). I begin with a pair of distinctions which have played a prominent part in the development of GIS itself: at the implementational level, the distinction between raster-based and vector-based data models, which reflects the distinction between field-based and object-based representations at a more abstract conceptual level. These distinctions are related to two distinctions with a long philosophical and technical pedigree, first the distinction between continuous and discrete phenomena, and second the distinction between absolute and relational concepts. These in turn are related to a distinction that is important for linguistics and the philosophy of language, between mass terms and count terms; at the ontological level, this distinction resurfaces as a distinction between matter (or 'stuff') and objects. I shall also discuss briefly the linguistic distinction between imperfective and perfective verbs, which has been recognised as having important parallels with that between mass and count nouns. At the conceptual level, this leads to a distinction between states (or more generally 'fluents') and events, which are important for any enterprise requiring an analysis of what goes on in time, and in particular for the integration of the time dimension in geography, an issue that is currently the focus of much attention in GIS.

As I have already indicated, and shall show in more detail in the rest of this paper, there is a complex net of interrelations amongst these various dichotomies, which can to some extent be explained on the supposition that they represent the same fundamental idea, surfacing in different contexts, e.g. at the level of language, concepts, or implementations, in relation to either space or time. It is perhaps not unreasonable to hope that by clarifying these interrelationships one might pave the way for the systematic development of spatio-temporal information systems which are both well grounded in theory and workable in practice.

This emphasis on dichotomies is natural in view of the universal predilection of human beings to divide the field of our experience into two, as for example, us and them, here and there, now and then, one and many, or, in more technical contexts qualitative and quantitative, continuous and discrete, analogue and digital, and relative and absolute. Everyone is familiar with the fact that uncritical adherence to this or that dichotomy, even in the face of abundant counter-indications, can result in immense harm, yet without any dichotomies at all human life—and certainly intellectual life—would become, if not impossible, then at least deeply impoverished. One well-known danger is that dichotomies which exhibit a certain degree of correlation have a tendency to become confused, as for example, the continuous/discrete distinction is often confused with analogue/digital, even though they are distinct. While bearing in mind the importance of avoiding this pitfall, we must be equally wary of neglecting genuine similarities and parallelisms where these do exist. In this paper, I have attempted to steer a middle course between these two extremes.

Properties of Spatial Representations

The Raster/Vector Distinction in GIS

This well-known distinction may be briefly summarised as follows. In a raster data model, the objects of interest are a set of locations within some geographical area, and for each attribute of interest (e.g. elevation, land use, rainfall), the value of the attribute is given for each location. In a vector data model, the objects of interest are various geographical entities (e.g. towns, rivers, railways), which are characterised in terms of various attributes, including locations specified by means of coordinates (hence 'vector').

Although it is usual to speak of a distinction between 'raster' and 'vector' data models, Peuquet (1984) pointed out that the term 'raster' is unduly restrictive, suggesting as it does a rectangular or square mesh. The important contrast, for Peuquet, is between *tessellation* models and vector models. These types of models are, in Peuquet's words, 'logical duals' of one another; as she puts it,

> Individual entities become the basic data units for which spatial information is explicitly recorded in vector models. With tessellation models, on the other hand, the basic data unit is a unit of space for which entity information is explicitly recorded. (Peuquet 1984, p. 85)

Sometimes, the distinction is regarded as a purely technical matter of little fundamental significance, and debates concerning which approach is more appropriate are sometimes dismissed with the observation that raster models can be converted to data models and vice versa. In fact this observation misses the mark in two ways: first, it is not strictly true, since in general the conversion between one or other format must involve loss of information; and second, as Helen Couclelis put it,

> the technical question of the most appropriate data structure for the representation of geographical phenomena begs the philosophical question of the most appropriate conceptualization of the geographic world. (Couclelis 1992, p. 65)

The debate, if such it is, can be conducted at several different levels. Worboys (1995) describes raster and vector as *GIS implementation approaches*, and distinguishes these from a more fundamental distinction between field-based and object-based *model types*. These inhabit a world more akin to the conceptualization level adverted to by Couclelis. In fact, by generalising from strict raster models (based on rectangular grids) to tessellation models in which the underlying spatial units may take almost any form, so long as they exhaustively partition the space of interest, Peuquet was already moving towards a consideration of conceptual, as distinct from purely technical issues.

Field-Based Versus Object-Based Models of Geographical Data

A useful way of characterising the distinction between field-based and object-based data models is in terms of the three key notions of *locations*, *attributes* and *objects*. In a pure field-based model, we start off with a spatial framework consisting of a set of locations related to each other by such relationships as distance, direction and contiguity. (This could, of course, be a rectangular grid, but many other ways of dividing up space are possible, and the notion of a field does not make any presuppositions on this score. Indeed, the locations could be dimensionless points densely packed in a continuous space.) We then introduce various attributes for the locations. At each location, the value of each attribute is specified. Attributes may have numerical value ranges, which may be continuous (e.g. height above sea level) or discrete (e.g. population), or non-numerical value ranges (e.g. land use, vegetation type, ownership). When the values of an attribute are determined for each location, we have a *field*, and a field-based description of a particular geographical area simply consists of a set of fields, which may be thought of as layers placed one on top of another over the basic grid of locations. The complete description of a given location is then provided by the attribute values lying vertically above the location in the layered structure.

In an object-based model, we start off with a set of *objects*. These may be of many different types (e.g. countries, towns, buildings, roads, railways, lakes, mountains, forests). Associated with each type of object is a set of attributes characteristic of that type (e.g. a town might have a name, population, administrative status), and each individual object of the type will have specific values for all the relevant attributes. Amongst the attributes, we can single out the location of the object. This will typically be specified by means of spatial coordinates for a set of points which together determine the location: a single point in the case of a point object (i.e. an object conceptualised as extensionless), a sequence of points in the case of a line object (conceptualised as having length but no area) and as a sequence of points returning to its starting point (thereby defining a polygon) in the case of an object conceptualised as extending over an area. Setting aside these implementational details, we may say that the attribute of location is a function from objects to their locations.

The diagram below gives a highly schematic representation of these two approaches.

$$\begin{array}{ccc} \textit{locations} & \qquad & \textit{objects} \\ \downarrow & \swarrow & \searrow \\ \textit{attributes} & \textit{attributes} & \textit{locations} \end{array}$$

The distinction between field-based and object-based representations is often associated with two other important distinctions: that between continuous and discrete representations and that between absolute and relational conceptions of space. I shall discuss these in turn.

Continuous Versus Discrete Representations

I shall not discuss in detail the mathematical and logical definitions of continuity and discreteness; discussion may be found in Galton (2000). I shall simply follow standard practice and say that discrete data is data presented in the form of integers or non-numerical symbols such as letters or words, while continuous data is data presented as real numbers, the most important feature of such data being that given any two continuous data values there is the possibility of interpolating arbitrarily many further data values of intermediate size. How does the continuous/discrete distinction align with the raster/vector distinction, or with the field/object distinction? This depends on what aspects of these distinctions we look at!

If we consider how space is represented, i.e. the model for the key term *location*, then in a raster-based representation, the set of locations effectively forms a discrete space, e.g. the set of cells of a rectangular grid, in which each cell has several 'next-door neighbours'. In a vector-based representation, on the other hand, locations are specified by means of coordinates which can in principle take any real-number value, and thus the set of possible locations forms a continuous space. (In practice, of course, there is a limit to the precision with which real-number values can be represented in any given system, and so the *implementation* of these values could be thought of as providing only a discrete space of values; but this is not a design decision so much as an inevitable consequence of our digital modes of representation.) Thus from the point of view of locations, we might say that raster/vector aligns with discrete/continuous.

On the other hand, if we look, not at locations, but at what is located there, the objects and attributes, we obtain a rather different picture. A vector model presents the items of interest in the form of objects each of which is sharply distinct from the others. The objects form, we might say, a discrete collection. A raster model, on the other hand, conceptualises the field of interest in terms of the variation of attributes over space, and for many of the attributes this variation may be essentially continuous (that is, the available values for the attribute may form a continuous range of variation, like the real numbers). Thus from another point of view, there is reason to align raster/vector with continuous/discrete, and this is the view which is more generally taken.[1] The above discussion made use of the raster/vector distinction rather than the field-based/object-based distinction. In fact, the discussion would have gone rather differently if I had taken the latter distinction as our starting point instead. In practice, a raster-based model takes the spatial framework as discrete, but as already noted, this is not essential to the field-based conception that it may be regarded as implementing. We can take as our spatial framework a continuous space in which each point constitutes a separate location at which each of the attributes under considerations assumes some value. Seen in this way, a field-based model can be entirely continuous. I shall have more to say about this issue below, in the section on the mass/count distinction.

Absolute Versus Relational Views of Space

The field-based approach can be said to start with a spatial framework and then populate it with values which are regarded as attributes of locations. The view of space implicit here is that space somehow exists independently of what exists in space (of course, this does not mean that one cannot operate with the field-based approach if one does not subscribe to this particular philosophical view of the nature of space). The object-based approach starts with objects, and regards location as an attribute of objects, somewhat on a par with their other, non-locational attributes. The view of space implicit here is that space does not exist independently of the objects which occupy it, that space is a logical construct from those attributes and relations of objects that are singled out as having a certain special character which may be labelled 'spatial'. In this way the debate concerning spatial data type keys into the classic debates about absolute versus relational views of space and time.[2]

Can we meaningfully talk of space and time in isolation from spatial and temporal phenomena such as objects and events? This is the problem of absolute versus relational conceptualisations of space and time.

The Newtonian conception of space and time is that they are frameworks which can be specified, and conceived of existing, independently of any objects or events to occupy them. On this view, the structure of these frameworks, both in the large and in the small (e.g. whether bounded or unbounded, discrete or continuous, and for time whether linear or branching, for space whether Euclidean or non-Euclidean, etc)

[1]Couclelis (1992) refers to the process of creating a vector model as that of 'carving up the continuous landscape into discrete objects'. The continuous/discrete contrast is here aligned not with raster/vector so much as with reality/model, the model being understood to be vector-based.

[2]Compare (Peuquet 1994), where the parallel is drawn between the raster/vector dichotomy and 'objective' versus 'subjective' views of space. Note that objective/subjective and absolute/relative form another pair of dichotomies which, though quite distinct, are closely enough related to be often confused.

can be discussed independently of any discussion of what the world and its history contains.

The opposite, Leibnizian view disclaims all such discussion as vacuous: first, we must talk about objects and events, then we can define various relationships amongst these as temporal (e.g. simultaneity and succession) or spatial (e.g. containment and contiguity), and on the basis of these define purely spatial or temporal entities (regions, points, intervals, instants) as logical constructs. For example, a time instant might be defined as a simultaneity class, that is, a maximal set of mutually simultaneous instantaneous events, where an instantaneous event is not defined as an event occurring at an instant but, for example, as an event with no internal parts; and a region of space may be defined by its occupancy by a specified object at a specified time.

How should we choose between using an absolute (Newtonian) or relational (Leibnizian) view of space and time? This depends on our purpose. Philosophical purposes provide different criteria from technical purposes. Philosophers are interested in establishing ultimate conceptual priorities, e.g. what set of concepts provides an adequate basis for establishing all the rest of them, what 'things' really exist, etc. A philosopher might hope (not necessarily with justification!) that there is a unique right answer to such questions. Technical purposes are more pragmatic and expedient, and different technical contexts will demand different treatments.

A bold generalisation is that philosophy generally prefers a relational view of space and time, in which objects and events are primary, places and times logically dependent on them, whereas for many technical (and indeed everyday) purposes it is more convenient to adopt an absolute view: space and time provide a pre-existing framework within which objects and events can be located. A pure object-based data model might therefore be regarded as being better motivated from a philosophical point of view than a pure field-based model, even though the latter may be for many technical purposes more technically convenient.

In practice, we might have some difficulty in deciding whether a model, be it object or field-based, is 'pure'. On the one hand, one might say that the locations which are assigned to objects are determined by a pre-existing spatial framework defined by the coordinate system being used; on the other hand, if, say, these coordinates are latitude and longitude values on the Earth's surface, then we might say that so far from constituting a pre-existing framework they are thoroughly dependent on a pre-existing object, namely, the Earth itself (although typically this will not be one of the objects considered by the model). And again, if one therefore regards latitude and longitude values as object-dependent, then similarly one might so regard almost any system of locations that is used as the basis for a field-based model. The point is that although one has an intuitive feeling that the field/object distinction aligns neatly with the absolute/relational distinction, the apparent neatness here has a tendency to dissolve on closer examination.

Note incidentally that the distinction between classical and relativistic physics cuts across this dichotomy. Special relativity replaces the classical space and time continua with a unified spacetime continuum within which the rules for transformation of coordinates, addition of velocities, etc. are different, but it leaves open the issue of whether event objects are primary, with spacetime points and regions logically dependent on them (relational view) or whether the spacetime continuum is a pre-existing framework within which event objects can be located (absolute view). Note, however, that in general relativity, the fabric of spacetime is more closely integrated with the distribution of object events, e.g. a mass 'is' the focus of a certain pattern of disruption to the curvature of space in the vicinity of a point.

Mass Terms Versus Count Terms, Matter Versus Objects

The distinction between fields and objects is related to, though (at least at first sight) distinct from another important distinction, that between *matter* and objects. This in turn aligns closely to the linguistic distinction between mass nouns and count nouns.

Count nouns are nouns which paradigmatically refer to discrete objects, e.g. 'book', 'table', 'man' and 'star'. Grammatically, the general mark of a count noun is that it can occur in the singular and plural, and more specifically, in English, count nouns co-occur with adjectives such as 'many', 'few' and the cardinal numbers. Mass nouns, by contrast, refer to types of matter or stuff, with no implication as to discrete chunking, e.g. 'water', 'wool', 'mud' and 'salt'. Grammatically, mass nouns are characterised by their not taking the plural and, in English, co-occurring with adjectives such as 'much' and 'little', as well as with measure phrases such as 'a litre of' or 'a kilogram of'. Many nouns can be used as both mass and count nouns, typically with different though related senses. Quine (1960) has a nice example of this: 'Mary had a little lamb'. If we are talking about Mary's pets, then 'lamb' here is a count noun; but if we are talking about her Sunday dinner, it is a mass noun.

What appears at the linguistic level as the distinction between mass nouns and count nouns resurfaces at the ontological level as a distinction between matter and objects. Thus a mass noun typically denotes a type of matter, whereas a count noun refers to a type of object. Compare 'stone', used as a mass noun to denote a type of matter (as in 'This table is made of stone'), with the same word used as a count noun to denote a type of object (as in 'He threw a stone through the window'). We refer to and measure stone matter quite differently from stone objects. For example, if John's bucket contains three large chunks of rock while Mary's bucket contains ten small chunks, we might say that John has more stone but fewer stones.

The matter/object distinction is important and familiar in philosophy and is deeply integrated into our language via the distinction between mass and count nouns (Pelletier 1979). Its importance to knowledge representation has been highlighted by Hayes (1985) and Bunt (1985). Here are some ways in which it manifests itself in more specialist contexts:

- In geography, the distinction between terrains and features (e.g. desert, moorland, marshland, urban areas, versus individual deserts, moors, marshes, towns).

- In anatomy, the distinction between tissues and organs. We may talk about muscle tissue in general, and individual muscles, each of which is a particular sample of muscle tissue, with a well-defined form, function and location.
- In engineering, the distinction between materials and structures.[3]

The logic here is that an object typically consists of portions of various kinds of matter disposed in a particular way (and, e.g. in a particular position, thus the femur is a bone, so it has a particular pattern of internal organisation of the matter it contains, but also it occupies one particular position with respect to the other bones in the body). Sometimes, any more or less clearly delineated portion of matter of some kind can be singled out as an object (e.g. a moor is an area of moorland identified by its location with respect to the surrounding land). The sharpness of its boundaries may vary from case to case but this is usually not very important.

We may say

Object O is made of matter of type M
Object O consists of such and such homogeneous parts

Note the importance of granularity here. A material is a texture of spatial occupancy by matter. At fine enough granularity this texture might reveal itself as an agglomeration of individual objects (e.g. woodland is characterised by the presence of many individual trees, sand is an aggregation of sand grains). The important thing is that matter is conceptualised as homogeneous, thus ignoring any intrinsic low-level heterogeneity, whereas objects are conceptualised as units.[4]

How are matter/objects related to space? We may say

Region R is filled with matter of type M
At point P there is matter of type M
Within region R there is some matter of type M
Region R contains a subregion filled with matter of type M
Object O occupies the whole of region R
Object O is located within region R
Object O occupies the whole of some subregion of region R

It is crucial to realise that we are not dealing here with distinctions pre-existing in the world but rather with distinctions between different ways of describing what we find in the world. I may use the count noun 'lake' to refer to the same portion of the world's matter as what another time I refer to using the mass noun 'water'; in so doing I am using a different grammar and a different logic (I am referring to

[3]Notwithstanding Gordon's assertion that 'the distinction between a material and a structure is never very clear' (Gordon 1968). But Gordon's otherwise excellent discussion shows little sensitivity to the mass/count distinction, e.g. between material (mass) and a piece of material (count), and between a structure (count) and what might be called 'structured material', characterised by the indefinite repetition of structural units. Greater sensitivity is shown by Vogel (1988): 'Bone is made of several materials in an orderly array; but is bone therefore a structure or merely a composite material; or do we call bone generically a material but a particular bone, with shape, a structure?'

[4]More generally, Bunt (1985) speaks of 'phenomena that we perceive not as consisting of discrete elements, but as having a more or less homogeneous, continuous structure'.

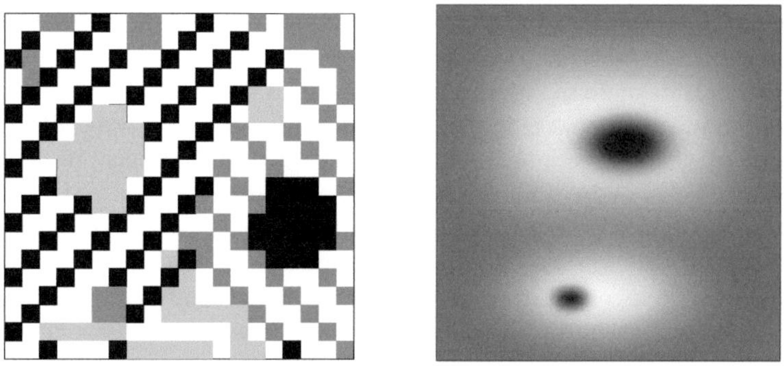

Fig. 4.1 'Matter' and 'objects' in discrete and continuous space

that portion of matter in different ways) but these differences engage not with what the world itself contains but with two different conceptual apparatuses I have at my disposal.

We can tie in the mass/count distinction with the field-based and object-based approaches to GIS. Consider the idea of 'forest', for example. In English, 'forest' can be either a mass noun or a count noun, as in

- The northern part of the country is mostly covered with forest
- There are three forests of area in excess of 10,000 sq km to the north of the city.

How might these statements be represented in a GIS? For the first statement, we might expect to see a Vegetation field, mapping locations onto a set of vegetation types such as forest, grassland and desert. We would find that a high percentage of the locations in the northern part of the country in question have forest as their vegetation type. For the second statement, we would expect to find three (and only three) distinct area objects of type forest, furnished with spatial coordinates from which one could infer that they are situated to the north of the city and each has an area in excess of 10,000 km.

Any matter type could in principle be represented in the form of a field, but the converse does not hold unless we are prepared to extend the notion of 'matter' to cover non-material properties like temperature and elevation.

It is tempting to align the mass/count distinction with continuous/discrete, and to an extent this alignment is well-motivated. But we must be careful; to see why, consider the following (Fig. 4.1):

- Imagine a discrete world, say a two-dimensional grid like an extended chessboard. Each square can take one out of some finite list of distinct colours (say black, white, red, yellow, green, blue). One area might be chequered, another striped, yet another uniform. These are textures defined by the pattern of distribution of colours on the cells in a small area.[5] A texture has the properties of a mass term, is conceptualised

[5] See Johansson (1998) for a discussion of the notion of *pattern* as an important ontological category. Johansson's patterns are not far removed from the notion of texture used here. Johansson hints at

as homogeneous (even though it exhibits heterogeneity at a fine granularity), is indefinitely extensible, etc. We may also pick out objects, for example, a roughly circular area of uniform colour surrounded by an expanse of diagonal striping (there are two such objects in the left-hand illustration of Fig. 4.1). We can ask how many such areas are found in a given region. The mass/count distinction is fully operational in this setting, but everything is discrete.

- Now imagine a continuous world, say the real Cartesian plane, and suppose that each point is assigned a colour out of a continuous range (e.g. the spectrum of visible light). Suppose also (so as to avoid the charge that I might be smuggling discreteness in through the back door, as it were) that the function assigning colours to points is mathematically continuous. Qualitatively, the same kinds of phenomena can occur as in the previous example (of course, additional phenomena, e.g. fractal boundaries, are also possible here). We can define textures and objects, e.g. a black area wholly surrounded by white—these are well individuated and can be counted (there are two of them in the right-hand illustration of Fig. 4.1). Once again the mass/count distinction is fully operational, yet everything is continuous.

The reason for the confusion is that the continuous/discrete distinction is much easier to explain and describe (and hence more familiar) than the mass/count distinction. We might conjecture that this is because the latter distinction tends to be built in to our language, whereas the former is external to it and therefore more readily perceived as a target for linguistic description.

Spatio-temporal Analogies

The drawing of analogies between time and space is a venerable pastime. In particular, the various distinctions discussed above in relation to spatial phenomena also appear in the temporal domain. The parallels between the spatial and temporal versions of these distinctions can be handled at several levels: at the linguistic level, the distinction between mass nouns and count nouns is paralleled by the distinction between imperfective and perfective verbs (Comrie 1976; Tedeschi and Zaenen 1981; Galton 1984; Verkuyl 1993); at the conceptual level, the distinction between matter and objects is paralleled by that between fluents and events; and at the implementational level, we can find temporal analogues to the raster/vector distinction—already noted by, for example, Worboys (1994).

I shall not discuss the linguistic level here; at the conceptual level, the natural distinction to make is between states and processes on the one hand, and events on the other. An event may be thought of as a single unitary happening, with a definite beginning and end, as for example, my cycling to the university this morning. By 'event' we often mean, in fact , an event *type*, under which any number of individual

---

a distinction between a count interpretation of pattern as a bounded patterned unity and a mass interpretation as an indefinitely extensible texture, but he does not explicitly acknowledge the importance of the mass/count distinction here.

occurrences can be described, e.g. the generic event consisting of my cycling to the university, which typically happens five times a week, most weeks of the year. The important thing is that event occurrences can be counted, and in this respect events are analogous to objects. A state or process, on the other hand, such as the state of my being at the university, or the process of my cycling (with no destination specified), is something ongoing, capable of continuation. The distinction has to do with the way these different types occupy time: an event takes up some interval of time, and does so as a whole: the time of the event is the whole interval, not any proper subinterval of it. A state or process occupies time in a more homogeneous way: I am at the university from one moment to the next, and if I am at the university over a particular period of eight hours, then this fact is a result of the fact that I am at the university at each moment during those eight hours. (Whereas I cannot say that if I cycle to the university over a particular period then this is because I cycle to the university at each moment during that period—for of course I cannot cycle to the university at a moment.)

In Artificial Intelligence, it is normal to handle states (and, to some extent, processes) by means of *fluents*. In the original treatment (McCarthy and Hayes 1969), a fluent was defined as a function from situations to values from some specified range, where a *situation* is to be understood as a complete snapshot of the world at a given time (either actual or possible). Thus, for example, the fluent 'Antony is on his bicycle' is a function which maps situations on to the value range $\{true, false\}$ (this is a *Boolean fluent*, which we may identify with a state). Other fluents can have more complex value ranges, e.g. the 'The population of London' can be regarded as fluent mapping situations onto non-negative integers. In order to determine what is the case at a given time, one needs to know what situations hold at what times. This is accomplished by a further mapping from times to situations, specifying one possible history of the world. It is not uncommon to by-pass the situations entirely and think of a fluent as a mapping directly from times to values, thus assuming one history as fixed.

Fluents are to be contrasted with actions or events, many of which can be thought of in terms of *transitions* between situations. Such an event type (or action type) is specified by means of a function which determines, for each situation, what the resulting situation would be if the event were to occur (or the action be performed) in that situation. For example, if we have a situation in which the fluent 'The door is shut' holds, then the action of opening the door transforms this into a situation in which 'The door is open' holds. This way of thinking of events is useful for the purposes of reasoning about how the world changes as a result of some given sequence of events, or conversely, what sequence of events is required in order to bring about a specified change in the world.

Fluents and events provide us with two different ways of conceptualising what goes on in time. The history of the world can be presented as a sequence of sets of fluent values: given a set of fluents, then at each moment in history, the state of the world (the current situation) can be described by means of the values assumed by those fluents at that moment. Or it can be presented as a collection of event occurrences, each assigned to a particular time, the interval during which it occurred (or, if we allow instantaneous events also, the instant at which it occurred).

There is a strong analogy here with the field-based versus object-based approaches to conceptualising the spatial domain. The analogies are made explicit in the following diagram.

SPACE	TIME
field: locations → values	fluent: times → values
object ↙ ↘ locations attributes	event ↙ ↘ times attributes

By analogy with what I noted earlier in the spatial case, it might be said that to think of history in terms of fluents is to presuppose an antecedently given temporal framework within which the fluents take their values, whereas it is possible to think of events, tied together by a network of such relations as precedence, overlap and simultaneity, as having conceptual priority, the flow of time then being *defined* in terms of this network. Thus the fluent/event distinction can be seen as aligning, at least loosely, with the absolute/relational distinction.

At the implementational level, the analogue of the raster/vector distinction familiar from the spatial domain leads to what may be called temporal raster and vector representations. This distinction can be illustrated by considering two ways of describing a simple domain consisting of the key events in a typical working day. In the upper diagram of Fig. 4.2, the day is characterised in terms of what state holds when. There are four states, *at home*, *travelling*, *at work*, and *at lunch*, and for each hour from 4 a.m. to 11 p.m. the graph shows which of the four states obtained during that hour. This is a raster representation. In the lower diagram, the same timeline is used, but instead a number of events are portrayed, namely, *leave home*, *travel to work*, *arrive at work*, *have lunch*, *leave work* and *arrive home*. Each of these events

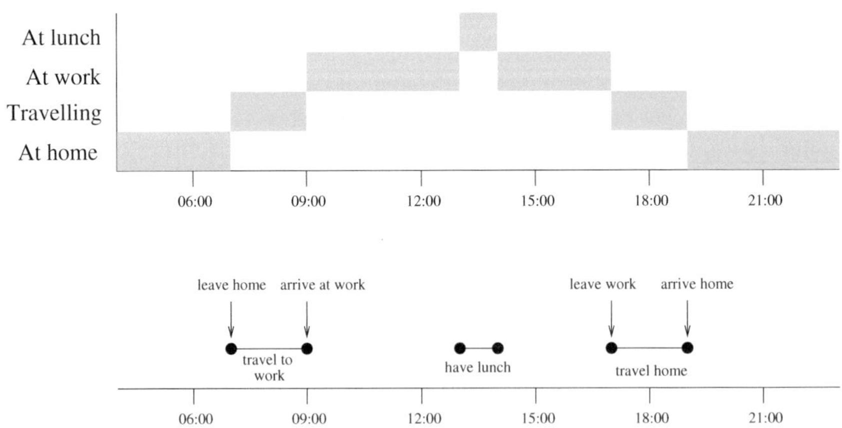

Fig. 4.2 Two portrayals of a working day

is represented by a dot or a line segment, depending on whether it is instantaneous or durative, and these elements are then placed in appropriate relation to the timeline. This is a vector representation.

To some extent, just as in the spatial case, these raster and vector representations can be considered to be interconvertible. For example, there is an *arrive at work* event whenever a period over which the state *at work* does not hold meets a period over which it holds (conversion from raster to vector). Conversely, the state *at lunch* holds at any time during the interval over which a *have lunch* event occurs (conversion from vector to raster). And as in the spatial case, it is only in simple cases that the interconversion can be accomplished without loss of information.

Integrating Spatial and Temporal Information

Integration of the spatial with the temporal can take a number of forms. At the simplest level of complexity, this can be a matter of *unifying* our treatments of space and time, that is, developing a single abstract model which can do duty for both, exploiting to the full the analogies between them. A more advanced enterprise would be to develop a scheme for spatio-temporal representation which subsumes (and therefore could replace) separate models of space and time. This is what the Special Theory of Relativity did for space and time in physics; but that theory does not in itself show how to accomplish such integration in earth-bound contexts such as geography.

Unification of Spatial and Temporal Theories

What one would like, and people have often set themselves to describe, is a unified abstract theory which covers both matter/objects and fluents/events as special cases. There is an important obstacle to doing this properly, which arises from a certain asymmetry in our relations to time and space.

Suppose we adopt a 'block world' view, i.e. a four-dimensional spacetime assigning four coordinates (x, y, z, t) to each point. Then it is natural for us to slice this block world at right angles to the t-axis, giving a succession of (x, y, z)-hyperplanes indexed by different values of t. Along this succession we identify portions of the (x, y, z)-hyperplane ('space') at one t value ('time') with portions at other times, e.g. we pick out an object O_1 existing in the (x, y, z) space at time t_1 and then for each time t' in some interval (t_1, t_2) we pick out an object O' which we call the 'same object' as O_1—the criteria for this identification being determined by the type of object to which O_1 is assigned. In other words, we think of the occupiers of space—objects and matter—as persisting through time. It is far less natural to do this slicing in any other way.[6] In particular, we do not think of the occupiers of time (fluents and events) as 'persisting' across space in the same way, although in principle we could do so. We could choose a particular point in London, say, and identify a particular episode in its history as 'The Great Fire of London', and then move to neighbouring points and pick out similar episodes from their histories and identify those as different 'spatial phases' of the Great Fire of London. But we very seldom actually do this. Occasionally, we do so, however: for example, we are reminded that an expression such as 'The Bronze Age', referring to a period of history, must be assigned to different dates in different geographical areas, just as the expression 'Poland', referring to a nation state, must be assigned to different sets of spatial coordinates at different times during its history.[7]

We conceptualise our world as being full of mobiles, i.e. objects and matter occupying different positions at different times. By contrast, we do not conceptualise the history of the world as full of events occupying different times in different places. Once again this is a statement about how we conceptualise the world; as such, it makes no claims about how the world is structured independently of any conceptualisation, though it is always tempting to suppose that we conceptualise the world in the way we do because the world itself is so structured as to readily lend itself to such conceptualisations (this is the view of the epistemological optimist), rather than because we are so constituted that we have no other way of conceptualising it, like the man who looks for his keys under the lamp post because that is where the light is rather than because he has any reason to suppose that the keys are really there (the epistemological pessimist).

In order to neutralise this asymmetry, we need to consider an area of discourse in which mobiles are not prominent. Geographical space is a very good example of this. At geographical scales (as portrayed in maps and atlases), change is mostly much slower than at the scale of individual humans, which is why we are able to publish meaningful maps at all, for if geographical features such as landforms, roads, countries and cities changed their shape or moved about at the same rate as humans do, maps would be largely useless. Thus to a first approximation geographical space is static. Of course, the recent drive to incorporate time in GIS belies this—but the very fact that this drive is recent precisely indicates that it was natural to begin GIS by taking a static view.

So, comparing time with static space, we find:

At each instant of time, some states obtain
At each point of space, some material properties hold

---

[6]Muller (1998) advocates a thoroughgoing four-dimensionalism in which the objects of enquiry are chunks of spacetime with no preferential form of slicing assumed; but within this framework, he accords privileged status to the notion of a 'temporal slice'. Note also that there is no discussion of how the mass/count distinction is manifested in the four-dimensional world.

[7]Zemach (1970) distinguishes four possible ontological schemes by whether they treat their entities as persisting across space, over time, both, or neither, and claims that all four are used in everyday thinking, though not with the same frequency.

Some event types may be defined as a certain localised bounded pattern of states, and each occurrence of an event type occurs on a certain interval.
Some object types may be defined as a certain localised bounded pattern of material properties; each instance of an object type occupies a certain region.
A derived state may be defined in terms of some pattern of occurrence of events; in particular the repeated occurrence of an event type may be regarded as a state.
A derived material property may be defined in terms of some pattern of location of objects; in particular the occurrence of many objects of a given type may be regarded as a material property.

The last two pairs allow recursive construction of ever more complicated descriptions, for example, a wood is an object defined as an area of woodland, woodland is defined as the sufficiently dense population of an area by trees, a tree is an object defined by a particular pattern of living tissue.

The distinction may be characterised in more abstract terms as follows. Given a *locational framework* F (which could be over space or time, we leave it undetermined),

- a *mass type* is a function f mapping points of F to values from some pre-assigned value range. For example, if F is time, f might give the temperature in London at each moment, or whether or not it is snowing in Paris, or the population of Asia—these are fluents. If F is geographical space on the Earth's surface, f might give the mean annual precipitation at each point, or the vegetation type, or the elevation, etc.—these are fields.
- a *count type* is a bundle of attributes with an identity, i.e. a set of functions mapping individuals to attribute ranges. For example, movement event types have attributes such as source, goal, path, subject, manner (a particular type of movement event type might be John cycling from his home to the station, which has source = John's house, goal = the station, path unspecified, subject = John, and manner = cycling). The object-type town has attributes such as population, name, and administrative status.
- a *count object* is a fully realised count type together with a location, i.e. an element or set of elements of F. A particular occurrence of John's cycling from his home to the station would have every attribute fully specified and would be assigned to a definite interval of time. A particular town has a definite name, population, etc., and is assigned to a particular geographical location.

In GIS, mass types are *fields*, whereas count objects are *objects*. As we have seen, the former are associated with raster-based representations, in which the data are presented in the form of functions assigning values to each location in some predefined tessellation (or raster); the latter with vector-based representations in which objects are modelled by points, lines, polygons, etc, defined in terms of vectors.

In temporal representations, we see a distinction between state-based (or fluent-based) schemes in which various states/fluents are defined and it is determined for each time instant which states hold (or what the value of each fluent is); and event-based schemes where individual events are specified and assigned to locations in time. For example, in a game we could divide the duration of the game into a sequence of short intervals and for each interval say what the score is (a fluent); or we could simply specify the events consisting of one side's scoring a point (events). Or we could model the history of a company by taking a series of snapshots, at each saying who the employees were, what posts they held, their salaries, etc.; or we could specify a sequence of appointment, promotion, demotion, firing and resignation events.[8]

The choice between the two types of representation is not uniquely determined. But some data are more appropriately represented using one rather than the other. A particularly interesting case is provided by land divisions.[9] A landmass is divided into countries, and countries are divided into states, departments, provinces or counties, as the case may be. At each level, the division is exhaustive and exclusive, that is, the divisions partition the land in a way that mountains, forests, rivers, lakes, towns, roads, etc. do not. In England, each point lies in one and only one county (or, exceptionally, on a county boundary). Thus the set of county locations is tightly constrained. Should we represent counties as objects or fields? There are reasons for each choice:

- As objects. A county has identity, location and other attributes such as administrative centre, and a county council having a particular constitution and membership. In a vector-based GIS, it is natural to represent it as a data object with its location specified by means of a polygon. Note how grammar points this way too: 'county' is a count noun, not a mass noun. But the object model does not in itself give us a natural way of encoding the partitioning role of the set of counties as a whole.
- Instead, we can regard 'county' as a field, a function from points to county names. At each point, just as we can assign an elevation and a vegetation type, so we can assign a county name. (There is, of course, a problem as to how boundary points should be handled.) For mapping purposes, a county name could be assigned a colour, then each point gets coloured in the colour of its county and we have a nicely coloured county map.

This example shows that there need be no hard and fast answer to the question whether a particular spatial concept should be represented by means of an object type or a field.

The temporal analogue of this is the division of time into years, months and days. We can regard January as an event (which occurs once a year and has duration 31 days), but then we need some special mechanism to ensure that all of time is partitioned into periods by successive month events. Or we can regard January as the value of a fluent, 'month', mapping instants onto the set January, February, ..., December (as when we say 'It's January now', on a par with 'It's raining now').

---

[8]This latter is the point of view of the Event Calculus (Kowalski and Sergot 1986) which has been highly influential in the development of temporal reasoning schemes for Artificial Intelligence.

[9]In the terminology of Peuquet (1984), this is 'adjacent polygon' data.

Combination of Spatial and Temporal Models

The Triad Framework of Peuquet (1994) is based on the three key notions of *locations* (understood in the spatial sense), *times* and *objects* ('where–when–what'). This framework supports location-based, time-based and object-based representations. As presented, the framework appears to be purely schematic, and does not address the issue of how the data should be stored so as to facilitate all three views without introducing unacceptable duplication of information.

In the light of our previous discussion, it might seem more reasonable to adopt a 'tetrad' framework based on the four key notions of *locations*, *times*, *objects*, and *events*. Just as an object may be thought of as primarily existing in space, but with a temporal extent, so an event may be thought of as primarily existing in time, but with a spatial extent. From a point of view in which the distinction between time and space is regarded as unimportant, it would be natural to subsume these under a single category of 'spatio-temporal entity', whose location is a chunk of spacetime, without any privileged status accorded to the segregation of the temporal dimension from the spatial dimensions. But this mode of presentation is entirely alien to our normal ways of conceptualising the framework of the world. Objects are primarily spatial because we think of the object as existing as a whole at each time during its history: it has spatial parts but not temporal parts; whereas events are primarily temporal because we *do* regard them as having temporal parts (though it is less clear that we complementarily deny them spatial parts, although we certainly sometimes do—another asymmetry between space and time).

The cardinal difficulty is how to sort out the complex interrelationships amongst the four key notions to provide an efficient and usable set of representations. The problem is that, on the one hand, there are many mutual dependencies amongst the elements of different representation types, but on the other hand, no subset of the types can be regarded as forming a primitive basis in terms of which the others can be defined. Thus, it seems that an effective spatio-temporal representation system should be able to handle locations, times, objects and events as primitive entities, to assign attributes to any of these, and to keep track of the interdependencies amongst the various attributes so assigned.

To illustrate, consider the sequence of maps in Fig. 4.3, representing the same area in 1900, 1950 and 2000. The grey shading represents built-up areas, the solid lines represent roads, the dashed lines county boundaries (the example is fictitious). Similar maps could be made for the intervening years. In this example, we may observe all of the following:

1. *Objects*. The towns A, B and C are objects with a distinct identity at 1900 and 1950. By 2000, they have coalesced to form a single entity which bears the name of A while B and C are relegated to the status of suburbs. In the earlier maps, A, B and C are represented as point objects (indicated by the white squares), whereas in the 2000 map this is no longer appropriate. The counties X, Y and Z exist at all three stages, but sometime between 1950 and 2000 the boundaries were redrawn so that the new city of A falls wholly within county X. There are

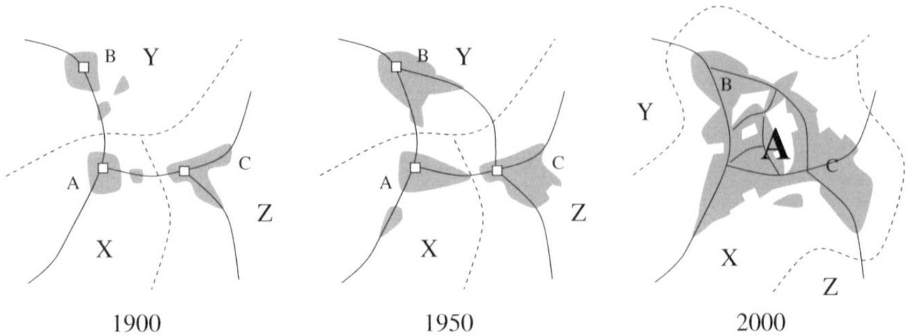

Fig. 4.3 Three stages in the development of a city

roads at all three stages, with new roads being created between 1900 and 1950 and between 1950 and 2000.

2. *Field attributes*. The field attribute portrayed here is land use, for which the two values *built-up* and *undeveloped* are represented by the presence and absence of grey shading. Note that it is not possible to identify the towns with any particular patches of built-up area. The outlier that has appeared on the road south of A in the 1950 map may or may not be regarded as part of A: this depends on political and administrative factors and cannot be determined on the basis of land use alone. The undeveloped enclaves within city A in the 2000 map most likely are to be regarded as belonging within the city, but again this interpretation is not enforced by anything in the map itself.
3. *Events*. The events are not represented explicitly but can be inferred from the map sequence. At some point between 1900 and 1950, a road was built connecting B directly to C. This was a durative event in the sense that its construction must have lasted some time, perhaps several months, although at a sufficiently coarse temporal granularity it could be regarded as a point (instantaneous) event, located at, say, the time the road was first opened for use. Although the material changes on the ground were all continuous (since they involved nothing more than the moving about of matter), from a conceptual point of view the transition between the condition in which there is no direct route from B to C to the condition in which there is one might be thought of as discontinuous. Two changes which presumably were discontinuous, since they were brought about not by the movement of matter on the ground but by fiat, were the redefining of the county boundaries between 1950 and 2000, and the redesignation of the conurbation consisting of the former towns of A, B and C as a unitary city A.
4. *Fluents*. The extent of built-up area steadily increases from 1900 to 2000. The maps only provide three instantaneous snapshots, but we know that in reality there must have been changes from year to year, if not from day to day. At the granularity represented by the map scale, the spreading outwards of developed

area is effectively continuous.[10] Ideally, one would like a sufficiently fine-grained representation of this ongoing process to enable one to determine for each location and each time whether or not that location was built on at that time.

There are many dependencies and interactions amongst these phenomena. For example, from a nature conservation point of view, the distribution of built-up areas is of far more significance than the socio-political designation of counties, towns and cities. From this point of view, the objects of interest include connected chunks of built-up area, and 'islands' of undeveloped land surrounded by built-up land. Although these can be named and counted like any other type of object, they are completely determined by the field attributes. Again, an event of some significance from the point of view of nature conservation occurred sometime between 1950 and 2000, with the disappearance of the corridor of undeveloped land separating the growing town of B from the A–C axis. Events of this kind can be named and counted like any other type of event, but they are completely determined by the changing values of fluents. However, if we only know the fluent values at rather widely separated time points—as in the case of our three maps—then the exact nature and timing of these events cannot be inferred. For this reason, it might be desirable to represent the events explicitly in the database. If this is done, it must be possible to check for consistency with known fluent values.

Conclusion

Returning to the European road map with which I began, we can recognise that the blue, yellow, green and grey patches, representing the sea and lakes, built-up areas, woodland, and mountainous country, all reflect different underlying fields: a two-valued field with values *water* versus *dry-land*, discrete many-valued fields of land use and vegetation types, respectively, and a continuous field of elevation. By contrast, the circles representing individual towns, the triangles representing individual mountains and the variously coloured lines representing different categories of roads all portray objects. The dotted, dashed and otherwise broken lines representing administrative boundaries of different ranks can be thought of as objects in themselves, although the areas they circumscribe are perhaps objects of greater significance. All these features provide a static view of certain aspects of a particular area of land at a particular time. If we compare a sequence of such maps spanning several decades, then we cannot ignore the time dimension. We may observe the wooded area steadily shrinking as forests are cleared, the steady growth of cities; each individual map provides an instantaneous snapshot of what is in reality an ongoing process. On the other hand, the opening of a new motorway and the construction of a dam are most naturally thought of as discrete events: an object is present in later maps that was absent from earlier ones. The redrawing of administrative boundaries is also almost

[10] At a finer granularity it looks more discrete: the successive addition of new buildings, streets, etc. At a finer granularity still all these events consist of the continuous motion of matter.

invariably a discrete event. If we had a map for each year of the century, we could present them cinematographically as an animated sequence; the distinction between ongoing processes and discrete events would then become much more obvious.

In this paper, I have explored the interconnections amongst a cluster of closely aligned distinctions in the field of spatio-temporal representations, ranging from the technical raster/vector distinction of GIS, through philosophical issues concerning matter and objects and states and events, to the linguistic distinctions between mass and count nouns and between imperfective and perfective verbs. Such distinctions must inform the construction of a workable ontology for spatio-temporal sciences such as geography, but the technicalities of coordinating both 'mass'- and 'count'-type representations for both spatial and temporal entities within a single unified system appear formidable. The purpose of this paper is not to suggest a solution to this problem but rather to highlight its existence and to place it in a broader setting by revealing the interconnections between superficially dissimilar distinctions arising in different contexts.

References

Bunt H (1985) The formal representation of (quasi-)continuous concepts. In: Hobbs JR, Moore RC (eds) Formal theories of the commonsense world. Ablex Publishing Corporation, Norwood, NJ, pp 37–70

Chrisman NR (1998) Rethinking levels of measurement for cartography. Cartogr Geogr Inf Sci 25(4):231–242

Comrie B (1976) Aspect: an introduction to the study of verbal aspect and related problems. Cambridge University Press, Cambridge

Couclelis H (1992) People manipulate objects (but cultivate fields): beyond the raster-vector debate in GIS. In: Frank AU, Campari I, Formentini U (eds) Theories and methods of spatio-temporal reasoning in geographic space. Lecture notes in computer science, vol 639. Springer, Berlin, pp 65–77

Galton A (1984) The logic of aspect: an axiomatic approach. Clarendon Press, Oxford

Galton A (2000) Qualitative spatial change. Oxford University Press, Oxford

Galton A (2011) Time flies but space doesn't: limits to the spatialization of time. J Pragmat 43:695–703

Galton A (2018) Processes as patterns of occurrence. In: Stout R (ed) Process, action, and experience. Oxford University Press, Oxford, pp 41–57

Gordon JE (1968) The new science of strong materials. Penguin, Harmondsworth

Hayes PJ (1985) Naive physics I: ontology for liquids. In: Hobbs JR, Moore RC (eds) Formal theories of the commonsense world. Ablex Publishing Corporation, Norwood, NJ, pp 71–107

Johansson I (1998) Pattern as an ontological category. In: Guarino N (ed) Formal ontology in information systems. IOS Press, Amsterdam, pp 86–94

Kowalski RA, Sergot MJ (1986) A logic-based calculus of events. New Gener Comput 4:67–95

McCarthy J, Hayes PJ (1969) Some philosophical problems from the standpoint of artificial intelligence. In: Melzer B, Michie D (eds) Machine intelligence, vol 4. Edinburgh University Press, Edinburgh

Muller P (1998) Eléments d'une théorie du mouvement pour la formalisation du raisonnement spatio-temporel de sens commun. Ph.D. thesis, Université Paul Sabatier, Toulouse III

Pelletier FJ (ed) (1979) Mass terms: some philosophical problems. D. Reidel, Dordrecht

Peuquet DJ (1984) A conceptual framework and comparison of spatial data models. Cartographica 21(4):66–113
Peuquet DJ (1994) It's about time: a conceptual framework for the representation of temporal dynamics in geographis information systems. Ann Assoc Am Geogr 84(3):441–461
Quine WVO (1960) Word and object. MIT Press, Cambridge, Mass
Stevens SS (1946) On the theory of scales of measurement. Science 103(2684):677–680
Tedeschi P, Zaenen A (eds) (1981) Tense and aspect. Academic Press, New York
Verkuyl HJ (1993) A theory of aspectuality: the interaction between temporal and atemporal structure, vol 64. Cambridge studies in linguistics. Cambridge University Press, Cambridge
Vogel S (1988) Life's devices: the physical world of animals and plants. Princeon University Press, Princeton, NJ
Worboys MF (1994) Unifying the spatial and temporal components of geographic information. In: Waugh TC, Healey RG (eds) Advances in geographic information systems: proceedings of the sixth international symposium on spatial data handling, Edinburgh, Taylor and Francis, London, pp 505–17
Worboys MF (1995) GIS: a computing perspective. Taylor & Francis, London
Zemach EM (1970) Four ontologies. J Philos 67:213–247

Chapter 5
Mapping the Deep Blue Oceans

Rasmus Grønfeldt Winther

Abstract The ocean terrain spanning the globe is vast and complex—far from an immense flat plain of mud. To map these depths accurately and wisely, we must understand how cartographic abstraction and generalization work both in analog cartography and digital GIS. This chapter explores abstraction practices such as selection and exaggeration with respect to mapping the oceans, showing significant continuity in such practices across cartography and contemporary GIS. The role of measurement and abstraction—as well as of political and economic power, and sexual and personal bias—in these sciences is illustrated by the biographies of Marie Tharp and Bruce Heezen, whose mapping of the Mid-Atlantic Ridge precipitated a paradigm shift in geology.

Keywords Cartography · GIS · Abstraction · Simplification · Selection · Exaggeration · Oceanography · Bathymetry · Scale · Map projections · Marie Tharp · Bruce Heezen · Lamont–Doherty Earth Observatory · Heinrich Berann · Ocean charts · Physiographic diagrams · Panorama maps · Plate tectonics · Cold War · Women in science · Bias · Discrimination · Workplace harassment

Introduction

The cartographer and geologist Marie Tharp recounts meeting oceanographer Jacques Cousteau in person only once, sometime between August 31 and September 12, 1959, in a hotel ballroom in New York City at the inaugural International Oceanographic Congress. She attended the Congress but did not present a paper. She and Cousteau spoke after a historical film screening, a conversation Tharp said

---

R. G. Winther (✉)
University of California, 9, Laidley Street, San Francisco, CA 94131, USA
e-mail: info@rgwinther.com
URL: https://www.rgwinther.com

© Springer Nature Switzerland AG 2019

T. Tambassi (ed.), *The Philosophy of GIS*, Springer Geography,
https://doi.org/10.1007/978-3-030-16829-2_5

she enjoyed.[1] One imagines the conversation was filled with mutual admiration, and possibly curiosity about each other's eccentricities and achievements.

At a recent conference in France, Bruce Heezen, Tharp's long-term collaborator, had given Cousteau a copy of the epoch-making 1957 physiographic diagram of the North Atlantic's ocean floor (Fig. 5.1) that Tharp had drawn from Heezen's deep-sea sonar data. The map depicted a mountain ridge in the middle of the Atlantic. Cousteau was extremely skeptical that this Mid-Atlantic Ridge existed; even so, he had hung the map up in the mess hall of his famed *Calypso*, so that he and his crew could study it.

On the *Calypso*'s way to New York City and the conference at which he and Tharp would meet, Cousteau decided he would prove Tharp and Heezen wrong once and for all. There could not possibly be such a strange phenomenon, which seemed to corroborate the much-maligned theory of continental drift and plate tectonics—a topic of heated discussion at the International Oceanographic Congress. Kilometers above the supposed ridge, the *Calypso* lowered its submarine camera "sled", the *Troika*, into deep, cold Atlantic waters. Sure enough, as his film projected to a large, enraptured audience of scientists at the Congress, the *Troika*'s camera recorded a high mountain ahead in the distance; a climb up that mountain; a steep descent; a

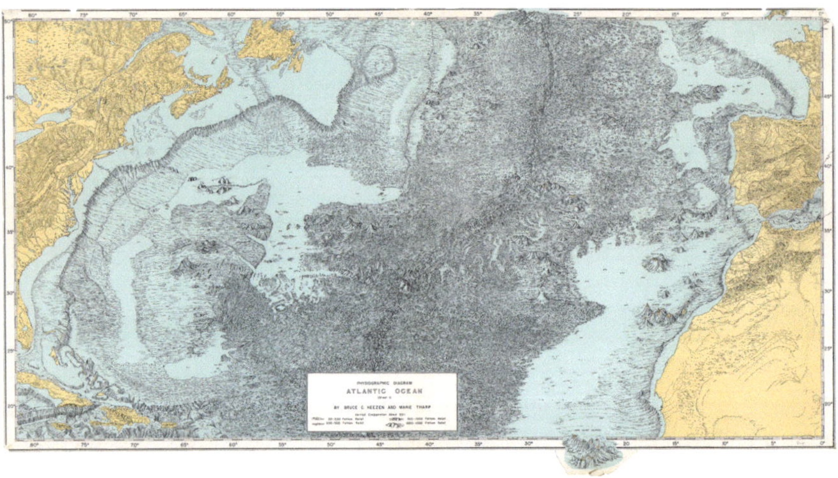

Fig. 5.1 North Atlantic physiographic diagram. (Published in 1957; map inset to Ewing, Heezen, Tharp 1959.) As indicated in the information box, the vertical exaggeration is 20:1. This box is located where it is because they did not have much data for that region of the ocean. In part Tharp and her collaborators chose to draw physiographic diagrams, because exact depth data need not be shown, and this information, while they had access to it, was classified by the US Military until at least the late 1960s. Reproduced by kind permission of Lamont–Doherty Earth Observatory and the Estate of Marie Tharp/©Marie Tharp Maps, LLC Fiona Yacopino, 8 Edward St. Sparkill, NY 10976

[1]Felt (2012), "enjoyed", Loc 2178.

trip across a plain filled with young lava; and a climb up another mountain. Cousteau and his team even turned the *Calypso* around and redid the whole exercise.

Tharp's map, Cousteau's film, and the Atlantic all agreed: The Atlantic ridge was real. *A map became the world through a film* (Winther 2020).

This chapter explores how mapping works, particularly with respect to abstraction practices of map-making, and with respect to the case of deep blue oceans. The oceans are not an immense, flat plain of mud. To map them accurately and wisely, we must understand how cartographic abstraction and generalization work both in analog cartography and digital GIS.

I see significant continuity between classic cartography and GIS (Winther 2015). The emergence of GIS, in my view, signals not the classic map's nostalgic swansong or tragic funeral, but rather a retooling and enrichment of possibilities for visual geographic practices. Differently put, a map-based science of data collection, management, and abstraction shifted to a computer based science of database management, spatial analysis and statistics, expert systems, and modeling.[2] In this shift, the power of the map was neither lost nor forgotten, as can be seen below with contemporary efforts of ocean floor mapping via satellite altimetry remote sensing.[3]

Finally, and perhaps most concretely, the intertwined biographies of Marie Tharp and Bruce Heezen capture many empirical and conceptual—as well as social and political—themes associated with mapping. By interweaving history and philosophy, I hope to interest you in how and why maps of the oceans are drawn; what importance this has for questions about power, values, and bias in science; and the relevance that mapping has for the future of the oceans, especially in a time of foreboding climate change and generalized ecological collapse.

Abstraction in Cartography and GIS

To create an analogy for how maps generalize and abstract from the world, imagine yourself sitting on an airplane as it leaves the terminal. You stare out the window and see the runway. As the plane accelerates, you feel the movement in your limbs and your gut. Buildings, cars, and hills whiz by, faster and faster. The plane climbs. The level of magnification changes. Trees and cars disappear. Rivers and highways become generalized curves. A quilt of greens, blues, and browns emerges.

Soon you are above it all, looking down with sweeping vision. The teeming world on the ground has become simpler and more abstract—the general features of a *map*. Whenever we compare a map to its territory, we find this flip from everyday, human-scale perception to a generalized abstraction.

---

[2]On broadening the concept, methods, and purposes of cartographic generalization, see e.g., Abler (1987), Shea and McMaster (1989), Couclelis (1992), Goodchild (1992), Schuurman (2004), Lüscher et al. (2009).

[3]Smith and Sandwell (1997).

Maps are produced by practices of abstraction, to somewhat similar effect. Once data has been collected—size, position, boundaries, landscape features, and so forth—abstraction must be performed in order to produce maps, some of which are highly dynamic and complex. Cartographic abstraction is akin to scientific theorizing. Whether a map is made via classic analog cartography or a geographic information system (GIS), a standard, classic set of abstraction protocols is used, including *selection*, *classification*, *simplification*, *symbolization*, and *exaggeration*—to which I add *perspectivizing* and *partitioning* in Chap. 3 of my forthcoming *When Maps Become the World* (2020). Here I will focus on selection, simplification, and exaggeration, with examples from mapping the deep blue oceans.

Cartography is the study of principles and rules of map making and map use (Winther 2020). An important question shaping this discipline was how to engage in abstraction and generalization when creating maps.[4] These practices are similar across both classic analog cartography and digital GIS.[5] Even in the digital, computational age, map abstraction remains very much that switch from human-scale perception and navigation to graphic representations at extreme scales.

As an emerging discipline, an important phase for GIS in the early 1990s was, as Nadine Schuurman plausibly suggested, a "switch" from "a map to model-oriented approach to generalization" (1999, 83). In North America, the "culture of cartography" was dominant; the map as such was the focus. Conversely, "Europeans had developed a *landscape model* [the database] that is based on derived data" (ibid.). The key shift was from "working with mental models of maps" to committing to "the database" as generative of "map objects" (2004, 48–49). Schuurman highlighted Brassel and Weibel's (1988) article on automated map generalization as instrumental to this shift. Here Brassel and Weibel characterized generalization "as an intellectual process, [which] structures experienced reality into a number of individual entities, then selects important entities and represents them in a new form" (1988, 230–1). The authors then distinguished two kinds of "objectives for spatial modeling", corresponding to two kinds of generalization: "spatial modeling for purposes of data compaction, spatial analysis and the like [i.e.,] *statistical generalization*" and "*cartographic generalization*", which, "in contrast, aims to modify local structure and is non-statistical" (232). By identifying a broader set of modeling strategies and purposes—beyond visual display and map-making—Brassel and Weibel prompted the emerging GIS community to, I believe, transform cartography and the map. Yet, the map remains.

Let us now turn to specific practices of map abstraction.

---

[4]I prefer the term "abstraction" for the process of inferring general features from the particulars of the world or our experience. Although most cartographers prefer to use "generalization", "abstraction" is the more appropriate, flexible, and general term. On my pragmatic account of abstraction and its shadow side, pernicious reification, see Winther (2014). Cartographic abstraction is structurally and substantively related to scientific abstraction (see Winther 2020, Chap. 3).

[5]The cartographic framework, and its take on abstraction, can be gleaned from close study of work such as Wright (1942), Koláčný (1969), Muehrcke (1969, 1972, 1974a, b), Wolter (1975), Robinson and Petchenik (1976), Wood (1992), MacEachren (1995), Harley (2001), Montello (2002). See also Winther (2015, 2020).

Selection: Scale

Selection in cartography is the intentional reduction of content, particularly as a consequence of choosing map scale and map projection. Scale sets a map's representational scope and granularity of detail,[6] while a map projection is a flat, two-dimensional geometric representation of a curved, two-dimensional surface of a globe, ellipsoid, or geoid. These are practices of abstraction because they involve the detachment of certain information from its context, generally emphasizing some features at the expense of others. The selection of scale and projection are also significant in that they constrain myriad other representational features of the map.

Scale is a ratio or proportion between features of the representation and properties of the world depicted. Depending on the map or model, the scale might be given in terms of time passage, the intensity of features, distances and sizes, or other parameters. Map scale can be shown visually (for example, with a graduated line representing 10 km), quantitatively (for example, 1:10,000,000), or in words. Scale affects all other abstraction practices.[7]

Scale should be selected based on how much area one desires to cover (*scope*), and at what level of detail (*grain*), while taking presentation constraints into account (for example, a book, a poster, or a screen with zooming capacities). The larger the scale, the more fine-grained, detailed, and concrete the map can be. At one logical limit is the famed one-to-one map of the world, a concept that is poetically and humorously exploded by authors such as Lewis Carroll, Mark Twain, and Jorge Luis Borges.[8]

Some authors classify maps according to scale.[9] World maps are small-scale; a map fitting on two leaves of an atlas could have a scale of 1–60 million (1:60,000,000).[10] In contrast, city maps have a larger scale, typically varying between 1:10,000 and 1:25,000 (see Footnote 9). Tharp and Heezen's maps (Figs. 5.1 and 5.2) represent at different scales—1:5,000,000 and 1:30,412,800 (480 miles:1 inch), respectively.

In general, many of the same considerations about the purpose-dependency and limits of scale from analog cartography pertain to digital maps. Selecting map scale is as necessary for digital maps as it is for analog maps. Digital maps such as Google

---

[6]The sciences are distinguished by differences in scale. The boundaries of particle physics, biochemistry, neuroscience, anthropology, or cosmology, etc., are set (if permeably) by the minimum or maximum spatial scale of the objects and processes of its domain, from the tiny to the enormous. Temporal scales also vary across the sciences. For instance, quantum mechanics and quantum chemistry trade in extremely short time scales, developmental biology in days, weeks, and months, geology in millennia and millions of years, cosmology in billions of years (Winther 2020).

[7]For a rigorous, mathematical treatment of map scales, see Bugayevskiy and Snyder (1995, 17–20).

[8]Carroll ([1893] 2010, 162–163), Twain (1894, Chap. 3, 57), Borges ([1946] 1975, 325). With humor and irony, Eco ([1992] 1994) playfully deconstructs the very concept of a one-to-one map.

[9]Greenhood (1964, 48–49), Muehrcke and Muehrcke (1998, 13, 537–546), Kimerling et al. (2009, 22–33), and Krygier and Wood (2011, 94–95).

[10]ESRI (n.d.) provides a list of common map scales.

Fig. 5.2 An absolute panorama map of the Atlantic Ocean floor as painted by Heinrich Berann, under close collaboration with Tharp and Heezen. Berann painted many panorama maps for *National Geographic*, also of the Himalayas and the Alps. This map appeared in the June 1968 issue of *National Geographic*. Notably, this image also graces the cover of Naomi Oreskes' well-respected and excellent 1999 book on continental drift. But Tharp herself is mentioned on just two pages of the main text of Oreskes' book. Heinrich Berann/National Geographic Creative/National Geographic Image Collection

Maps are often zoomable,[11] but the grain can thus increase only because the system adds new information as we increase the scale—or else it would be like visually blowing up a photograph to reveal its basic pixels.

Selection: Projection

As for projections, the Mercator projection remains favored in the mapping of the oceans, including by Tharp and Heezen and ocean mappers and coauthors Walter Smith (of the National Oceanic and Atmospheric Administration) and David Sandwell (of the Scripps Institution of Oceanography).[12] Well known for its use in marine charts, the Mercator conformal projection projects the world onto a cylinder such that lines of constant bearing on Earth (i.e., *rhumb lines*) are transformed to straight lines on the map.

While still taught, the study of map projections, which filled geography and cartography classes and textbooks before the rise of GIS, has massively declined in importance. As I show in Chap. 4 of *When Maps Become the World*, part of the reason lies in the triumphant biography of the Mercator projection, in its various guises, including Johann Heinrich Lambert's 1772 "transverse Mercator", in which the cylindrical developable surface is oriented not around the equator, but along a meridian. In cartographic argot, this projection has a *transverse* rather than an *equatorial* aspect (orientation). The transverse Mercator became central to the ellipsoid datum's coordinate system (i.e., WGS 84) in the mid-twentieth century.[13]

Furthermore, for various cognitive and social reasons, such as familiarity and historical inertia, GIS and online mapping services such as Google Maps, Bing Maps, and ArcGIS Online employ a "Web Mercator". These equations render Earth in a near-conformal, cylindrical projection.[14] Perhaps Web Mercator has become the online and digital cartographic representation standard because "north is always the same direction"; it simply "look[s] right"; it "allows for simpler (and therefore quicker) calculations […] [and] continuous panning and zooming at any area, at any location, and at any scale"; and it "allows close-ups (street level) to appear more like reality."[15] But these are not sufficient explanations for Web Mercator's dominance, since computers could always retranslate projections, depending on which parts of the world one wishes to show.

---

[11] Since 2009, Google Earth shows the oceans based on, among other data sources, Marie Tharp as well as Smith and Sandwell, and collaborator's maps and data. See Jha (2009).

[12] Heezen et al. (1959, 3), Smith and Sandwell (1997), Sandwell et al. (2014, 66). The mathematics, visualizations, and quandaries involved in and with map projections are discussed extensively elsewhere (e.g., Snyder 1993; Winther 2020), so I shall set it aside here.

[13] See Rankin (2016).

[14] E.g., Brotton (2012, Chap. 12), Strebe (2012), Battersby et al. (2014).

[15] First two quotes from Strebe (2012); third quote from Battersby et al. (2014, 88–9); last quote from Google representative Joel H., August 4, 2009. https://productforums.google.com/forum/#!topic/maps/A2ygEJ5eG-o.

To be fair, map projection distortions become less important as scale increases—after all, a large-scale map shows a roughly flat area, with just a little bit of curvature. However, there is no reason not to be able to compare map projections for small-scale maps. Consider map aficionado Tobias Jung's *Compare Map Projections* website.[16] While Mercator's projection is useful for navigational purposes and also the standard projection Marie Tharp and Bruce Heezen used, there is nothing to stop it from being just one among multiple projections in a flexible, GIS integration platform, where the context-dependency and advantages and disadvantages of each map projection are indicated and discussed, as per Jung's website, and as further discussed in Winther (2020).

Simplification

Simplification is the omission and streamlining of information such that general features of a pattern or process are represented on the map, but unnecessary detail is abstracted away. We can emphasize or omit any number of patterns from a rich data set, representing only some aspects of the data. For instance, houses and roads can be removed, a meandering river straightened out, or a large number of trees aggregated into a small simple patch of green.[17] And there is much that cannot be represented on a map. The more simplified a map is, the more abstract it is (even if abstraction involves much more than simplification).

We might also simplify because we are privy to limited data, because of limited technologies or imperfect surveying opportunities, or even because a map was "born classified",[18] all of which were the case with Tharp and Heezen's maps.[19] In such conditions, we would wish only to perform the *minimal* amount of interpolation within, and extrapolation across, the available data. As Hali Felt quotes Marie Tharp in her creative biography of the oceanographic cartographer, "Deep sea soundings obtained along a ship's track were as a ribbon of light where all was darkness on either side."[20]

An early protocol of automated line simplification is the Ramer–Douglas–Peucker algorithm, which outputs a simplified zigzag line from a complex real-world line, while preserving the latter's basic properties (Fig. 5.3).[21] The algorithm first connects

[16]https://map-projections.net/index.php.

[17]An interesting *material* simplification strategy is described in *Hammond's Compact Peters World Atlas*: "Cartographers have struggled with the best way to create hillshading for hundreds of years. In this atlas the 3-D relief comes from photographing specially made plaster relief models and blending these photos with hand-rendered coloring" (Hardaker 2002, 7).

[18]Doel et al. (2006, 605).

[19]Tharp spoke thus: "The displacement of peaks and other topographic features [in physiographic diagrams] due to the vertical exaggeration blurred their actual positions as demanded by a classification regulation" (Felt 2012, Loc 1779).

[20]Felt (2012), Loc 1720.

[21]Ramer (1972), Douglas and Peucker (1973).

the two ends of the complex, real-world line and finds the real-world line bend point farthest from that connecting line. Releasing the first connection, the algorithm then connects the first end point and that farthest point, and the second end point and that farthest point. We now have two straight lines angled and embedded along the entire length of the real-world line. The algorithm subsequently operates recursively on each of these two lines, and so forth (Fig. 5.3). The recursion ends, overall, when *every* farthest point is within a set maximum *tolerance distance*.[22] The Ramer-Douglas-Peucker algorithm marks an important milestone in the development of the *digital*, computational map.

Exaggeration

Exaggeration is the technically inaccurate adjustment or reproportioning of the size and placement of map elements. The purpose of exaggeration is to increase legibility, comprehensibility, and communicative power. An example is expanding the width of streams or highways on a map to make them visible rather than razor-thin. More dramatically, Harry Beck's classic London Tube map sacrifices geographically accurate location of stations by exaggerating their relative location, fixing their placements into topologically accurate, user-friendly straight lines.[23]

Tharp exaggerated vertical cross section profiles of mountain ridges: "With a few exceptions all profiles are represented with a 40:1 vertical exaggeration"[24] (Fig. 5.5). She had to do this in order to show the Mid-Atlantic Ridge profile in a meaningful and memorable way. Otherwise, the profile would have nearly disappeared into a solid line barely crawling along the ocean bottom. The ocean is so wide that even towering mountains look small by comparison.

Map elements become exaggerated in various ways. When geographic features have to be shown at different scales of a GIS map, then the map elements often have to be exaggerated in distinct ratios. For instance, as we zoom in, that river need not become thicker in proportion to the scale. It could remain relatively thin and still be visually recognizable. However, the software platform will probably update

---

[22]More concretely, Ramer's code selects every anchoring point of what becomes an irregular polygon constructed from the target real-world line. [An anchoring point was a farthest orthogonal point or vertex, in the prior step ($N - 1$).] Vertices exceeding maximum distance (see: lower left hand column box of Fig. 5.3) "open" the polygon at each step, and are labeled as such in the program stack. The polygon becomes "closed" when the two new line segments from that point to the original anchoring points are constructed. This automated procedure is repeated, until no further vertices (orthogonal points) are greater than d_m (the maximum tolerance distance) and the polygon becomes fully closed. For a dynamic rendition of the Ramer–Douglas–Peucker algorithm, see: https://en.wikipedia.org/wiki/Ramer%E2%80%93Douglas%E2%80%93Peucker_algorithm#/media/File:Douglas-Peucker_animated.gif.

[23]See the "Harry Beck's Tube map" post on the website of London's transit agency, tfl.gov.uk.

[24]Heezen et al. (1959, 15).

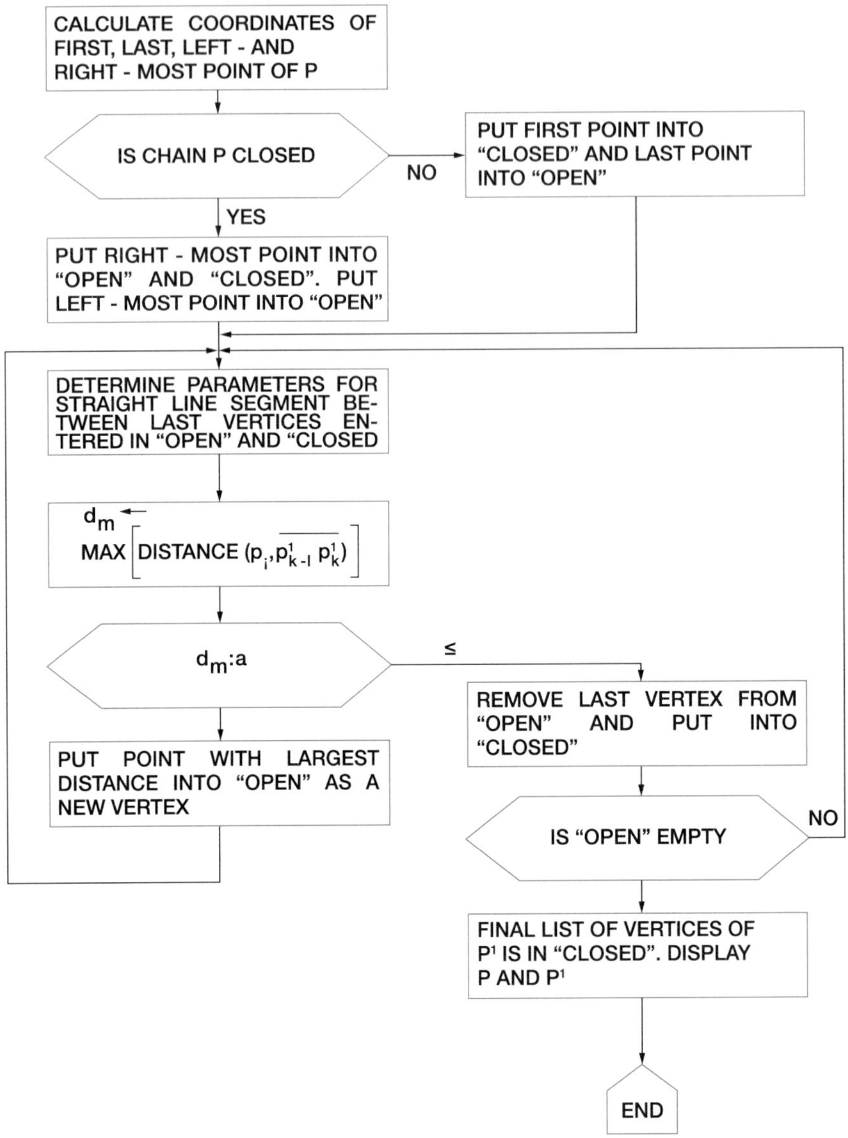

Fig. 5.3 Polygon generation flow diagram p. 248, Ramer (1972). Redrawn by cartographer and graphic designer Mats Wedin

the snakiness of the river, recalculating perhaps with the Ramer–Douglas–Peucker algorithm.

A digital map captures too much data and interpretation to represent in any split-second visualization on a screen or on paper. Software packages by Esri, for instance, store and sometimes compress data. Google Maps stores data elsewhere, far from users' computers. The digital map is more like an extended network, where the visualization is a tip of the iceberg, a hyperlocal mapping-as-you-go, rather than something you can hang on a wall.

Tharp-Heezen Maps

As a historical prolegomenon to a fuller story of abstraction in cartography and GIS, consider the case of Marie Tharp's maps of the deep. These maps changed the face of the Earth Sciences: "This physiographic map '*is* in some ways the ocean floor', former Heezen graduate William Ryan later mused: 'It's our only multi-dimensional picture of it… that map and every subsequent revision to it'."[25] Through her Mid-Atlantic Ridge *profiles*, her *physiographic diagrams* reminiscent of geographer Armin Lobeck's,[26] and her long-term collaboration with Bruce Heezen (and, to a lesser extent, Heinrich Berann) on *perspective* and *panorama* maps, Marie Tharp gave us the ocean floor. Tharp's representations also suggested a mechanism for explaining the ocean floor's features. Tharp's maps became the world.

Tharp shows the importance of characterizing a system *as a whole*, not merely as an aggregation of parts. My argument here resonates with Evelyn Fox Keller's analysis of Nobel Prize-winning corn geneticist Barbara McClintock in her *A Feeling for the Organism* (1984). There are clear indications that, just like McClintock, Marie Tharp possessed powerful capacities to see all the parts of a system in a holistic, dynamic, and interactive manner. She eschewed atomism and reductionism. She was also able to intuit hypothetical patterns via scientific interpolation and extrapolation. She actually *integrated the oceans* in her physiographic diagrams and in her coaching of Austrian painter Heinrich Berann's panoramas.

Both McClintock and Tharp had a perhaps more traditionally feminine (only weakly and statistically associated with actual sex) capacity to approach a set of complex biological or geological processes—genetic inheritance and ocean floor bottoms, respectively—with a broad vision. They investigate important scientific phenomena with their all-inclusive, embodied Gefühl.[27] Such a floodlight vision

[25] Doel et al. (2006, 620).

[26] Lobeck was hired as a full professor in Geology at Columbia University, home of Lamont, in 1948. Tharp had "devoured" his 1924 *Block Diagrams* book (Felt 2012, Loc 1715). Lobeck developed the physiographic diagram and was involved with the US military, especially during the two world wars. His "Physiographic Diagram of the United States" (1948) was influential. For a brief biography, see Smith (1959).

[27] For early work on the epistemology of gender, sex, and science, see Harding (1986), Keller and Longino (1996).

complements the sharp cutting, analytic spotlight vision typically permeating science. Effective research at the community level requires a commitment to a panoply of distinct research styles, each expressed by changing constellations of individual researchers and research groups.[28]

The role of political and economic power and of personal bias in contemporary GIS is illustrated by Tharp's and Heezen's biographies.

Tharp-Heezen Timeline

1947. While an undergraduate geology student at Iowa State University, Bruce Heezen heard a lecture by Maurice "Doc" Ewing and was enraptured. Ewing invited him to join an expedition of the Mid-Atlantic Ridge on the *Atlantis I*. Heezen accepted, and eventually became a graduate student at Columbia, receiving his doctorate under Ewing in 1955.[29]
1948. Marie Tharp had completed a bachelor's in English literature and music at Ohio University, a master's degree in geology at the University of Michigan, and a bachelor's in mathematics at the University of Tulsa in Oklahoma. In 1948 she was hired by Ewing as a research associate.[30] After a few years, she was working almost exclusively with Heezen on their shared interests in ocean mapping (Fig. 5.4).
1949. The Lamont Geological Observatory was officially established in Palisades, NY, associated with Columbia University. Ewing was its founder and source of energy.[31] While Ewing and Heezen had a close and productive collaborative effort in the first years of this institution, their relationship would sour. Heezen was associated with Lamont for the remainder of his life, even with a much-diminished role, starting in 1966. Tharp was treated unjustly by Lamont after Heezen's death.
1952. Tharp completed six *profile drawings* of the Mid-Atlantic Ridge (Fig. 5.5) primarily using Lamont survey data, much of it collected by Heezen on *Atlantis I*, but also with some data from the German ship *Meteor* and other sources. These profiles were based on sonar sounding data, as ships crossed what turned out to be the Mid-Atlantic Ridge at different latitudes. Particularly striking about these drawings—and what took Tharp initially by surprise, and then approximately a year to convince Heezen of—was the *valley* depicted inside the ridge. According to Tharp, Heezen "initially dismissed my [her] [rift valley and continental drift] interpretation of the profiles as 'girl talk'". Ironically, the rift valley V-shape indentation was indeed a form of girl talk, in a genuinely productive way.[32] This smelled of continental drift, because it meant the plates were coming apart, with lava oozing out from the wound.

[28]Longino (2001), Winther (2012, 2020).

[29]Tharp and Frankel (1986, 3).

[30]Tharp and Frankel (1986, 2–3), Barton (2002, 216–217). See Landa (2010) for discussion of Tharp's early biography, and her "ties" to her father, a soil surveyor.

[31]Consult Lamont–Doherty Earth Observatory (n.d.).

[32]Tharp (1999). Helen Longino provided constructive feedback.

Fig. 5.4 Marie Tharp in front of profiles and globes that she, Heezen, and their assistants used in preparing and drawing physiographic maps. Pictured here is her first 1957 physiographic diagram. Some of their globes were made with "acrylic applied to a basketball" (Doel et al. 2006, ftnt. 72, p. 625). These globes remained unpublished, but avoided any map projection distortions. Cf: Bressan (2018). Reproduced by kind permission of Lamont–Doherty Earth Observatory and the Estate of Marie Tharp

(Heezen defended an alternative theory: an expanding Earth coming apart at its seams.)

1953. Given the profile drawings and further sounding data, Tharp started her first sketches of *physiographic* diagrams. Her diagram of the North Atlantic was completed in 1956,[33] officially published in 1957,[34] and presented as a map inset to Heezen et al. (1959) (Figs. 5.1 and 5.4). To aid in these efforts Lamont secured the research ship *Vema*, which became one of the most influential oceanographic research ships, with over 1 million kilometers of total sailing during its lifetime as a research vessel.

1954. Tharp and indirect collaborator Howard Foster, a Ph.D. student who was drawing maps of *earthquake data* on maps of the same scale on the table adjacent to hers

[33]Felt (2012), "by the end of 1956" Loc 1880.

[34]It appeared as an addendum to Elmendorf and Heezen (1957). In the acknowledgments of that paper, Marie Tharp is thanked first and the last sentence reads "The encouragement and guidance of Dr. Maurice Ewing has been of great value" (1093).

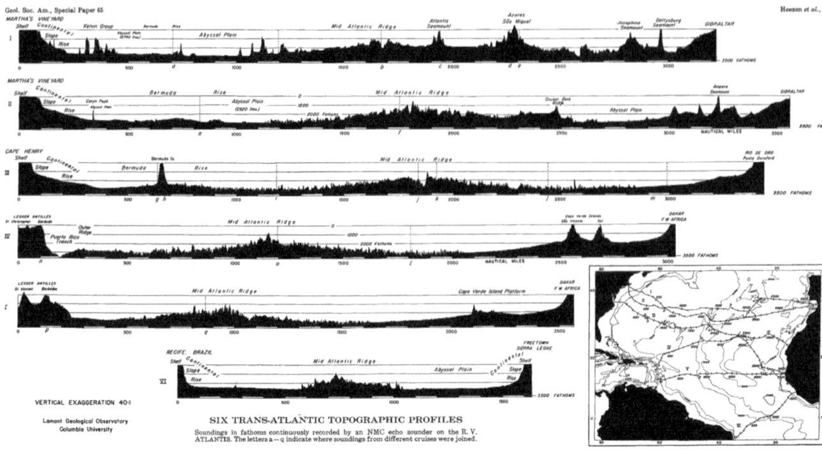

Fig. 5.5 "Six Trans-Atlantic Topographic Profiles" (with 40:1 vertical exaggeration). Bruce C. Heezen, Marie Tharp, and Maurice Ewing, (1959). The Floors of the Oceans: I. The North Atlantic: Geological Society of America Special Paper 65, https://doi.org/10.1130/SPE65. Heezen et al. (1959) (Lamont Geological Observatory, Columbia University), Plate 22

at Lamont, made an important discovery. Heezen had insisted that they draw their maps on the same scale.[35] While the exact date and circumstances are unclear, one (or both) of them, having superimposed the earthquake data map on the Mid-Atlantic Ridge valley map, noticed a strong concentration of earthquakes in the valleys and very few earthquakes beyond the ridge. This was of course further evidence for some kind of movement of the ocean floor. This earthquake data from Gutenberg and Richter (1954) and the USGS was shown as Plate 29 of Heezen et al. (1959).

1957. On March 26, 1957, Heezen gave a talk on the rift in the Mid-Atlantic Ridge to the Princeton Geology Department, at the end of which the influential geologist Harry Hess rose to his feet and declared, "Thank you, Bruce, for a lecture that shakes geology to its very foundations."[36] Some years prior, Hess had rejected a paper by Heezen on the very topic of the Mid-Atlantic Ridge, and its rift. Hess would become one of the key developers of modern plate tectonics.

1959. Publication of the monograph *The Floors of the Ocean: 1. The North Atlantic* by Heezen, Tharp, and Ewing. Choice passages about map abstraction include one where they discuss the difference between preparing a terrestrial and a marine physiographic diagram: "In the former the major problem is to select from more-detailed maps the features to be represented. [...] In contrast, the preparation of a marine physiographic diagram requires the author to postulate the patterns and trends of the relief on the basis of cross sections and then to portray this interpretation in the diagram."[37]

[35]Felt (2012), "same scale" Loc 1737.

[36]Meritt (1979), 273.

[37]Heezen et al. (1959, 3).

1966. The long-term episode Tharp and Heezen came to call "the harassment", which had already started to rumble due to their 1965 trip to India, Thailand, Taiwan, and Australia, if not before, intensified and came to a head for all involved parties as a consequence of a press conference at the 1966 2nd International Oceanographic Congress in Moscow. Heezen and Tharp shared information at the Congress that they had not strictly been authorized by Ewing, and Lamont more generally, to circulate. Furthermore, a paper Heezen had co-authored with other "Lamonters" (but not Ewing) was initially rejected by *Science*, but then accepted by *Nature*. Ewing was upset because Lamont policy was to have two senior scientists approve every paper before these were submitted to conferences, conference proceedings, or specialist journals. This protocol had not been followed when Heezen and co-author's paper was sent to *Nature*.[38] Of various descriptions of the harassment available, Marie Tharp puts it most directly and authoritatively:

> We had planned to study the Mediterranean Sea next, but we were diverted instead to the Indian Ocean [Fig. 5.6], because a diagram of it was urgently needed to help plan the International Indian Ocean Expedition. Now our efforts were [eventually] thwarted by a long-lasting falling out between Bruce and Doc. There are two sides to that story, but the result was that Doc banned Bruce from Lamont ships and denied Bruce access to Lamont data. He tried unsuccessfully to fire Bruce, who had a tenured faculty position at Columbia, but he did fire me. From then on, I was paid through research grants that Bruce received from the Navy, and I continued the mapping working at home.[39]

1967. First angled panorama map (Fig. 5.6) produced by Tharp, Heezen, and Berann. Tharp and Heezen would, on several occasions over the years, stay at Berann's house near Innsbruck, Austria, for long periods of time. According to Felt "The story of how the Indian Ocean map came into existence unfolds rather like the plot of a *Mission: Impossible* episode," and the interested reader is invited to consult Chap. 17 of her book for background on Fig. 5.6.[40]

1968. First absolute panorama map (Fig. 5.2) by Tharp, Heezen, and Berann. Moreover, Figs. 1 and 7 of W. J. Morgan's influential and classic 1968 *Journal of Geophysical Research* article "Rises, Trenches, Great Faults, and Crustal Blocks" were based on Heezen and Tharp maps.

1977. Heezen dies off the coast of Iceland. Tharp reports: "On June 21, 1977, Bruce Heezen died suddenly of a heart attack in a submarine [NR-1] near the Reykjanes Ridge. I was on the research ship *Discovery* studying the Ridge from above. We had recently completed work on our world ocean floor panorama and each had proofs with us on our respective boats."[41] *The New York Times* published an obituary two days after Heezen's death, which included this sentence: "The Heezen-Tharp physiographic maps, first of the North Atlantic and then of all major oceans of the world, were widely circulated by the National Geographic Society."[42]

---

[38] Felt (2012), "senior research scientists", Loc 2900.

[39] Tharp (1999).

[40] Felt (2012), Loc 2451.

[41] Felt (2012), Loc 3818.

[42] Sullivan (1977).

1977/1978. Appearance of World Ocean Floor Panorama Map.[43] The mid-oceanic rift system spanning the entire globe is now shown as a *single system*—Earth as a Frankenstein-monster patchwork of tectonic plates (Fig. 5.7).
1978. Tharp attends a session of the General Bathymetric Chart of the Oceans (GEBCO) guiding committee in Ottawa, Canada, where plans for future editions of the World Ocean Floor Panorama were being considered. GEBCO figuratively ripped her map out of her hands in an act that could appropriately be called "systemic piracy". An online article puts it dramatically, but accurately: "Marie Tharp [...] had to sit still while a roomful of men dismembered her legacy and divvied up the remnants among themselves in a frenzy of violent opportunism [...] She watched ocean after ocean snatched from her grasp, her prospects for future work

Fig. 5.6 The Indian Ocean Floor angled panorama map by Berann, Tharp, and Heezan was a supplemental, foldout map in the October 1967 issue of *National Geographic*. Subscriptions to that magazine numbered six million in the USA alone (Felt 2012, Loc 2810). Heinrich Berann/National Geographic Creative. National Geographic Image Collection

[43]Proofs completed in 1977. Felt (2012), Loc 4121: "The first copy of the World Ocean Floor Panorama—conceptualized by Marie Tharp and Bruce Heezen, painted by Heinrich Berann with assistance from Heinz Vielkind, and funded by the U.S. Office of Naval Research—rolled off the presses at about 7:00 p.m. on May 17, 1978." In the final stretch of producing the WOFP, Tharp had hired a Ukrainian cartographer, Luba Prokop. WOFP has since appeared in many places, in various avatars, and even in poster format.

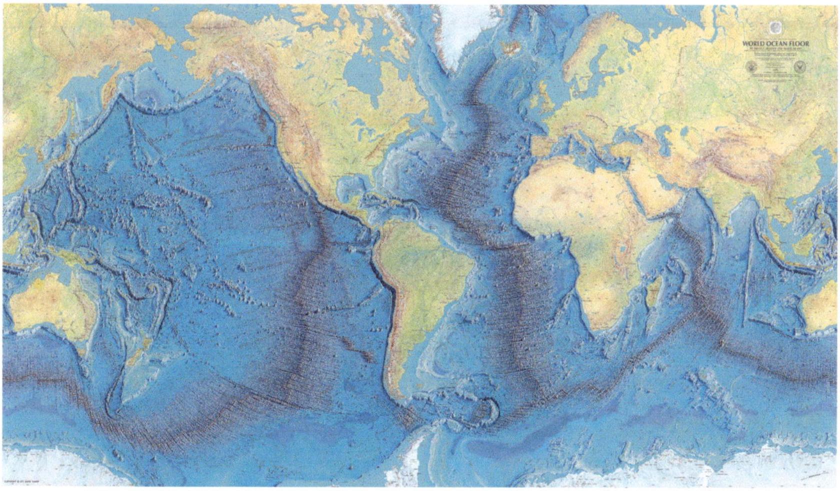

Fig. 5.7 World Ocean Floor Panorama, Bruce C. Heezen and Marie Tharp, 1977. Copyright by Marie Tharp 1977/2003. Reproduced by kind permission of Marie Tharp Maps LLC image provided by Lamont–Doherty Earth Observatory

chopped to a few sectors around Australia, hardly enough to sustain her financially or intellectually for more than a few months."[44]

1982. An official version of what Tharp calls her "Opus" appears in a commemorative volume on Heezen.[45] For the remainder of her life, she works and revises this autobiographical writing, which otherwise remains unpublished.

1997. Tharp is named one of the four greatest cartographers of the 20th century by the Library of Congress's Geography and Map Division's Philip Lee Society. That same year, her work is shown in a Library of Congress exhibition *American Treasures from the Library of Congress*, marking the 100th anniversary of the Jefferson Building. At the opening gala, for which President Clinton is present, she sees the original draft of the Declaration of Independence, maps drawn by George Washington, and the Emancipation Proclamation, among other treasures, from her wheelchair. The friend accompanying her recounts how she cried when her eyes finally fell on one of her ocean floor maps. She tells him, "I wish that Papa and Bruce could see it."[46]

2001. Tharp receives the first annual Lamont–Doherty Heritage Award.[47]

2006. Tharp dies of cancer in Nyack, New York.

[44]Debakcsy (2018).

[45]Tharp (1982).

[46]Felt (2012), Loc 4637.

[47]Bizzarro (2001).

Map Types

Tharp and Heezen produced five kinds of maps:

Physiographic diagrams provide a "45 degree view" from above, with stylized iconography and shading (Figs. 5.1 and 5.4).

Profiles are cross-sections of the ocean floor, with vertical heights exaggerated 40 times (Figs. 5.4 and 5.5).

"*Panorama maps*" painted by Berann, are of 3 kinds:

Perspective maps by Berann, under Tharp and Heezen's guidance, were similar to Richard Edes Harrison's World War II perspective maps, as if looking at Earth from a satellite some 40,000 km above Earth's surface (Northern Atlantic Ocean[48]; Winther 2020, Chap. 3).

Angled panorama maps are a kind of bird's-eye, abstracted view where the whole image is angled/curved, yet the horizon is flat (Fig. 5.6).[49] Berann painted the Himalayas and Alps in this manner, and – under the guidance of Tharp and Heezen – the ocean bottoms. Mapping the deep blue oceans indeed.

Absolute panorama maps are painted as an all-knowing view from an absolute vantage point – the Mercator projection is appropriate for this purpose, and was used (Figs. 5.2 and 5.7).

Cartopower and the Future of Mapping

The depths of the oceans are a mystery. No comprehensive, fine-grained bathymetric map exists. Not yet. Only 5% of the ocean bottoms have been fully mapped.[50]

Recent satellite technologies permit precise measurements of sea surface topography and gravitational anomalies across the planet. With satellite altimetry and gravitational potential measurements, new comprehensive, small-scale maps can be drawn (Fig. 5.8). Such maps do not, for better *and* worse, use interpolation and extrapolation. These coarse-grained maps precisely portray the data at the highest level of resolution the data permit. Yet, much work remains to be done.

Whatever our future mapping will look like, one thing is certain: Like all forms of representation, they will exist within a power structure. I call the specific forms of power encoded in maps "cartopower". Cartopower is twofold: first, it is the political, economic, and social power structure, often invisible, behind a map; second, it is the power that a map exerts in the world via its ontological assumptions. Power scaffolds maps, and maps exert power—maps thereby build worlds.[51]

An anatomy of the cartopower of Tharp and Heezen's maps illuminates, yet again, the ubiquity and disproportionate importance of military and corporate interests in many scientific endeavors.

---

[48]*National Geographic*, June 1968. Found here: http://www.berann.com/panorama/archive/image/PN_W_10.jpg. Accessed November 8, 2018.

[49]Patterson (2000) explores this kind of panorama map.

[50]See Copley (2014) for a clear exposition of what this actually means.

[51]I discuss cartopower in terms of my multiple representations account of ontologizing in Chap. 5 of Winther (2020). See Harley (2001) and Wood (1992) for related views.

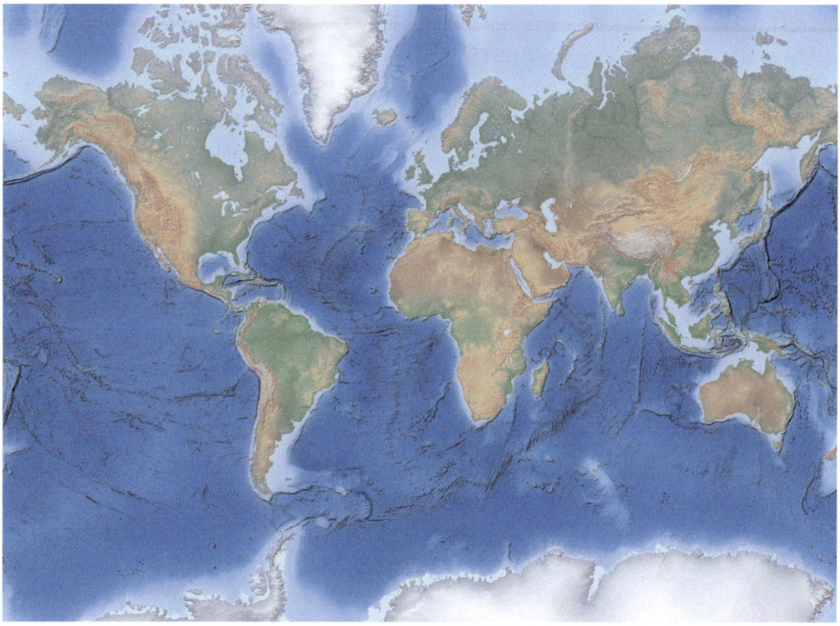

Fig. 5.8 Bathymetric map using gravitational anomalies and satellite altimetry from data provided at the Scripps Institution of Oceanography (UCSD), and originally explained in Smith and Sandwell (1997). Data available here: https://topex.ucsd.edu/WWW_html/mar_topo.html Drawn by cartographer and graphic designer Mats Wedin

Consider the power structures that enabled Tharp and Heezen's groundbreaking research. A historical article by Doel et al. (2006) explores how underwater bathymetry became very secretive during and after WWII, after an initial global free information/open source period immediately following WWI. The US Navy now wished to collect secret information about where their submarines could hide, the location of seamounts and mountains into which submarines could crash, and the whereabouts of enemy submarines.[52] To these ends, the Navy was busy developing an underwater "SOund SUrveillance System" (SOSUS)—listening devices that could detect Soviet submarines. In all of this, the Pentagon decided that "creating a comprehensive map of the ocean floor" was essential.[53] Thus, "Lamont Geological Observatory was a quintessential Cold War institution, largely dependent on military contracts to support its research programs."[54] Heezen's and Tharp's research was funded by heavy military interests.

Corporate interests did not take a backseat. In the early 1950s, AT&T Bell Labs was busy trying to create the first trans-Atlantic commercial marine telephone lines.

---

[52]Doel et al. (2006, 608).

[53]Doel et al. (2006, 608).

[54]Doel et al. (2006, 609).

Interestingly, these labs also worked closely with the American military on its classified SOSUS project. Heezen's direct (and Tharp's indirect) collaboration with AT&T Bell Labs provided them with two crucial resources: "a rich, vastly expanded source of seafloor data" and "invaluable financial resources."[55] I would also go so far as to agree with Doel et al's statement that "Bell Labs funding made the Heezen-Tharp North Atlantic physiographic map *possible*."[56]

Much like the Cold War growth of physics, space technology, and computer science, the emergence of maps of the ocean floor by Tharp and Heezen was suffused with cartopower. This was a high-stakes mapping project. As precise, beautiful, scientific, and creative as Marie Tharp's maps were, they were also buffeted about in a perfect storm of political, culture, military, and economic power.

Just as there is a continuity of positivist, scientific abstraction practices across cartography and GIS, so there are ongoing concerns with bias, discrimination, and moneyed interests.[57] For instance, in a milestone article, feminist GIS'er Mei-Po Kwan asks: "is GIS an inherently masculinist technology or social practice? How are particular subjectivities or gendered identities constituted through routine interaction with GIS technology? Do women and men interact with or use GIS technology differently? [...] How may GIS technology perpetuate gender inequality or occupational segregation in the information technology labor market and women's status in geography?"[58] These questions are clearly important to any perspective on critical issues in GIS, and the ongoing nature and deployment of cartopower. They also point to issues of sexual and other forms of bias that researchers may suffer, shaping the way that their work becomes available or not.

Consider the problem of what I will call "personal style harassment", which is when creative spirits, with independent flair, find themselves moving around restlessly—both in their minds and in the world—unable to fit into the power structures at their institutional home. In the case explored in this chapter, Heezen's creative personal style conflicted strongly with the power structure of Director Ewing's Lamont. The institution issues rules, which are open to interpretation. Even when there is some modicum of clarity about such rules, there are many of them, and a reasonable and overworked human being is simply not able to follow all of them. Such limitations are tacitly accepted by the upper administration, which ignores small infractions or suppresses, to some extent, enforcement of narrow rule-following. They do this *until* a creative thinker comes upon the scene, trying to contribute on her own terms, in her own tempo, sometimes shaking the foundations of her field. She then gets every rule thrown at her. Tharp and Heezen were both subject to such a personal

[55] Doel et al. (2006, 610, 611).

[56] Doel et al. (2006, 611).

[57] Discussions of the simultaneous empirical and technological *and* social and political facets of GIS can be found in, e.g., Kwan (2002), Schuurman (2004), Pavloskaya (2006), St. Martin and Wing (2007), Cope and Elwood (2009), Crampton (2010), Dodge et al. (2009, 2011).

[58] Kwan (2002, 275, footnotes suppressed).

style harassment, with Tharp experiencing further bias or harassment in the form of sexism.[59]

Science journalist Stephen S. Hall made the point clearly and forcefully in his obituary of Tharp:

> Maurice (Doc) Ewing, the brilliant and autocratic director of what is now the Lamont-Doherty Earth Observatory at Columbia, remained famously unpersuaded by the growing evidence of continental drift and began to clash with Heezen over both ideas and ego. Heezen had become a tenured professor, but Ewing did what he could to thwart the mapping project. He refused to share important data about the sea floor with the map makers — data that Heezen's graduate students sometimes surreptitiously "exported" to Tharp and her assistants. He stripped Heezen of his departmental responsibilities, took away his space, drilled the locks out of his office door and dumped his files in a hallway. Most important, Ewing blocked Heezen's grant requests and, as [Paul J.] Fox said, "was essentially trying to ruin Bruce's career."[60]

We must address scientific and technical features of cartography and GIS *as well as* these complex and interrelated fields' social, political and economic aspects. This includes social conditions at institutional as well as interpersonal scales. Science is data, theory, and knowledge; but science is also politics, economics, and ethics. Whatever the fate of the map within GIS, its conceptual framework developed within cartography has much to teach us, even those of us working on GIS.

Conclusion: Ocean Mapping and Gratitude

The first part of this chapter reviewed some basic map-making abstraction practices. Whenever we compare a map to its territory, we shift from everyday, human-scale perception to something more detached and abstract. I tried to show the continuity between analog and digital cartography in strategies of abstraction. As abstraction practices, selection, simplification, and exaggeration apply as much to old-school cartographic maps as to cutting-edge GIS efforts.

I also surveyed the tremendous careers of Marie Tharp and Bruce Heezen. In addition to carefully studying Tharp's maps, if we also learn about how she turned her house, and later Heezen's house, into a cartographic data management center, training zone, gourmet kitchen, studio, and library, we are left with no doubt about how remarkable these researchers were. In the end, Tharp looked back at her life with gratitude:

---

[59]Doel et al. (2006, 609) proclaim: "Their early careers offer a snapshot of the divergent opportunities for men and women in the earth sciences in mid-twentieth century America. One of the very few female researchers at Lamont during its first decades, Tharp had limited financial security and few opportunities to attend scientific meetings. Typical for this period, her contributions often remained invisible." Moreover, recall the 1978 GEBCO affair above, where Marie Tharp's work was forcefully removed from her—in my moral universe, this was an act of piracy against Tharp. For a discussion of the "climate and consciousness" (9) of women in geography (not geology) see Monk (2004).

[60]Hall (2006).

> Not too many people can say this about their lives: The whole world was spread out before me (or at least, the 70 percent of it covered by oceans). I had a blank canvas to fill with extraordinary possibilities, a fascinating jigsaw puzzle to piece together: mapping the world's vast hidden seafloor. It was a once in a lifetime—a once in the history of the world—opportunity for anyone, but especially for a woman in the 1940s. The nature of the times, the state of the science, and events large and small, logical and illogical, combined to make it all happen.[61]

The stories of Tharp and Heezen also remind us, however, that politics, greed, and discrimination die not. We have much to do not only on environmental and ecological matters but also on social equity. In gratitude for what we have today and with hope for a genuinely sustainable future let us please get to work.

Acknowledgements Kristine Sergejeva provided expert project coordination. Laura Laine secured figure permissions. Mats Wedin processed and finalized figures. Lucas McGranahan and Gloria Sturzenacker provided outstanding editorial commentary. Heidi Svenningsen Kajita and Mette Bannergaard Johansen were sounding boards. Gratitude to all.

References

Abler RF (1987) The National Science Foundation National Center for Geographic Information and Analysis. Int J Geogr Inf Sci 1(4):303–326

Barton C (2002) Marie Tharp, oceanographic cartographer, and her contributions to the revolution in the Earth sciences. Geol Soc Lond Spec Publ 192(1):215–228

Battersby SE, Finn MP, Usery EL, Yamamoto KH (2014) Implications of the Web Mercator and its use in online mapping. Cartographica 49(2):85–101

Bizzarro D (2001) Lamont-Doherty Earth Observatory bestows Heritage Award on Marie Tharp, pioneer of modern oceanography. Columbia News. http://www.columbia.edu/cu/news/01/07/marieTharp.html. Last modified 18 September 2002

Borges JL ([1946] 1975) On exactitude in science. In: Collected fictions (trans: Hurley A). Penguin, London, p 325

Brassel KE, Weibel R (1988) A review and conceptual framework of automated map generalization. Int J Geogr Inf Sci 2(3):229–244

Bressan D (2018) Marie Tharp, the woman who discovered the backbone of Earth. Forbes (Online, Accessed 30 July 2018.)

Brotton J (2012) A history of the world in 12 maps. Viking, New York

Bugayevskiy LM, Snyder JP (1995) Map projections: a reference manual. CRC Press, Boca Raton, FL

Carroll L ([1893] 2010) Sylvie and Bruno concluded. Nabu Press, Charleston, SC

Cope M, Elwood S (2009) Qualitative GIS. A mixed methods approach. Sage, London

Copley J (2014) Just how little do we know about the ocean floor? The Conversation (website). https://theconversation.com/just-how-little-do-we-know-about-the-ocean-floor-32751. Accessed 2 Oct 2018

Couclelis H (1992) People manipulate objects (but cultivate fields): beyond the raster-vector debate in GIS. In: Frank AU, Campari I, Formentini U (eds) Theories and methods of spatial–temporal reasoning in geographic space. Springer, Berlin, pp 65–77

Crampton JW (2010) Mapping: a critical introduction to cartography and GIS. Wiley-Blackwell, Chichester, UK

[61] Tharp (1999).

Debakcsy D (2018) Making continents move: the ocean cartography of Marie Tharp. Women You Should Know (website). https://womenyoushouldknow.net/cartography-marie-tharp/. Accessed 18 Oct 2018
Dodge M, Kitchin R, Perkins C (eds) (2011) The map reader: theories of mapping practice and cartographic representation. Wiley-Blackwell, Chichester, UK
Dodge M, Perkins C, Kitchin R (2009) Mapping modes, methods, and moments: a manifesto for map studies. In: Dodge M, Kitchin R, Perkins C (eds) Rethinking maps: new frontiers in cartographic theory. Routledge, London, pp 220–243
Doel RE, Levin TJ, Marker MK (2006) Extending modern cartography to the ocean depths: military patronage, Cold War priorities, and the Heezen-Tharp mapping project, 1952–1959. J Hist Geogr 32:605–626
Douglas DH, Peucker TK (1973) Algorithms for the reduction of the number of points required to represent a digitized line or its caricature. Can Cartogr 10:112–122
Eco U ([1992] 1994) On the impossibility of drawing a map of the Empire on a scale of 1 to 1. In: How to travel with a Salmon and other essays. Harcourt, New York, pp 95–106
Elmendorf CH, Heezen CH (1957) Oceanographic information for engineering submarine cable systems. Bell Syst Tech J 36(5):1047–1093
Esri (n.d.) FAQ: what are some common map scales? https://support.esri.com/en/technical-article/000010532. Accessed 20 Aug 2018
Felt H (2012) Soundings: the story of the remarkable woman who mapped the ocean floor, Kindle edn. Henry Holt, New York
Goodchild MF (1992) Geographical information science. Int J Geogr Inf Sci 6(1):31–45
Greenhood D (1964) Mapping. University of Chicago Press, Chicago
Gutenberg B, Richter CF (1954) Seismicity of the Earth. Princeton University Press, Princeton, NJ (As cited in Heezen et al. 1959, 110.)
Hall SS (2006) The contrary map maker. The New York Times, Dec 31. https://www.nytimes.com/2006/12/31/magazine/31Tharp.t.html. Accessed 6 Nov 2018
Hardaker T (2002) Cartographic introduction. In: Hammond. Compact Peters World Atlas. Hammond, Union, NJ
Harding S (1986) The science question in feminism. Cornell University Press, Ithaca, NY
Harley JB (2001) The new nature of maps: essays in the history of cartography. Johns Hopkins University Press, Baltimore, Maryland
Heezen B, Tharp M, Ewing M (1959) The floors of the oceans. I. The North Atlantic. Text to accompany the physiographic diagram of the North Atlantic. Special paper 65. The Geological Society of America, New York
Jha A (2009) Google Earth drops into the oceans. The Guardian, February 2. https://www.theguardian.com/environment/2009/feb/02/google-earth-oceans. Accessed 7 Nov 2018
Keller EF (1984) A feeling for the organism: the life and work of Barbara McClintock, 10th anniversary ed. Henry Holt, New York
Keller EF, Longino HE (1996) Feminism and science. Oxford University Press, Oxford, UK
Kimerling AJ, Buckley AR, Muehrcke PC, Muehrcke JO (2009) Map use: reading and analysis, 6th edn. ESRI Press, Redlands, CA
Koláčný A (1969) Cartographic information: a fundamental concept and term in modern cartography. Cartogr J 6:47–49
Krygier J, Wood D (2011) Making maps: a visual guide to map design for GIS, 2nd edn. Guilford Press, New York
Kwan M-P (2002) Is GIS for women? Reflections on the critical discourse in the 1990s. Gend Place Cult: J Fem Geogr 9(3):271–279
Lamont-Doherty Earth Observatory (n.d.) History of Lamont. https://www.ldeo.columbia.edu/about-ldeo/history-lamont-0. Accessed 12 Oct 12 2018
Landa ER (2010) The ties that bind: soil surveyor William Edgar Tharp and oceanographic cartographer Marie Tharp. Phys Chem Earth 35:868–880

Lobeck AK (1948) A physiographic diagram of the United States. (Map.) A. J. Nystrom, Chicago. https://www.davidrumsey.com/luna/servlet/detail/RUMSEY~8~1~234507~5510052:United-States----Physiographic-Diag. Accessed Oct 2018
Longino HE (2001) The fate of knowledge. Princeton University Press, Princeton, NJ
Lüscher P, Weibel R, Burghardt D (2009) Integrating ontological modelling and Bayesian inference for pattern classification in topographic vector data. Comput Environ Urban Syst 33:363–374
MacEachren A (1995) How maps work: representation, visualization, and design. Guilford Press, New York
Merritt JL (1979) Hess's geological revolution. In: The best of PAW: (100 Years of the Princeton Alumni Weekly), pp 273–277. https://www.princeton.edu/geosciences/about/publications/Harry_Hess_bop.pdf Accessed 24 Aug 2018
Monk J (2004) Women, gender, and the histories of American geography. Ann Assoc Am Geogr 94(1):1–22
Montello DR (2002) Cognitive map-design research in the twentieth century: theoretical and empirical approaches. Cartogr Geogr Inf Sci 29(3):283–304
Morgan WJ (1968) Rises, trenches, great faults, and crustal blocks. J Geophys Res 73(6):1959–1982
Muehrcke PC (1969) Visual pattern analysis: a look at maps. PhD dissertation, University of Michigan
Muehrcke PC (1972) Thematic cartography. Resource paper no. 19, Commission on College Geography. Association of American Geographers, Washington DC
Muehrcke PC (1974a) Map reading and abuse. J Geogr 73(5):11–23
Muehrcke PC (1974b) Beyond abstract map symbols. J Geogr 73(8):35–52
Muehrcke PC, Muehrcke JO (1998) Map use: reading, analysis and interpretation, 4th edn. ESRI Press, Redlands, California
Oreskes N (1999) The rejection of continental drift: theory and method in American earth science. Oxford University Press, New York
Patterson T (2000) A view from on high: Heinrich Berann's Panoramas and landscape visualization techniques for the U.S. National Park Service. Cartogr Perspect 36:38–65
Pavloskaya M (2006) Theorizing with GIS: a tool for critical geographies? Environ Plan A 38:2003–2020
Ramer U (1972) An iterative procedure for the polygonal approximation of plane curves. Comput Graph Image Process 1:244–256
Rankin W (2016) After the map: cartography, navigation, and the transformation of territory in the twentieth century. University of Chicago Press, Chicago, IL
Robinson AH, Petchenik BB (1976) The nature of maps, essays toward understanding maps and mapping. University of Chicago Press, Chicago
Sandwell DT, Müller RD, Smith WHF, Garcia E, Francis R (2014) New global marine gravity model from CryoSat-2 and Jason-1 reveals buried tectonic structure. Science 346(6205):65–67
Schuurman N (1999) Reconciling social constructivism and realism in GIS. ACME: Int E-J Crit Geogr 1(1):75–90
Schuurman N (2004) GIS: a short introduction. Blackwell Publishing, Malden, MA
Shea KS, McMaster RB (1989) Cartographic generalization in a digital environment: when and how to generalize. In: AUTO–CARTO 9, ninth international symposium on computer-assisted cartography, Baltimore, Maryland, pp 56–67
Smith G-H (1959) Armin Kohl Lobeck, geomorphologist and landscape artist, 1886–1958. Ann Assoc Am Geogr 49(1):83–87. https://www.jstor.org/stable/2561290
Smith WHF, Sandwell DT (1997) Global sea floor topography from satellite altimetry and ship depth soundings. Science 277(5334):1956–1962
Snyder JP (1993) Flattening the Earth. University of Chicago Press, Chicago, IL
St. Martin K, Wing J (2007) The discourse and discipline of GIS. Cartographica 42(3):235–248
Strebe D (2012) Why Mercator for the web? Isn't the Mercator bad? (Posted 15 March 2012.) https://www.mapthematics.com/forums/viewtopic.php?f=8&t=251

Sullivan W (1977) Dr. Bruce C. Heezen, 53, dies; mapped ocean floors. The New York Times, June 23. https://www.nytimes.com/1977/06/23/archives/dr-bruce-c-heezen-53-dies-mapped-ocean-floors.html. Accessed 20 Oct 2018

Tharp M (1982) Mapping the ocean floor: 1947–1977. In: Scrutton RA, Talwani M (eds) The ocean floor: Bruce Heezen commemorative volume. Wiley, New York, pp 19–31

Tharp M (1999) Connect the dots: mapping the seafloor and discovering the mid-ocean ridge. In: Lippsett L (ed) Lamont-Doherty Earth Observatory: twelve perspectives on the first fifty years, 1949–1999. Lamont-Doherty Earth Observatory of Columbia University, Palisades, NY. Parts available at: http://www.whoi.edu/page.do?pid=7500&tid=282&cid=23306. Accessed 26 Oct 2018

Tharp M, Frankel H (1986) Mappers of the deep. Am Hist 95(10):48–62. http://www.faculty.umb.edu/anamarija.frankic/files/Oceanography%20Spring%2008/MidAtlantic%20ridge%20discovery.pdf.gz. Accessed 14 July 2018

Twain M (1894) Tom Sawyer Abroad. Charles Webster & Co

Winther RG (2012) Interweaving categories: styles, paradigms, and models. Stud Hist Philos Sci Part A 43(4):628–639

Winther RG (2014) James and Dewey on abstraction. The Pluralist 9(2):1–28

Winther RG (2015) Mapping kinds in GIS and cartography. In: Kendig C (ed) Natural kinds and classification in scientific practice. Routledge, New York, pp 197–216

Winther RG (2020) When maps become the world. University of Chicago Press, Chicago, IL

Wolter JA (1975) Cartography—an emerging discipline. Can Cartogr 12(2):210–216

Wood D (1992) The power of maps. Guilford Press, New York

Wright JK (1942) Map makers are human: comments on the subjective in maps. Geogr Rev 32(1):527–544

Chapter 6
On the Distinction Between Classical and Nonclassical Geographies: Some Critical Remarks

Timothy Tambassi

Abstract In *Ontological Tools for Geographic Representation*, Roberto Casati, Barry Smith, and Achille Varzi have formalized and introduced the (geo-ontological) distinction between classical and nonclassical geographies. Although that distinction makes no essential reference to maps, the authors have pointed out that the dichotomy can be useful to specify the kind of geography that is implied in the spatial representation. Thus, the aim of this paper is to showcase the main assumptions behind the distinction between classical and nonclassical geography and to present some possible issues arising from its application to cartographic representation. Accordingly, the first two sections offer a short introduction to the scopes of the ontology of geography, and to the main theoretical tools needed for advancing a (formal) theory of spatial representation. The third section shows some issues emerging from the application of the distinction between classical and nonclassical geographies to the cartographic representation, by discussing (and expanding) the list of examples provided by the three authors.

Keywords Ontology of geography · Classical geography · Nonclassical geographies · Cartographic representation · Geographical entities

Ontology of Geography and Spatial Representation

In some recent publications (Tambassi 2016, 2017a, b), I define the ontology of geography as that part of the philosophical ontology concerned with the mesoscopic world of geographical partitions, and aimed at the following:

1. arguing whether and how the geographical descriptions of reality emerging from common sense can be combined with descriptions proposed by professional geographers (Geus and Thiering 2014);

T. Tambassi (✉)
Department of Humanities, University of Eastern Piedmont, Vercelli, Italy
e-mail: timothy.tambassi@gmail.com

ICUB, University of Bucharest, Bucharest, Romania

© Springer Nature Switzerland AG 2019

T. Tambassi (ed.), *The Philosophy of GIS*, Springer Geography,
https://doi.org/10.1007/978-3-030-16829-2_6

2. establishing what kinds of geographical entities exist (and might exist) and how they can be defined and classified in an ontological system that gathers them together (Smith and Mark 2001);
3. developing a formal theory of spatial representation, with special reference to spatial phenomena on the geographic scale (Casati et al. 1998).

Here, I do not intend to deny such a definition on the aims of the ontology of geography, even if some clarifications might be useful. The first one concerns the fact that, in the analytic area, the development of *an ontological reflection on geography* has been mainly focused on the theoretical and technical needs of GIS and geo-ontologies, rather than on making explicit assumptions and commitments of geography as a discipline. In this sense, it might be more precise to talk about this area of research in terms of "ontology of GIS and geo-ontologies",[1] and to underline its distance from those ontological reflections on geography coming from the "continental philosophy" (Hacking 2002; Elden 2003; Schatzki 2003; Escobar 2007; Harvey 2007; Dean 2010; Law and Lien 2012; Joronen 2013; Kirsch 2013; Shaw and Meehan 2013; Springer 2013; Whatmore 2013; Blaser 2014; Bryant 2014; Roberts 2014; Joronen and Häkli 2017). The second clarification concerns the geographical debate, within which there are many different (geo-)ontological approaches, other than the perspective taken in these pages (Vallega 1995; Berque 2000; Raffestin 2012; Boria 2013). Accordingly, the analysis of the relationship between ontology and geography shows a multifaceted nature, for which it would probably be more correct talking of ontologies in the plural and, then, of ontologies of geography (Tambassi 2018a).

With these thoughts in mind, in the following pages, I intend to maintain the definition above, by highlighting that points 2 and 3 are strictly interconnected: in particular, the latter may be thought as dependent upon the former. Indeed, as Casati and Varzi (1999) remark, the advance of a theory of spatial representation should be combined with (if not grounded on) an account of the kinds of entity that can be located or take place in space. In short, it means to provide a definition of what may be collected under the rubric of *spatial entities* and to outline how to distinguish them from purely *spatial items* (such as points, lines, regions and so forth). Moreover, developing such a formal theory also implies choosing between absolutist and relational theories of space. The former maintains that the space exists as an independently subsistent individual (a sort of container) over and above its inhabitants (objects, events, and spatial relations between objects and events, or without all these entities). Conversely, the relational theory considers that spatial entities are cognitively and metaphysically prior to space. Thus, there is no way to identify a region of space except by reference to what is or could be located at that region (Casati and Varzi 1999).

[1] However, according to Pesaresi (2017), the ontology behind GIS could be more extensive than the three points presented here.

Classical and Nonclassical Geographies

In order to enhance a theory of spatial representation, the (analytical) ontology of geography has developed three main theoretical tools strictly interconnected and mutually interacting:

- mereology, which is the theory of parthood relations (Simons 1987; Smith and Mark 1998; Casati and Varzi 1999; Mark et al. 1999);
- topology that is the theory of qualitative spatial relations such as continuity, contiguity, overlapping, and so forth (Smith 1994, 1995, 1996; Smith and Varzi 2000);
- theory of spatial location, which deals explicitly with the relationship between an entity and the spatial region it occupies or in which it is located. In a strict geographical sense, this relation is not the one of identity—a geographical entity is not identical with the spatial region it occupies, besides two or more different geographical entities can share the same location at the same time—and it does not imply that any single geographical entity is located somewhere, or that any spatial region is the location of a geographical entity (Casati et al. 1998; Varzi 2007).

In addition to these tools, Casati et al. (1998) also introduce the distinction between classical and nonclassical geographies that, according to the authors, can be useful for the specification of the (kind of) geography behind spatial representations.

Starting from the fact that there is no single universally recognized formulation that precisely indicates what classical geography is, the authors characterize a geography on a region *R* as a way of assigning (via the location relation) geographic objects of given types to parts or subregions of *R*. Then, they propose to put forward some principles for a minimal characterization of geographic representation, which are such that the violation of one of those principles produces intuitively incomplete representations.

Under these assumptions, the three authors sustain that the term classical geography[2] (CG) does not carry any normative claim. It simply describes a rather robust way of *tiling* regions in the presence of certain general axioms, which specify that:

CG1. every single geographic entity (nations, lakes, rivers, islands, etc., but also mereological combinations of these entities) is located at some unique spatial region;
CG2. every spatial region has a unique geographic entity located at it;
CG3. if two (or more) entities are located are at the same spatial region, then they are the same entity.

Consequently, a geography can be considered as nonclassical (NCG) if it:

NCG1. drops one or more of the axioms of CG;
NCG2. (and/or) adds axioms to those of CG.

[2]For an analysis of the notion of "classical geography" in a geographical context, see Lukermann (1961), Geus and Thiering (2014).

According to the second condition (NCG2), we could, for example, add an axiom to obtain the effect that all geographic units are connected, or consider how the properties of geographic boundaries relate themselves to the axioms of classical geography. Instead, for what concerns the first alternative (NCG1), we might observe that it licenses nonspatial geographic units as well as maps with gaps and gluts. To be more precise, denying that every geographical entity is located at some unique region allows to include also nonspatial geographical entities, entities with multiple location or duplicates of the same geographical entity. Again, to discard that every spatial region has a unique geographic entity located at it enables us to consider maps with regions that are assigned no entity, or two or more competing units.

Issues from Cartographic Representation

Now, although the distinction between CG and NCG makes no (essential) reference to maps, Casati, Smith, and Varzi maintain that a model of CG may also be visualized as a set of instructions for coloring maps, according to which, once that a set of colors is fixed:

1. every subregion of the map has some unique color;
2. (and) every color is the color of a unique region of the map.

CG1 would consequently be satisfied by point 2, CG2 by point 1 and CG3 may be thought as a logical consequence of 1 and 2. Analogously, we could easily generate a CG via an act of tiling which, for example, divides the Earth's surface:

ECG1. into land and water;
ECG2. (or) among nations (including quasi-nations such as Antarctica), national waters, and international waters.

But how to imagine a model of NCG? By circumscribing the analysis to NCG1, it is interesting to underline that Casati, Smith, and Varzi purpose, in their examples, at least four different models of NCG.

The Capital of Singapore

In the first case, a model of NCG is obtained by dropping CG3, according to which if two or more entities share the same location (are located at the same spatial region), then they are the same entity. According to the authors, such a drop would allow maps with spatial regions that are assigned two or more competing for geographical entities. As a consequence, this would also permit the representation of disputed lands, on which, for example, two (or more) different nations could concurrently declare their sovereignty. The resulting nonclassical map could be easily rethought also in terms that preserve the axioms of CG, for example by considering all such spatial regions as occupied by geographical entities of the *Disputed Land* type.

However, in my opinion, a more controversial situation might arise by means of another example: that is, by adding to ECG2—the map that divides the Earth's surface into nations, national waters, and international waters—some points that locate, on that map, the capitals of all the nations. In this context, on the one hand, we could think that CG3 is not respected. For example, we could consider the points that locate the capitals, as points where two different geographical entities are located at the same time: the nations themselves and their capitals. Otherwise, on the other hand, we may not have difficulties also in considering CG3 as respected. Indeed, we could take all those points in the map as occupied by the (geographical entities) capitals exclusively. As an alternative, we could also show how the different conditions of existence and the criteria of identity of nations and (their) capitals do not prevent them from sharing the same spatial location, without creating overlaps—that, in terms of CG, would lead them to be considered as the same entity.

Looking for No One's Land

The second model of NCG is achieved by dropping CG2, according to which every spatial region has a unique geographic entity located at it. Besides having maps with two or more competing geographical entities located at the same spatial region, such a drop licenses maps with spatial regions that are assigned no geographical entity. In this context, Casati, Smith, and Varzi remark that a default assignment that preserves the axioms of CG would consist in considering all spatial regions of the kind as occupied by an object of the *No Man's Land* type.

However, all the clarifications offered by the authors may not exhaust the issues of the cartographic representation related to the drop of CG2. Let's take, for example, ECG2 and remove the geographical entity *Suriname* from the spatial region occupied by Suriname itself. According to the authors, in this case, we could preserve the axioms of CG by assigning to the spatial region no longer occupied by the geographic entity *Suriname* an entity of the *No Man's Land* type. Despite this, in my opinion, the same result could be also obtained by placing no entity at all on that spatial region, neither *No Man's Land* entity. Indeed, if we keep the distinction between spatial regions and geographical entities, we could do without *Suriname* and *No Man's Land* (entities), only by thinking of *something* whose conditions of existence and identity are simply defined by the boundaries of the neighboring geographical entities—in this case, by French Guyana to the East, Brazil to the South, Guyana to the West and Atlantic Ocean to the North. At this point, we might further ask whether the boundaries of those geographic entities define a geographic entity or a spatial region. Consequently, it is reasonable to infer that in the former case we might have to do with a CG, whereas in the latter with an NCG.

Sailing to Thule

The last two models of NCG are the outcome of the drop of CG1, according to which every single geographic entity is located at some unique spatial region. In the first of the two cases, Casati, Smith, and Varzi consider the possibility of duplicates of the same geographical entity, which would contradict CG, because the entity in question would be located at (at least) two different spatial regions.

The example given by the authors is the People's Republic of China (located in the Mainland China) and the Republic of China (located on the Island of Taiwan): both claim to be *only* China, but we cannot accept neither of the claims since we assume CG1. Despite the double connotation of China, are we really faced with a duplicate of the same geographical entity, given that People's Republic of China and Republic of China have different conditions of existence and identity? In other words, are we dealing with a model of NCG that shows duplicates of the same geographical entity? To preserve the axioms of CG, the authors have suggested considering *China* (entity) as the mereological sum of the competing spatial regions, which correspond to Mainland China and the Island of Taiwan. In line with this, may we actually consider (the whole) China as the result of such a mereological sum? To be more precise, could we really define China as the sum of spatial regions currently occupied by the People's Republic of China and Republic of China?

Just to add further hurdles, we might also consider the puzzling case of Thule and the several theories about its possible location, which include, among others, the coastline of Norway, Iceland, Greenland, Orkney, Shetland, Faroe Islands, and Saaremaa. Now, if we imagine a map that shows all these locations, then, in my opinion, we would be hardly inclined to consider Thule as the mereological sum of all the locations ascribed to it. At the same time, it would be unlikely to consider the various Thule represented on the map (with different conditions of identity) as duplicates of the same geographical entity. Perhaps, we could take the various points that locate Thule on that map as indicating different geographical entities, to which different authors have attributed the same connotation. But, in that case, how might we interpret that map? As a model of CG or NCG?

Poland into Exile

The last possibility of providing a model of NCG that excludes CG1 considers the inclusion of nonspatial geographic entities. The example provided by Casati, Smith, and Varzi is that of Poland during the Era of Partition—namely, during the era in which the entity in question did not have any territory to call as its own land. According to the authors, such a model of NCG might be converted into a model of CG by naming a certain, arbitrarily chosen, region as *Ersatz*-Poland—for example, the headquarters of the Government in Exile in London—so to preserve its spatial location.

But, on the basis of these assumptions, how could we talk about geographical entities such as Kosovo, Holy Roman Empire, cities, villages or Benelux in ECG2 that, as we said, divides the Earth's surface into nations, national waters, and international waters? Considering the nonunanimous recognition by all UN member states, the case of Kosovo is interesting, since its inclusion in ECG2 could depend on its being accepted as an independent nation or as a region that belongs to Serbia. Consequently, if ECG2 did not include Kosovo, such a map could be a model of:

- CG for Serbia, which does not recognize Kosovo as a nation;
- NCG for Italy, which recognizes Kosovo as a nation.[3]

Conversely, if ECG2 included Kosovo, ECG2 could be considered as a model of CG for Italy, but not for Serbia that could see in ECG2 the drop of CG3 for the presence, in the spatial region occupied by Kosovo, of two different competing entities: Kosovo and Serbia. Such a situation may be further complicated by the Holy Roman Empire. Indeed, if we can hardly discard it as a geographical entity, does its exclusion from ECG2 make that map a model of NCG—given the existence of a geographical entity that is actually nonspatial? Or, eventually, should we think about models of CG and NCG only by reference to geographical entities that actually have or can have a spatial location? And if it would be the case, should we also include nations such as Sahrawi Arab Democratic Republic, Pridnestrovian Moldavian Republic and Republic of Somaliland (nations that have a limited recognition)[4] in order to consider ECG2 as a model of CG? And what about geographical entities such as cities and villages? Does the absence of these geographical entities in ECG2 make the entities nonspatial in this context and, consequently, turn ECG2 into a model of NCG? The same question can be extended to geographical entities such as Benelux or imaginary geographical entities, but also to entities that are the result of a mereological sum of other geographical entities (for example, the mereological sum of New Zealand, Prussia, and Normandy). In all these cases, could the absence of imaginary or arbitrary geographical entities make ECG2 becoming a model of NCG?

Conclusion

The purpose of these pages has been that of sketching out the distinction between classical and nonclassical geographies, which, according to Casati, Smith, and Varzi, can be useful for specifying the kind of geography that is implied in the spatial representation. Presuppositions and axioms of these sorts of geographies have not been criticized. However, some possible ambiguities related to the application of this

[3]Accordingly, Italy might also consider Kosovo as a nonspatial geographical entity, at least in this specific context.

[4]To be more precise, Sahrawi Arab Democratic Republic is a non-UN member state recognized only by few UN member states; the Pridnestrovian Moldavian Republic is a non-UN member state recognized only by non-UN member states; the Republic of Somaliland is a non-UN member state not recognized by any state.

distinction to the cartographic representation has been remarked. Those ambiguities, in my opinion, make the distinction not entirely clear cut and open up the possibility of different interpretations, which (in turn) could subordinate the distinction to the subjects (whoever they are) involved in its application.

I find a recurring *leitmotiv* that might be identified in the absence of a (shared) definition of geographical entity. Such a *leitmotiv* arises from the four models of NCG and makes the individuation of the kind of geography implied in different spatial representations a difficult task to pursue. Thus, by circumscribing the possible issues related to the geo-ontological debate, it is only with a shared definition of geographical entity that we might lay the foundation for an unambiguous application of the distinction between classical and nonclassical geographies to cartographic representation.

So, what is a geographical entity? What are its conditions of existence and identity? Should we include in our rubric of geographical entities only entities that could be portrayed on a map or also nonspatial and/or abstract entities? Should we consider only those entities that actually have (or could have) a spatial location or also those entities that actually do not have, have never had and/or will have? How to deal with the geographical entities the location of which is (or was) vague or indeterminate? In Chap. 9, we try to answer these (and other) questions, dealing with the ontological conundrums represented by the absence of a (shared) definition of a geographical entity.

Acknowledgements This article is a partial translation of Tambassi (2018b). Ontologia della geografia e rappresentazioni cartografiche: uno spunto critico. Bollettino della Società Geografica Italiana, 1(1), 19–26. Doi:https://doi.org/10.13128/bsgi.v1i1.86. Online: https://riviste.fupress.net/index.php/bsgi/article/view/86, with some minor revisions. I am grateful to the publisher (Firenze University Press) and the editor (Margherita Azzari) for permission to use the material in question. Sections 2 has also some overlaps with Tambassi (2017a). Thanks are due to Giulia Lasagni and Paola Zamperlin. They know why.

References

Berque A (2000) Ecoumène: introduction à l'étude des milieux humains. Belin, Paris
Blaser M (2014) Ontology and indigeneity: on the political ontology of heterogenous assemblages. Cult Geogr 21(1):49–58
Boria E (2013) Genealogie intellettuali e discontinuità nazionali nella storia della cartografia. Bollettino della Società Geografica italiana 6(3):443–460
Bryant LR (2014) Onto-cartography. an ontology of machines and media. Edinburgh University Press, Edinburgh
Casati R, Smith B, Varzi AC (1998) Ontological tools for geographic representation. In: Guarino N (ed) Formal ontology in information systems. IOS Press, Amsterdam, pp 77–85
Casati R, Varzi AC (1999) Parts and places. MIT Press, Cambridge (MA)
Dean M (2010) Governmentality: power and rule in modern society. SAGE, London
Elden S (2003) Reading genealogy as historical ontology. In: Milchman A, Rosenberg A (eds) Foucault and heidegger: critical encounters. University of Minnesota Press, Minneapolis
Escobar A (2007) The 'ontological turn' in social theory. Trans Inst Br Geogr 32(1):106–111

Geus K, Thiering M (2014) Common sense geography and mental modelling: setting the stage. In: Geus K, Thiering M (eds) Features of common sense geography. implicit knowledge structures in ancient geographical texts. LIT, Wien
Hacking I (2002) Historical ontology. Harvard University Press, Cambridge (MA)
Harvey D (2007) The Kantian roots of Foucault's dilemmas. In: Crampton JW, Elden S (eds) Space, knowledge and power: Foucault and geography. Ashgate, Hampshire
Joronen M (2013) Heidegger, event and the ontological politics of the site. Trans Inst Br Geogr 38(4):627–638
Joronen M, Häkli J (2017) Politicizing ontology. Prog Hum Geogr 41(5):567–579
Kirsch S (2013) Cultural geography I Materialist turns. Prog Hum Geogr 37(3):433–441
Law J, Lien ME (2012) Slippery: field notes in empirical ontology. Soc Stud Sci 43(3):363–378
Luckermann F (1961) The Concept of Location in Classical Geography. Ann Assoc Am Geogr 51(2):194–210
Mark DM, Smith B, Tversky B (1999) Ontology and geographic objects: an empirical study of cognitive categorization. In: Freksa C, Mark DM (eds) Proceedings of spatial information theory. cognitive and computational foundations of geographic information science. international conference COSIT'99 stade, Germany, Springer, Berlin, Heidelberg, pp 283–298
Pesaresi C (2017) Applicazioni GIS. Principi metodologici e linee di ricerca. Esercitazioni ed esemplificazioni guida. Torino, Utet
Raffestin C (2012) Space, territory, and territoriality. Environ. Plan. D: Soc. Space 30(1):121–141
Roberts T (2014) From things to events: whitehead and the materiality. Environ and Plan D: Soc Space 32:968–983
Schatzki T (2003) A Nnew societist social ontology. Philosophy of the Social Sciences 33(2):174–202
Shaw IG, Meehan K (2013) Force-full: power, politics and object-oriented philosophy. Area 45(2):216–222
Simons P (1987) Parts: an essay in ontology. Clarendon Press, Oxford
Smith B (1994) Fiat objects. In: Guarino N, Pribbenow S, Vieu L (eds) Parts and wholes: conceptual part-whole relations and formal mereology. proceedings of the ECAI94 workshop, ECCAI, Amsterdam, pp 15–23
Smith B (1995) On drawing lines on a map. In: Frank A, Kuhn W (eds) Spatial information theory —a theoretical basis for GIS. proceedings, international conference Cosit'95, semmering, Austria (Lecture notes in computer science), vol 988. Springer, Berlin, pp 475–484
Smith B (1996) Mereotopology: a theory of parts and boundaries. Data Knowl Eng 20:287–303
Smith B, Mark DM (1998) Ontology and geographic kinds. In: Poiker TK, Chrisman N (eds) Proceedings of the eighth international symposium on spatial data handling. International Geographical Union, Burnaby, British Columbia, pp 308–320
Smith B, Mark DM (2001) Geographical categories: an ontological investigation. Int J Geogr Inf Sci 15(7), 591–612. http://idwebhost-202-147.ethz.ch/Courses/geog231/SmithMark_GeographicalCategories_IJGIS2001@2005-10-19T07%3B30%3B52.pdf
Smith B, Varzi AC (2000) Fiat and bona fide boundaries. Philos Phenomenol Res 60(2):401–420
Springer S (2013) Human geography without hierarchy. Prog Hum Geogr 38(3):402–419
Tambassi T (2016) Rethinking geo-ontologies from a philosophical point of view. J Res Didact Geogr (J-READING), 2(5), 51–62. https://doi.org/10.4458/7800-04
Tambassi T (2017a) The philosophy of geo-ontologies. Springer, Cham
Tambassi T (2017b) A geographical taxonomy for geo-ontologies. Axiomathes 27(4):355–374. https://doi.org/10.1007/s10516-016-9309-z
Tambassi T (2018a) Introduzione. Filosofia della geografia. Semestrale di Studi e Ricerche. Geografia 30(2):9–14
Tambassi T (2018b) Ontologia della geografia e rappresentazioni cartografiche: uno spunto critico. Bollettino della Società Geografica Italiana, serie 14, 1(1), 19–26. https://doi.org/10.13128/bsgi.v1i1.86

Vallega A (1995) La regione, sistema territoriale sostenibile. Compendio di geografia regionale sistematica. Mursia, Milano

Varzi AC (2007) Spatial reasoning and ontology: parts, wholes and location. In: Aiello M, Pratt-Hartmann I, van Benthem J (eds) Handbook of spatial logics. Springer, Berlin, pp 945–1038

Whatmore S (2013) Earthly powers and affective environments: an ontological politics of flood risk. Theory Cult Soc 30(7–8):33–50

Part III
Geo-ontological Perspectives

Chapter 7
Drawing Boundaries

Barry Smith

Abstract In "On Drawing Lines on a Map" by Smith (*Spatial information theory. A theoretical basis for GIS*, Springer, Berlin/Heidelberg/New York, 1995), I suggested that the different ways we have of drawing lines on maps open up a new perspective on ontology, resting on a distinction between two sorts of boundaries: *fiat* and *bona fide*. "Fiat" means, roughly: human-demarcation-induced. "Bona fide" means, again roughly: a boundary constituted by some real physical discontinuity. I presented a general typology of boundaries based on this opposition and showed how it generates a corresponding typology of the different sorts of objects which boundaries determine or demarcate. In this paper, I describe how the theory of fiat boundaries has evolved since 1995, how it has been applied in areas such as property law and political geography, and how it is being used in contemporary work in formal and applied ontology, especially within the framework of Basic Formal Ontology.

Keywords Ontology · Geospatial information science · Spatial boundaries · Fiat boundaries · Fiat objects · Truthmakers

Introduction

In "On Drawing Lines on a Map" (Smith 1995), I described an approach to the ontology of reality resting on the thesis that extended entities can have boundaries of two different sorts. On the one hand, there are what we might call *bona fide* boundaries, which correspond to physical discontinuities of the sort illustrated by coastlines or the surface of your skin.[1] On the other hand, are *fiat* boundaries, which are boundaries introduced in the absence of physical discontinuities, for example,

[1] I am grateful to Thomas Bittner, Berit Brogaard, Roberto Casati, David Mark, Alan Ruttenberg, Andrew Turk, Achille Varzi, Laure Vieu, Lars Vogt, and Wojciech Zelaniec for many helpful contributions.

B. Smith (✉)
Department of Philosophy, University at Buffalo, Buffalo, NY 14260, USA
e-mail: phismith@buffalo.edu

© Springer Nature Switzerland AG 2019

T. Tambassi (ed.), *The Philosophy of GIS*, Springer Geography,
https://doi.org/10.1007/978-3-030-16829-2_7

the boundary of Utah or the boundary through your body separating your upper from your lower torso.

The idea of such a distinction was inspired by the theory of boundaries and the continuum sketched in Brentano (1988) and in Chisholm (1989). Both Brentano and Chisholm recognized that there is something problematic in treating a continuum as a collection (set, class, aggregate) of noncontinuous points or elements. Each continuum is, rather, ontologically prior to the boundaries or cuts that we may make within its interior. Such cuts are, by definition, not present in the continuum itself. Rather, they are added, for example, through an act of demarcation. A formal theory of boundaries and the continuum was developed in this light in my paper in the Chisholm volume of the *Library of Living Philosophers* (Smith 1997), and this led in turn to the formal theory of bona fide and fiat boundaries presented in Smith and Varzi (1997, 2000).

Fiat boundaries may lie entirely skew to all boundaries of the bona fide sort (as in the case of the boundaries of Utah and Wyoming). They may also (as in the case of the boundaries of Indiana and Pennsylvania) involve a combination of fiat and bona fide portions; or they may be constructed entirely out of bona fide portions which because they are not themselves intrinsically connected, must be conjoined via a fiat boundary (Smith 2007).

Once fiat boundaries have been recognized, we can apply the fiat–bona fide dichotomy also to the corresponding (bounded) *entities*.[2] Fiat entities—as for example in the case of parcels of real estate—are distinguished by the fact that they exist only because certain fiat boundaries exist. In some cases, this will reflect some specific human decision. In other cases, a fiat boundary will exist not in virtue of some specific human decision but rather in the reflection of the physical properties of the object itself.[3] That fiat part of a mountain which is above 500 feet above sea level exists independently of any specific contour map, and independently of the institution of contour maps. But it is a fiat part, nonetheless.

We can draw fiat boundaries also in the temporal realm to yield fiat *processes*: the Renaissance, the Millennium, the Second World War, the Reagan Years, my childhood, and so forth. All of these are perfectly objective sub-totalities within the totality of all processes making up the history of the universe, even though the spatial reach, as well as the initial and terminal boundaries of, for example, the Second World War, were decided (in different ways) by fiat.

The examples of fiat entities mentioned above are all cases where proper parts are delineated or carved out (by fiat) within the interiors of larger bona fide wholes. They are examples of entities created by moving from the top (or middle) down. But we can also proceed from the bottom up, by constructing higher level fiat objects out of

---

[2]On the use of "fiat entity" rather than "fiat object" see the section on Basic Formal Ontology, below.

[3]In a series of papers, beginning with (Vogt et al. 2011), Lars Vogt and his collaborators have presented a powerful critique of the conception of fiat boundaries as originally formulated in "On Drawing Lines on a Map", emphasizing above all the need to modify the assumption that fiat boundaries go hand in hand in every case with human decision or demarcation (Vogt 2018a, b; Vogt et al. 2012a, b).

lower level bona fide objects as parts. This is because, while bona fide entities such as tables, apples, persons, planets are unitary, and thus connected, fiat entities may be scattered; they may be such as to include separate bona fide objects as parts (Smith 1999a). Polynesia is a geographical example of this sort; other examples might be: the Polish nobility, the constellation Orion, or the species *cat*. Such higher order fiat object aggregates may themselves be unified together into further fiat entities (say: the Union of Pacific Island Nations). The fiat boundaries to which higher order fiat entities owe their existence are the mereological sums of the (fiat and bona fide) outer boundaries of their respective lower order constituents.[4]

Fiats Perceptual and Ecological

Are entities of these fiat sorts of ontological significance? Can basic principles of metaphysics really turn on the rather elaborate beliefs and conventions that human beings have evolved in relation to place, space, and politico-administrative jurisdiction? To see why these questions must be answered in the positive, consider what happens when two political entities (nations, counties) or two parcels of real estate lie adjacent to one another. The entities in question are then said to share a common boundary. This sharing of a common boundary is a peculiarity of the fiat world. For when two bona fide objects converge upon each other, for example, people shaking hands, then what happens physically in the area of apparent contact has to do first of all with a compacting of molecules on either side, and ultimately with aggregates of subatomic particles whose location and whose belongingness to either one or other of the two bodies are only statistically specifiable. Genuine coincidence of bona fide boundaries is thus impossible, if "coincidence" means: identity of spatiotemporal location.[5] To see what is involved here we note first of all that in the geographical realm (or in other words on the geographic level of granularity)—for example, in the geographic region where a coastal territory meets the sea—we draw fiat boundaries, even though that are delineated not by *sharp* outer boundaries but rather by boundary-like regions which are to some degree indeterminate. The boundary between two hands in a handshake is a boundary of this sort.

We can draw in this connection on the work on visual perception of the ecological psychologist J. J. Gibson, who takes as his starting point not internal visual imagines nor retinal excitations but rather the entities out there in the world which are the targets of perception. As Gibson writes:

> We are tempted to assume … that we live in a physical world consisting of bodies in space and that what we perceive consists of objects in space. But this is very dubious. The terrestrial

[4]Compare Bittner and Smith (2001, 2003a, b) on the theory of granular partitions.

[5]Details are provided in Smith and Varzi (2000), which sets forth the formal differences between the coincidence of boundaries in the fiat realm and the mere proximity of boundaries which is achievable in the realm of physical bodies.

> environment is better described in terms of a medium, substances, and the surfaces that separate them (Gibson 1979: 16).

Here substances are stuffs—rock, soil, sand, oil, wood, the tissues of animals. They are all more or less resistant to deformation and penetration by solid bodies and more or less permanent in shape, and they are all generally opaque. Media, in contrast, such as air and water, are relatively insubstantial, and solid bodies can move through them without much resistance. Surfaces, finally, separate media from substances and they separate substances from each other where they come into contact (Stroll 1988: 126).

The combinations of medium, substances, and surfaces that we experience exhibit what Gibson calls "affordances", which he defines as "what the environment offers the animal, what it provides or furnishes, whether for good or ill" (Gibson 1979: 127), as a chair affords sitting, a staircase affords climbing, or an angry bear affords fleeing.

Affordances involved in every case a combination of bona fide and fiat entities. A fiat boundary is created when light casts a shadow across a part of your cave, or when an animal looks up as you cross into its territory. Affordances may involve what we might call negative entities—holes, cavities, openings (Casati and Varzi 1994). A tunnel, for example, is bounded physically by its walls, floor, and roof; at its entrance and exit, however, it must make do with fiat boundaries. There is a tunnel through your body that passes from the esophagus through the stomach and on to the small and large intestines. These various parts of the tunnel are separated in virtue of bona fide boundaries founded in the different microscopic structures of the different portions of the tunnel lining. But the tunnel itself is continuous, and so the boundaries separating the successive subtunnels within the tunnel are fiat in nature.

Varzi (2016) presents the ingredients of a view according to which reality is one single continuum, so that all boundaries are fiat in the sense that all boundaries are human-induced. A more extreme view would have it that *no* entities are fiat in nature. Rather, our talk of what we are here calling "fiat entities" is, precisely, talk, and thus of no further ontological significance. Some friends of what we might think of as "ultimate physics" hold that there are bona fide objects—ultimate atoms of reality—but that these exist on a level way beneath our everyday ken. They would thus reject candidate meso- and macroscale bona fide objects such as people and planets, viewing these, again, as a matter of mere *façons de parler*.[6]

The Linguistic Windowing of Reality

There are also fiat entities which arise specifically in virtue of the groupings and refinings which are involved in our use and understanding of language. This occurs in a two-fold process. On the one hand, linguistic entities such as spoken words and sentences are themselves processes demarcated in fiat fashion out of concrete

[6]See Davies (2018) on the metaphysical implications of the fiat /bona fide distinction.

sound-material that is in itself not cleanly separated out into linguistic units via discontinuities in the flow of sound. In addition, the flow of sound and of sound production is fused in one or other way with underlying mental and neurological accompaniments. On the other hand, external reality, too, is tailored, or pared down by fiat, in order that it will fit our linguistically generated windows of salience. Thus, if I say, truly, "John built mud pies in the sand", then the real-world target of my utterance is a certain portion of reality[7] involving John, some sand, a complex plurality whose constituent unitary parts are comprehended through the phrase "mud pie". If I say, truly, "John shocked Mary", then the real-world correlate of the verb of this sentence is a complex dynamic affair (a fiat process, or what is elsewhere called a "process profile" (Smith 2012a)) that is comprehended through the transitive verb "shock". Participants in the process involve John, Mary, some utterance or gesture on the part of John, and some mental process on the part of Mary that is caused by this utterance or gesture and has both Mary and John as its targets.

The way in which natural language contributes to the generation of fiat boundaries can be illustrated in relation to the correlated linguistic phenomena of *mass* versus *count* and *verbal aspect* (Galton 1984; Mourelatos 1981).

When I point toward a cattlefield and assert "there is cattle over there", then the target of my assertion differs from my target when I assert "there are cows over there". The underlying real bovine material is in both cases the same, but I impose different sorts of boundaries on this material in the two cases.

Verbal aspect has to do with the "internal temporal constituency" of the events toward which our empirical judgments are directed (Comrie 1976). Consider the concrete factual material that is John kissing Mary on a given occasion. In this factual material, here again, fiat boundaries can come to be drawn by language in a variety of ways. It can be comprehended as: "John is kissing Mary", "John is repeatedly kissing Mary", "Mary is constantly being kissed by John", "Mary has been being constantly kissed by John since 1884", and so on (Thomsen and Smith 2018).

Such carving out of linguistic fiat objects is in part a matter of simple grouping together, for example, through the use of plural referring expressions such as "Hannah and her sisters" or "Siouxsie and the Banshees" (Ojeda 1993). But it is in part also a matter of windowing or foregrounding (Talmy 1996). If I point to a group of irregularly shaped bumps in the sand and say "dunes", then the real-world correlate of my expression is a complex plurality (a higher order fiat object with non-crisp boundaries) divided, via the general type *dune*, into constituent non-crisply delineated parts (Smith 1987). Cognitive linguists such as Talmy (1995, 1996) and Langacker (1987/1991) have rightly emphasized the degree to which language effects subtle articulations of this sort.

Another variety of fiat boundary-creation is effected through our use of expressions such as "this" and "that" in relation to objects in space. This involves in each case the drawing of a transient, imaginary boundary, lying in the region in front of the speaker, which is such that the objects labeled "this" and "that" lie on opposing sides of the boundary. The use of "here", similarly, involves the creation of an

[7] Ceusters and Smith (2015).

ephemeral fiat boundary comprehending a roughly spherical volume of space around the speaker, a volume whose size, shape, and location, and sometimes also degree of crispness (Smith and Brogaard 2003a) are contextually determined. Transient boundary-creation of this sort is effected in the same way independently of order of magnitude, from the tiniest ("this flea") to the grossest ("that empire," "that galaxy").

As Talmy has shown (1995), boundaries are at work also in cases of the following sort:

I offered Agnes the book [creates a virtual sphere around the recipient].

She accepted the book [Agnes allows the sphere to be broken].

She rejected the book [Agnes maintains the sphere unbroken].

which involve the creation of nonphysical paths and boundaries of a range of different sorts.

It is important to realize that the fiat boundaries drawn in cases such as this are drawn in the world of bona fide objects. While it is true that all objects which we grasp linguistically are grasped through our linguistically expressed concepts, we should not move from there to all objects which we grasp linguistically exist only in virtue of our linguistically expressed concepts.[8]

Everyday objects and processes are described by cognitive linguists such as Talmy (1995) and Lakoff (1987) as existing in the "conceptual realm". Even space itself is often described by Talmy as a mere "conceptual domain" in a way that implies that, in the absence of concept-using subjects like ourselves, space would not exist. What I am proposing here, however, is that the fiat boundaries induced through natural language are of a piece with geographical fiat boundaries. This makes it clear how Talmy's position is to be corrected: the fiat boundaries to which reference is constantly made in our natural language utterances are not in any sense in our heads, or in some conceptual sphere. Rather, they are out there in the world. They are not, however, physical in nature. Rather, they are analogous to other ephemeral sociocultural formations—such as debts, claims, bank balances—entities which are parts of what Frege would call "objective" reality, yet not such as to fall within the domain of physical science (Frege 1884: Chap. 1).

And now, if some fiat boundaries—like the borders of nations or postal districts—are social entities in this sense, then like the latter they will be subject to legal regulations. When the legal system takes up into its orbit a vaguely bounded region (a wetland) or vaguely bounded processes (for example, the process of dying) then it characteristically adds a rule that is designed to make the relevant boundary precise. Private property in some jurisdictions extends to the mean low water mark, and for any coastal portion of the United States or Canada, there are some legal definition based on mean low, high, average, etc. tide level, as to where private property stops and a commons starts. Definitions are needed also as to how such determinations apply when boundaries cross the mouths of rivers. If the legal system needs to know where the shoreline is in order to regulate access, then it will need to pick some particular stage in the tidal cycle, such as mean low tide level; it thus creates a fiat

8 Stove calls the argument in favor of views of this "the gem". See Stove (1991).

shoreline that is fixed and reasonably crisp, and this exists as it were alongside the bona fide shoreline that moves with the tides. You cannot see or touch or trip over the fiat shoreline; but the fiat shoreline is there, nonetheless, as a part of reality: if you cross it, you will be fined.

Truthmakers as Fiat Objects

We can now expand our treatment of linguistically generated fiat boundaries to throw light on the notion of truth as classically understood in terms of a *correspondence* between a judgment or assertion on the one hand and a certain portion of reality on the other. Reality does not, of course, come ready-parceled into judgment-shaped portions that are predisposed to stand in relations of correspondence of the given sort. The discipline of logical semantics has thus tended to treat, not of truth as such (truth to reality), but rather of truth *in a model*, where the model is a specially constructed set-theoretic reality-surrogate. The theory of fiat boundaries can help us to avoid the need for this resort to surrogates by allowing us to treat judgment itself as a *sui generis* variety of drawing fiat boundaries.

True judgments effect a drawing of boundaries that is successful in the sense that it does not conflict with reality. The resultant boundaries themselves are drawn, as already described, in the extended world of genuine objects and associated processes. The fiat entities they circumscribe are typically many-sorted: they include both objects and processes (as the correspondent sentences include both nouns and verbs). Such entities are on the one hand autonomous: that region of reality through which the given boundary is drawn—for example, the complex of objects and processes which are involved in John's kissing Mary—exists in and of itself, regardless of our judging activity, and so do all its constituent subregions. The whole itself is, however, also in a certain sense dependent on our judgment. For in the absence of the judging activity through which the drawing of the fiat boundary is effected, a portion of reality of just this sort would in no way be demarcated from its surroundings. Judgment-shaped portions of reality can in this way exist in reality objectively, and be precisely tailored to make our judgments true, but they are fiat rather than bona fide in nature.

There is, as Talmy puts it, a windowing of reality that is effected by our uses of language, especially of those descriptive uses of language which are involved when we make true empirical judgments. The ephemeral fiat boundaries effected through declarative sentences are analogous to the ephemeral boundaries of the visual fields associated with our acts of visual perception as described in (Smith 1999c). Veridical judgments stand to those portions of reality which are their fiat judgment-correlates as acts of perception stand to their associated visual fields.

Each true empirical judgment can be seen as effecting a division of reality in fiat fashion in such a way as to mark out a certain truthmaking region consisting of those entities that are relevant to the truth of the judgment in question. Truth itself can then be defined as the relation of correspondence between judgment and its corresponding truthmaking region, and a true judgment is in this sense analogous to a map of the corresponding portion of geographic reality.[9] A view of truth along these lines rests on an account of the windowing of reality via language that is of a piece with the ecological account of perceptual windowing advanced by Gibson.

Fiat Boundaries in Feature Spaces

The fiat–bona fide opposition can be identified also in the realm of qualities. We distinguish first between determinate and determinable qualities, where the former—for example, *this specific shade of red*—are specifications of the latter—for example, the quality *red*. We can imagine the determinates of a determinable such as color arranged in a quasi-spatial way, as happens in accounts of color- or tone-space (Gärdenfors 2000; Guarino 2013; Johansson 1989). When an object changes its color continuously, for instance moving through the color spectrum from red to violet as a result of continuous heating, we then draw fiat divisions along this spectrum through our use of color terms, dividing it into *red*, *orange*, *yellow* and so forth. This process is subject to a certain degree of variation in determining where the boundaries are to be drawn, for example, between different cultures and different specialized areas (colors of wine, hair, and so forth).

We draw analogous fiat partitions also in spaces of variation along non-qualitative dimensions, for example in classifying geographic entities such as "strait" and "river" (Mark et al. 2001). The English language might have evolved with just one term, or three terms, comprehending the range of phenomena stretching between *strait* and *river* or, in French, between *détroit* and *fleuve*. For while the Straits of Gibraltar are certainly not a river, and the Mississippi River is certainly not a strait, things like the Detroit River, the Saint Claire River, the Dardanelles, and the Bosphorus are borderline cases. All are flat, narrow passages that ships can sail through between two larger waterbodies (lakes, seas), and all have net flow through them due to runoff. Is Lake Erie really a lake, or just a wide, deep part of the river-with-five-names that is called the St. Lawrence as it flows into the sea? Well, that depends on what you mean by "lake".

Quine has put forward a radical proposal according to which even classical conceptual distinctions drawn in metaphysics are distinctions of this fiat sort. Consider three scattered partitions of the world made up of rabbits, of rabbit stages, and of undetached rabbit parts, respectively. The results of each of these partitions, as Quine sees it, make up when taken collectively just the same scattered portion of the world. The only difference "is in how you slice it" (Quine 1969: 32). What he means, in our terms, is that the conceptual divisions between continuants, stages and undetached parts are mere products of fiat. Since the reference is behaviorally inscrutable as concerns such distinctions, Quine concludes that there is no fact of the matter that they might reflect—no fact of the matter on the side of the objects themselves as these

[9] A detailed formal theory of truthmaking along these lines is presented in Smith (1999b).

existed before they were addressed in our language. Continuants, parts, and stages do not differ from each other in virtue of any corresponding (bona fide) differences on the side of the corresponding entities in reality. Rather they differ from each other in the way in which, when asked to count the number of objects in the fruit bowl, you can say either: "one orange", or "two orange-halves", or "four orange-quarters", and so on—and you will give the right answer in each case. The distinctions in question are merely the products of our fiat partitions of one and the same reality.

But note that Quine is being too hasty when he affirms in defense of his thesis of "ontological relativity" that there is no ontological fact of the matter as concerns the reality to which we are related when using singular referring terms. For it follows from his own doctrine that it is a fact of the matter that this reality is intrinsically undifferentiated as far as the mentioned ontological distinctions are concerned. This is just the other side of the coin from the fact that the corresponding boundaries are on his view entirely fiat in nature. This putative ontological fact of the matter, however, faces problems. For it is itself a presupposition of the thesis of ontological relativity to the effect that there are no ontological facts of the matter.

Quine comes close to a view according to which *all* boundaries on the side of objects in reality are of the fiat sort. Objects of reference, for him, can comprise any content of some portion of spacetime, however, heterogeneous, disconnected and gerrymandered this may be. For us, on the other hand, there are some ways of referring to things and processes that track bona fide boundaries in reality and others that do not. It is the job of fundamental science to move us in the direction of such bona fide joints of reality, though even when science has completed this job there will of course still be room for delineations of the lesser sort, which track boundaries—for example of Quebec, of the 70 mph zone on the highway, or of the No Smoking Section of your favorite restaurant—which exist only as a result of our acts of fiat.

Jeffersonian Fiats

When Jefferson first drew his map of proposed states of the Northwest Territory in 1784 (see Fig. 7.1), drawing off 14 neat checkerboard squares between the boundaries of the Atlantic colonies and the Mississippi River, his result was sufficiently inaccurate that it did not even have the Great Lakes in the right place. In the end, 10 states were nonetheless created in this area, having boundaries that follow Jefferson's lines in large degree.

Delineations such as these can be effective in creating objects in the geospatial realm only if the pertinent boundaries are, in the jargon of topology, *Jordan curves* (broadly: the boundary of a geopolitical or administrative entity must be free of gaps and must nowhere intersect itself). They are effected from the top down in the sense that there are no units or elements from out of which the corresponding fiat entities could be seen as being constructed in analogy with the way in which sets are constructed out of their members.

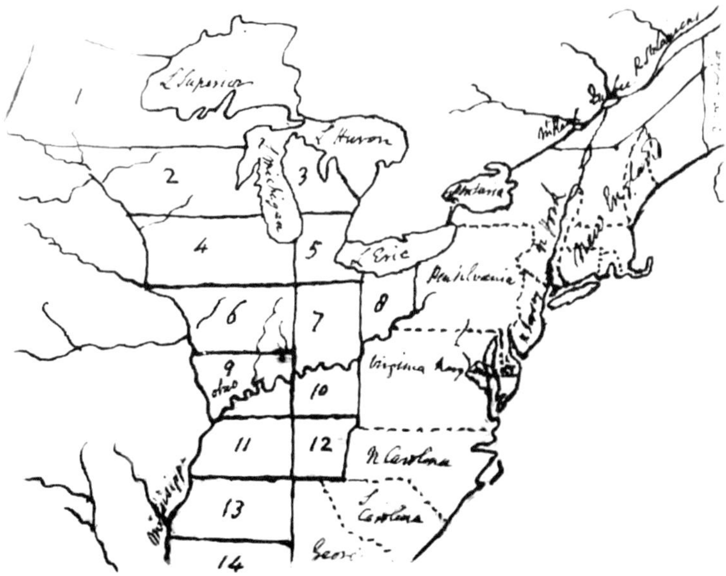

Fig. 7.1 Map of proposed states of the Northwest Territory, drawn by Thomas Jefferson in 1784

This is because geographers deal with regions of different shapes and sizes, and with sub-regions of these regions, and with the ways these regions and sub-regions overlap or fail to overlap (Casati and Varzi 1999). They deal, in other words, with a mereologically structured world. Some of Jefferson's delineations correspond to bona fide boundaries: river banks, coastlines, and the like. These are boundaries in the things themselves, and they would exist (and did indeed already exist) even in the absence of all delineating or conceptualizing activity on our or Jefferson's part. Many borders of political and administrative units in the North-American continent correspond to no genuine heterogeneity on the side of the bounded entities themselves.

Often, of course, such boundaries do in course of time come to involve boundary-markers: walls, barbed-wire-fences, border-posts, watch-towers, and the like, and these will sometimes replace what is initially a fiat boundary with something more tangible, something physical. Fiat and bona fide objects are interrelated also epistemologically. Thus, in cadastral practice certain objects, for example, surveyors' pegs placed to establish a boundary, enjoy a privileged status in determining at later times where the boundary lies. This means that there are laws governing the use of such objects, as also of posts, walls, fences and so forth, as evidence of boundary location, laws, for example, having the effect of limiting the degree to which walls may be moved when rebuilt. Such laws institute a new layer of fiat boundaries, attached to the primary layer and constituting surrounding fiat zones of tolerance.

But, however, arbitrary a given geospatial demarcation might be, there are reasons of a nonarbitrary sort why these and those fiat boundaries are created rather than others. Thus, it was a complex medley of considerations relating to shipping, trade winds, harbors, climate, markets, and so on, which led our ancestors to create the fiat region called "the North Sea" in a way which could not just as well have motivated them to create what would have been called, say, "the Middle Sea" stretching between the Bermudas, the Azores, and Gotland. As already noted, fiat boundaries, in general, owe their existence not merely to human fiat but also to associated real properties of the relevant factual material (they are functions of affordances, in Gibson's terms). As demarcated in mesoscopic (geographical) reality they are in every case linked to bona fide objects at various scales without which the relevant demarcations could not be effected at all. It is already for this reason a confusion to suppose that all objects (or all mesoscopic objects) might be in some sense of the fiat type. As the reports of boundary commissions make abundantly clear, the very possibility of fiat demarcation presupposes the existence of bona fide landmarks in relation to which fiat boundaries can be initially specified and subsequently relocated.

It is interesting in this respect to consider the question when an imaginary mathematical line (a fiat boundary) was first recognized as a political limit separating two territories. In his *The Renaissance Rebirth of Linear Perspective* (Edgerton 1975: 115), Edgerton describes how, during the wars of 1420, a longitudinal line was proposed as the boundary between the two states of Milan and Florence. The reference is to the treaty between Filippo Maria and Florence dated February 8, 1420, which designated the ideal line connecting Magra and Panaro as the limit of their respective spheres of influence (which themselves referred back to another treaty, from 1353, where Milan and Florence each agreed to stay out of the affairs of Tuscany and Lombardy). It is, however, very unlikely that this line was a true boundary between the two territories. Thus, the question as to the first genuine geopolitical fiat boundary remains unresolved.

Vagueness, Gluts, and Intervolvements

As already pointed out, geographical fiat objects will, in general, have boundaries that involve a combination of bona fide and fiat elements. The shores of the North Sea are bona fide boundaries, but we conceive the North Sea as a fiat object nonetheless, because where it abuts the Atlantic it has a boundary of a non-bona fide sort. The status of the latter is noteworthy in that there seem to be few practical consequences that turn on the issue as to where, precisely, it lies. Political boundaries were once themselves standardly created in places (mountain ridges, middles of rivers, marshes, swamps, deserts) where there is little human activity and thus little chance or occasion to look into their exact location.

Something similar holds also in regard to many geographical boundaries of a nonpolitical sort—for example, the boundary between a hill and an associated valley. The treatment of such cases requires a further opposition between crisp and indeterminate boundaries. Spatial entities such as deserts, valleys, mountains, noses, tails are delineated not by crisp outer boundaries but rather (on some sides at least) by boundary-like regions which are to some degree indeterminate. We here leave open the question whether bona fide reality involves both crisp and scruffy (fuzzy, hazy, indeterminate) entities as part of its ultimate furniture. Here, vagueness will be seen as a matter of semantics. If you point to an irregularly shaped protuberance in the sand and say "dune", then the correlate of your expression is a fiat object whose constituent unitary parts are comprehended through your idea of what a dune is. The vagueness of this idea is responsible for the vagueness with which the referent of your expression is picked out. And what this means is that each one of a variety of overlapping determinate portions of reality has an equal claim to being such a referent.

The above corresponds to the so-called supervaluationist account of vagueness (Fine 1975; McGee 1997; Varzi 2001), which sees vagueness in terms of precisification so that to say that a demarcation line is *vague* it to say that there is a multiplicity of acceptable ways of *making it precise*. A view along these lines can be sustained only if account is taken of the fact that the assignment of a range of candidate-precisified referents to a given expression is dependent on the context in which that expression is used. This is because the degree of vagueness we can comfortably allow in our delineations varies inversely with the degree to which a given boundary is of practical relevance—and what is and is not of practical relevance is of course such as to vary from one context to another (Bittner and Smith 2001, 2003a, b; Smith and Brogaard 2003a).

When you have a map, and it has a shoreline with ins and outs, and on the water adjacent to one of the ins is a label saying "Baie d'Ecaigrain", it is fairly easy for a human to see where the bay is. The outer boundary of the bay (seaward) is in most contexts irrelevant to action or practice, and thus a wide range of precisifications is allowed. In a context in which regulators have ceded all the islands (or oil) in the bay to some other country, however, a quite different and much narrower range will be required. Human beings can cope quite well with such vagueness of reference and with contextually determined reference shifts, and with different sorts of vagueness along different dimensions—as for instance where a bay is recorded as extending from there to there on the coastline, but as just fading off to seaward. The well-known indeterminations involved for example in establishing the number of lakes in Finland or the length of the coastline of Norway (Sarjakoski 1996) illustrate the phenomenon whereby the range of admissible precisifications can vary widely—according to the purpose the measurements are being made, the governing regulations or the definitions or the measuring instruments or protocols employed.

We can all agree that mountains, hills, ridges, capes, points, necks, brows, shoulders, heads, knees, shanks, rumps, pockets, fronts, backs, pits are real; and that it is obvious where the top of a mountain or the end of a cape is to be found. The crisply determined features of such entities—for example, the heights of mountains—can be looked up in reference books. But where is the boundary of Cape Flattery on the inland side? Where is the boundary of Mont Blanc (we mean the base of the mountain) on the French and Italian sides? (Smith and Mark 2003).

Modern geopolitical boundaries are distinguished in being infinitely thin (crisp, determinate, precise). Political and legal boundaries must, it seems, enjoy at least ideally and in the long run a geometrical perfection of this sort, which is to say that they must take up no space. For otherwise, disputes would threaten to arise in relation to the no-man's-lands that the boundaries themselves would then occupy. If a wall or river separates two distinct portions of land, then either the wall or the river must be split equally down the middle, or it must be assigned as a whole to one or other of the two parties, or it must be declared common property (and then there will exist two infinitely thin boundaries separating each of the two distinct parcels of land from the commonly owned region which divides them).

Each adjacent pair of geopolitical boundaries (say: on the Franco–German border) manifests, in addition, the phenomenon of coincidence of boundaries which are yet not identical. The boundary of France along this border is not also the boundary of Germany: each points inwards toward its own respective territory. Contrast, in this respect, the Western boundary of the old German Democratic Republic or the southern border of the present Turkish Republic of Northern Cyprus: here, exceptionally, no coincident twin was established, since the relevant neighbors did not see fit to institute a boundary of their own.[10] Moreover, as the case of Texas and the U.S.A. makes clear, distinct geopolitical boundaries may also coincide from within. That is, they may coincide for a part of their length along which they serve as boundaries on the same side. As a map of the states of the continental U.S.A. makes clear, the modern geopolitical ideal is a world of boundaries which form an irregular tessellation of each geopolitical area, until the boundary of each cell in the tessellation ends up having the topology of a Jordan curve.

There are departures from this ideal of a sort not to be cataloged under the heading of vagueness. First, there are gaps which we have discussed already above in dealing with the no-man's-lands that have not yet been assigned to one jurisdiction or another. Today, however, almost all gaps have been eliminated via treaties. Gluts are a more intriguing matter. Consider the border between Germany and Luxemburg: where borders between states usually run down the middle of water bodies, the bed and banks of the rivers Mosel, Sauer and Our belong to *both* Germany and Luxemburg, which hold them in a condominium, a status which has been shared by all the water bodies forming the boundary between these two countries since 1816, which is the year of the first written agreement on the boundary separating the United Netherlands from Prussia.

An ontological status that is still more problematic is enjoyed by Lake Constance, which forms part of the boundary between Austria, Germany, and Switzerland. Lake Constance is an ontological black hole in the heart of Europe, whose territorial status is in seemingly unresolvable limbo. While one part of the lake, Lake Überlingen (which is not truly a lake), belongs completely to Germany, the course of the border in the rest of Lake Constance has not been laid down. For while Switzerland holds the view that the border runs through the middle of the Lake, Austria and Germany are of the opinion (albeit on different grounds) that the lake stands in condominium of all the states on its banks. Hence no international treaty establishes where the borders of Switzerland, Germany, and Austria in or around Lake Constance lie. If you buy a

[10]Compare the treatment of boundaries of different plerosis in Brentano (1988).

ticket to cross the Lake in a Swiss railway station, your ticket will be valid only to the point in the middle of the Lake where, as the Swiss see it, their jurisdiction ends.

Scattered Fiat Objects

The drawing of fiat boundaries can create fiat parts within larger bona fide wholes. But it can also, in the manner of Micronesia or Polynesia, create fiat wholes out of smaller bona fide parts. And then, while bona fide objects are in general connected, the fiat objects that are circumcluded by fiat boundaries in this way are scattered entities.

There are also cases where the two distinguished factors—on the one hand the carving out of fiat parts, and on the other hand the gluing together of fiat wholes—-operate in tandem, so that geographical objects are created via the fiat unification of disconnected parts within larger wholes, for example in coastal nations in whose territory islands, or portions of islands, are included.

The Holy Roman Empire of German Nations in around 1500 serves as a nice example in this regard. Here "German Nation" signifies one or other of some hundreds of kingdoms, principalities, duchies, counties (*Grafschaften*), prince-bishoprics, free cities, and so forth. These were often scattered, which means that they included parts disconnected from each other and embedded in the interiors of other German Nations. Scattered fiat objects of this sort may be intervolved—intercalated inside each other—in a variety of ways. Consider the case of the Belgian enclave of Baarle-Hertog, which is depicted, together with its neighbor, the Dutch community of Baarle-Nassau, in Fig. 7.2.

This figure represents an area of roughly three square kilometers situated some five kilometers from the Dutch–Belgian border near Turnhout. The lighter shaded areas in the figure represent the community of Baarle-Hertog. The small darker shaded areas depict the tiny Dutch enclaves of Baarle-Nassau. Each such enclave is surrounded by a portion of Belgian territory, which is in its turn surrounded once more by territory that is Dutch. This peculiar arrangement arose as a consequence of Dutch independence from Spain in 1648 when the Dutch border was defined on the basis of a long-standing feudal provincial boundary, which in turn featured numerous enclaves and exclaves. A strong religious divide between the Netherlands and Spain in 1648, coupled with rural conservatism favoring the status quo, together stymied all governmental attempts to exchange or cede the enclaved lands. The two families of enclaves around Baarle were briefly merged in 1815 with the formation of the United Netherlands at the Congress of Vienna. But with the independence of Belgium in 1830, the old situation was resurrected, and once again ancient provincial limits were used as the international border. Being unable to determine a more rational boundary than those involved in negotiating the 1843 Treaty of Maastricht was forced to resort to the individual determination of national ownership of each of 5732 plots in the two communes, yielding a delineation of the border that survived until 1995, when modern administration, infrastructure, and legal systems necessitated an exacting

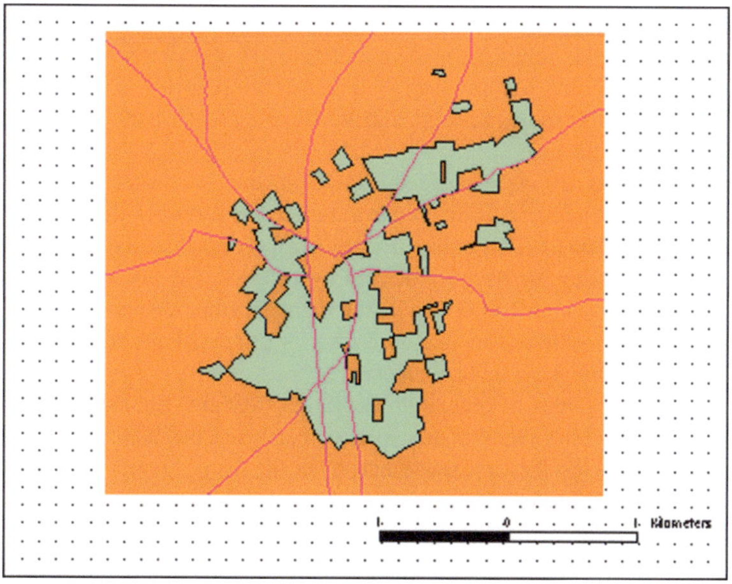

Fig. 7.2 The Enclaves of Baarle-Hertog and Baarle-Nassau

survey which has cemented the existence of the enclaves in the arrangement depicted above.[11]

Fiat and Bona Fide Boundaries in the Material Realm

Organisms and cells are marked as fundamental units of biology by the coverings or membranes which extend continuously across their surfaces, allbeit with small apertures—such as pores, mouths, nostrils—which allow interchange of substances such as air and food between interior and exterior (Smith and Brogaard 2003b). Objects of these sorts are thus bounded by bona fide coverings, which are parts of the objects which they bound.[12] For organisms in early phases of their lives, complex layered bona fide coverings have evolved with the function of protection, for instance, against bacterial invasion, toxins, and damage through physical force. The mammalian embryo is protected by the successive layers of (starting with the outermost layer): maternal epithelial covering (the outer protective layer formed by the mother's skin), the uterine wall, the placenta, chorion, and amnion, and ultimately by the outer layer of cells of the embryo itself. The eggshell of a chicken similarly protects the developing chick through an outermost inorganic layer called the cuticle

[11] Details are presented in Whyte (2002). See also Vinokurov (2007).

[12] They are "fiat object parts" in the terminology of BFO.

of the egg, inside which is a succession of organic layers including the vertical crystal, and the palisade layers followed by the mammillary cone, external shell membrane, internal shell membrane, and limiting shell membrane (Hinckel et al. 2012). At the same time some 17,000 pores allow air and moisture to penetrate through these layers into the interior of the egg.

Many other kinds of bona fide objects—including our own planet, starting with the Earth's crust—have exteriors structured in a similar way by bona fide external layers, coverings or membranes. This holds of many artifacts for instance of automobiles, whose aluminum alloy panels are protected by successive layers of paint designed to protect against weather and UV radiation damage, stone-chipping, and so forth. Layered structures of these sorts are used also for protective purposes in roofs and walls, and also in highways and pavements.[13]

Objects such as lumps of granite do not have coverings or membranes of these sorts. But this does not mean that they lose their status as bona fide objects. Indeed, they are bona fide in just the same sense that physically distinct layers of an epithelium or roadway are bona fide, namely in virtue of the physical discontinuity between them and the substances or media that surround them.

For all the mentioned kinds of objects manifest on their outermost surfaces what we shall call an interface. There is an interface between a block of granite and the surrounding air. There is an interface between any two adjacent layers in a multilayered structure. And there is an interface between an outermost layer and the surrounding medium and between the innermost layer in a protective layered structure and the enclosed medium surrounding the protected entity. Stroll, in his book on *Surfaces* (1988: 44f.), calls such interfaces "Leonardo surfaces"—drawing on the discussion in Leonardo's *Notebooks* of a surface as "the common boundary of two things that are in contact", for example, the boundary between the air and sea. For Stroll, as for Leonardo, surfaces are without bulk—as contrasted with the associated bulky portions of matter which they are the surfaces of. For us here, however, interfaces are thin zones at the exteriors of bona fide objects where microparticles may be such that their belongingness to one or other of the interfacing objects is only statistically specifiable.

A general theory of such matters is provided by J. J. Gibson in his *Ecological Approach to Visual Perception* (1979), in terms of a trichotomy of substance, surface and medium. As he puts it, where any two of solid, liquid and gas come into contact there is constituted a surface (Gibson 1979: 16). Portions of liquid and gas serve as media with four characteristics that are of relevance to animal motion and perception:

1. detached solid bodies can move through them without resistance, and they thus afford locomotion,
2. they are generally transparent, transmitting light, and thus afford vision,
3. they transmit vibration or pressure, and thus afford auditory perception,
4. they allow rapid chemical diffusion, thus affording olfactory perception (detection of a substance from a distance).

---

[13]Up to 8 such layers of timber, chalk, stone and other materials were employed already in the construction of Roman roads (Flaherty 2002: 226).

Light, sound, and smell together guide and control motion, and as the animal moves it is tuned to the information contained in its environment, about things that reflect light, vibrate, or are volatile. They allow the animal to detect places that afford eating, to sniff out allies and predators, and so forth. More generally the animal is endowed with the ability to perceive objects, persons, and places (for example water holes) as persisting entities that can be detected and tracked. The animal is attuned to its surrounding environment along salient dimensions; its perceptual system thereby becomes immediately sensitized to salient differences in its environment. As Gibson puts it, "it resonates to the invariant structure"—it is able to simply extract salient invariants from the flowing array (Gibson 1979: 249, 255): to recognize a facial gesture immediately and spontaneously as welcoming or antagonistic, to gauge the movement of vehicles approaching an intersection in such a way as to assess immediately and spontaneously whether it is safe for you to cross the intersection with your own vehicle.

The environment, and the affordances, of a dragonfly or a water strider are of course very different from those of fish or human beings. On some readings of Gibson, this is taken to imply a relativistic view on Gibson's part, according to which organisms of different species *live in different worlds* (Katz 1987; compare Smith 2009). Water, for example, is a substance in one world and a medium in another. Katz infers from this that one could never say what water is, without saying for whom it is, and conversely (Katz 1987: 120). Gibson himself expresses the matter as follows:

> The natural environment offers many ways of life, and different animals have different ways of life. The niche implies a kind of animal, and the animal implies a kind of niche. Note the complementarity of the two. But note also that the environment as a whole with its unlimited possibilities existed prior to animals. The physical, chemical, meteorological, and geological conditions of the surface of the earth and the pre-existence of plant life are what make animal life possible. They had to be invariant for animals to evolve. (Gibson 1979: 128)

Here, therefore, Gibson embraces a realist perspective, according to which there is a common world, and a common space, and a common set of feature-spaces to which all species-specific niches and all associated collections of affordances belong. This common space (as we may here assume) is a continuum, and like all continua it can be partitioned in a range of different ways. From this perspective the various conflicting "perceptual spaces" are compatible; they reflect distinct partitions, roughly: partitions at different levels of granularity, of one and the same reality (Bittner and Smith 2003a). With each of these partitions there is associated a family of affordances which ground relational dispositions linking animals to their external environments and to each other:

> What the male affords the female is reciprocal to what the female affords the male; what the infant affords the mother is reciprocal to what the mother affords the infant; what the prey affords the predator goes along with what the predator affords the prey; what the buyer affords the seller cannot be separated from what the seller affords the buyser, and so on. The perceiving of these mutual affordances is enormously complex, but it is nonetheless lawful, and it is based on the pickup of the information in touch, sound, odor, taste, and ambient light (Gibson 1979: 135).

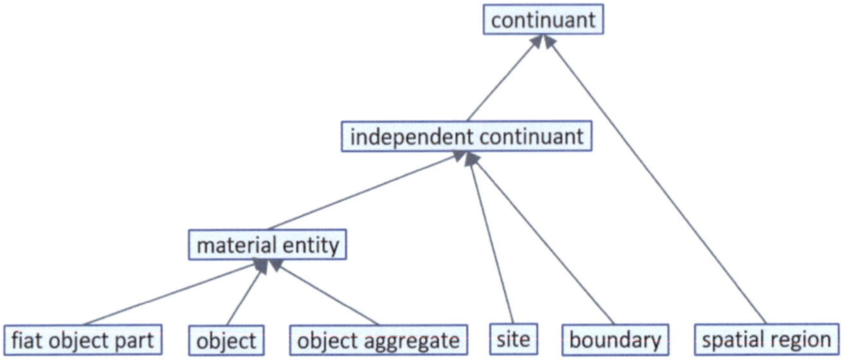

Fig. 7.3 *Independent continuant* and its subcategories in BFO 1.1

The demarcations associated with such mutual affordances bring into being zones of different sorts (Smith and Varzi 2002), for example, bubble-like zones around each person in a public area, forming what is called their *personal space*, which other persons (for examples persons of one or other sex) may or may not be allowed to penetrate; zones created where persons interact sexually in given environments which demarcate, for example, those parts of the body for which touching is permissible, from those parts of the body that are not permissible to touch; zones in which an infant is allowed to play freely; zones in which a potential prey can feel itself secure from encroachments by its predators; zones in a department store which demarcate areas where only very expensive products are for sale (sometimes including locked cabinets) from other zones with cheaper products, and so forth.

Fiat Objects in BFO

One application of the fiat/bona fide opposition is in the field of applied ontology, where it serves as one pillar of the treatment of spatial entities in Basic Formal Ontology (BFO), a top-level ontology that forms the shared architecture of some three hundred ontology initiatives in a range of different domains (Arp et al. 2015).

The spatial ontology in early versions of BFO is illustrated in Fig. 7.3.[14] Here continuant entities are divided into *spatial regions* on the one hand and *independent continuants* on the other, the latter being divided further into five subcategories, as follows:

The term "*Object*", in BFO, means bona fide object, in other words: mind-independent, material unit, whether natural or artefactual. Objects in this sense include organisms, artifacts such as laptops or screwdrivers, and unitary portions of matter such as molecules, planets and lumps of stone. Some objects (for example,

[14] Arp and Smith 2001; see also http://www.ifomis.org:80/bfo/owl, last accessed January 1, 2019.

multicellular organisms) may have other objects (for example, cells) as parts. Objects do not merge continuously into each other. Two objects may be contingently adjacent to each other, as for instance in the case of adjacent cells in your body; they may also be contingently connected, as for instance in the case of a lamp connected to a wall by means of an electric cable, or of a neuron connected to a second neuron through a synapse. Objects behave to a large degree independently, and have their own types of causal unity—for example of the sorts characteristic of organisms, artifacts, and solid portions of matter (Smith 2012b). BFO does not have a special term for "bona fide object", since all objects are bona fide. But their status as such derives not from any special character of their boundary but rather from their type of causal unity as defined in this paper.[15]

Fiat object parts are entities carved out within the interiors of objects by means of fiat boundaries. *Object aggregates* are collections of objects, for example, teams, committees, populations, products in a warehouse. *Sites* are for instance cavities, trenches, tunnels—entities within which objects can be situated.

The spatial ontology in the current version of BFO, (see Fig. 7.4), includes in addition to the existing spatial region subcategories also three categories of what are called "continuant fiat boundaries", namely: *fiat point*, *fiat line*, and *fiat surface*, which together replace the term "boundary" that was used in earlier versions. This is not because the new version of BFO rests on a presupposition to the effect that all continuant boundaries are fiat in nature. Rather it reflects the recognition of the fact that the apparent boundaries of, for example, organisms are not simple, static two-dimensional surfaces but rather complex cloud-like formations of moving microparticles. The boundaries we assign to organisms and the other three-dimensional entities that we perceive in everyday experience and whose dimensions we record in measurements are, therefore, fiat boundaries in the originally intended sense; they are boundaries that we create by a process analogous to one of precisification. This does not mean that the objects that they bound are fiat also. The fact that we cannot determine precisely, for example, the location of the boundary of the planet Earth, does not mean that the planet Earth is not a separate, mind-independent material entity.

The BFO treatment of the external boundaries of objects as viewed under the new dispensation reads as follows. First, when we measure the boundaries of such objects, then we impute to them fiat boundaries (in ways to be described below), and all spatial measurement data pertain to fiat boundaries of this sort. Whether such boundaries correspond to bona fide boundaries—to what we might think of as joints in reality—is a complex question the answer to which must be decided anew for each sort of case.

Certainly, the boundaries between planets, or between apples and oranges, or between free oxygen molecules, do reflect joints in reality. But other cases are more difficult. Consider, for example, the boundaries for example between the various layers of seal-coating and asphalt on a highway. Where microscopic examination

[15]Note that the account, there, is deliberately open-ended: thus further sorts of causal unity might be documented by BFO in due course.

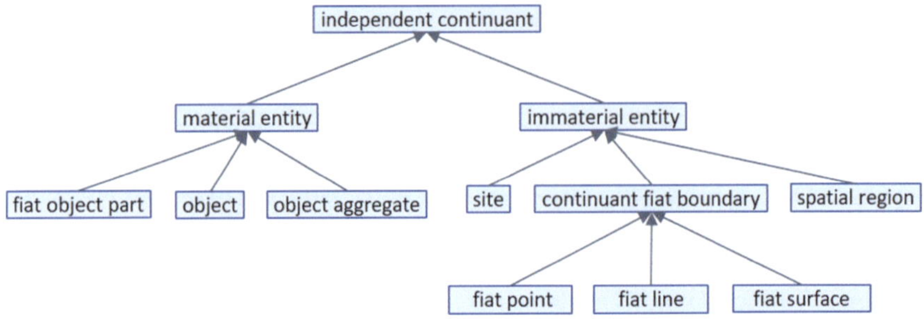

Fig. 7.4 Spatial categories in BFO-ISO

of such layers reveals that they are bonded together chemically, then the boundary between them is fiat in nature; where, in contrast, two steel plates are bolted together, then the boundary between them is bona fide, and something similar holds at the exteriors of solid bodies, for example at the interface between a lump of granite and the surrounding mass of air (Fig. 7.4).

The paradigmatic examples of continuant fiat boundaries in BFO are the North Pole (a fiat point), the Equator (a fiat line), and the plane separating the Northern from the Southern hemisphere (a fiat surface). Other examples are the center of mass of a solid body; isobars, isotherms, and isohyets; Utah. When a land surveyor draws lines on a map he projects these lines onto reality—treats reality as if it contained fiat lines of a corresponding sort. When we use a ruler to measure the distance between two points then we create (roughly: in our imagination) a fiat line connecting these points, and we position the edge of the ruler to coincide with this fiat line. For practical purposes, we substitute for the physical edge of the ruler (again in our imagination) a fiat boundary (we imagine the edge of the ruler as a fiat line). Similarly, when we observe the meniscus of a mercury column in a thermometer, and compare what we imagine to be its point of maximal height to the scale of the thermometer, then we treat the latter as consisting of a series of fiat lines of infinite thinness, and when the meniscus level falls between two such lines then we imagine a further such fiat line to be interpolated at the appropriate distance between them. In all such cases our perception, imagination and interpolation are subject to the constraints of our visual acuity, the resolution of our measuring apparatus, and so forth.

References

Arp R, Smith B, Spear A (2015) Building ontologies with Basic Formal Ontology. MIT Press, Cambridge, MA

Bittner T, Smith B (2001) Granular partitions and vagueness. In: Welty C, Smith B (eds) Formal ontology and information systems. ACM Press, New York, pp 309–321

Bittner T, Smith B (2003a) A unified theory of truth and reference. Log Anal 169–170:49–93

Bittner T, Smith B (2003b) A theory of granular partitions. In: Duckham M, Goodchild MF, Worboys MF (eds) Foundations of geographic information science. Taylor & Francis Books, London, pp 117–151
Brentano F (1988) Philosophical investigations on space, time and the continuum. English translation by B. Smith. Croom Helm, London and Sydney
Casati R, Varzi AC (1994) Holes and other superficialities. MIT Press, Cambridge, MA
Casati R, Varzi AC (1999) Parts and places. The structures of spatial representation. MIT Press, Cambridge, MA
Ceusters W, Smith B (2015) Aboutness: towards foundations for the information artifact ontology. In: Proceedings of the sixth International conference on biomedical ontology (ICBO), Lisbon, Portugal (CEUR 1515), pp 1–5
Chisholm RM (1989) On metaphysics. University of Minnesota Press, Minneapolis
Comrie B (1976) Aspect. An introduction to the study of verbal aspect and related problems. Cambridge University Press, London and New York
Davies R (ed) (2018) Natural and artifactual objects in contemporary metaphysics: exercises in analytic ontology. Bloomsbury, London
Edgerton S (1975) The renaissance rebirth of linear perspective. Basic Books, New York
Fine K (1975) Vagueness, truth and logic. Synthese 30:265–300
Flaherty CA (ed) (2002) Highways. The location, design, construction and maintenance of road pavements, 4th edn. Elsevier, New York
Frege G (1884) Die Grundlagen der Arithmetik. Koebner, Breslau. English edition: Austin JL (1959) Foundations of arithmetic. Blackwell, Oxford
Galton A (1984) The logic of aspect: an axiomatic approach. Clarendon Press, Oxford
Gärdenfors P (2000) Conceptual spaces. The geometry of thought. MIT Press, Cambridge, MA
Gibson JJ (1979) The ecological approach to visual perception. Houghton-Mifflin, Boston
Guarino N (2013) Local qualities, quality fields, and quality patterns: a preliminary investigation. In: Proceedings of SHAPES 2013 (CEUR 1007), pp 75–81
Hinckel MT et al (2012) The eggshell: structure, composition and mineralization. Front Biosci 17:1266–1280
Johansson I (1989) Ontological investigations. Routledge, New York
Katz S (1987) Is Gibson a relativist? In: Costall A, Still A (eds) Cognitive psychology in question. Harvester, Brighton, pp 115–127
Lakoff G (1987) Women, fire, and dangerous things. University of Chicago Press, Chicago
Langacker R (1987/1991) Foundations of cognitive grammar, 2 vols. Stanford University Press, Stanford
Mark D, Skupin A, Smith, B (2001) Features, objects, and other things: ontological distinctions in the geographic domain. In: Spatial information theory (COSIT 2001), pp 489–502
McGee V (1997) Kilimanjaro. Can J Philos 23(Suppl.):141–195
Mourelatos APD (1981) Events, processes and states. In: Tedeschi PJ, Zaenen A (eds) Tense and aspect (syntax and semantics), vol 14. Academic Press, New York, pp 191–212
Ojeda A (1993) Linguistic individuals. CSLI Publications, Stanford
Quine WVO (1969) Ontological relativity. In: Quine WVO (ed) Ontological relativity and other essays. Columbia University Press, New York
Sarjakoski T (1996) How many lakes, islands and rivers are there in Finland? A case study of fuzziness in the extent and identity of geographic objects. In: Burrough PA, Frank AU (eds) Geographic objects with indeterminate boundaries. Taylor & Francis, London, pp 299–312
Smith B (1987) On the cognition of states of affairs. In: Mulligan K (ed) Speech act and Sachverhalt: Reinach and the foundations of realist phenomenology. Nijhoff, Dordrecht/Boston/Lancaster
Smith B (1995) On drawing lines on a map. In: Frank AU, Kuhn W (eds) Spatial information theory. A theoretical basis for GIS, vol 988. Lecture notes in computer science. Springer, Berlin/Heidelberg/New York, etc, pp 475–484
Smith B (1997) Boundaries. In: Hahn LH (ed) The philosophy of Roderick Chisholm. Open Court, LaSalle, IL, pp 534–561

Smith B (1999a) Agglomerations. In Freksa C, Mark DM (eds) Spatial information theory. Cognitive and computational foundations of geographic information science. Proceedings of COSIT '99. Springer, Berlin, pp 267–282
Smith B (1999b) Truthmaker realism. Australas J Philos 77:274–291
Smith B (1999c) Truth and the visual field. In: Petitot J, Varela FJ, Pachoud B, Roy J-M (eds) Naturalizing phenomenology: issues in contemporary phenomenology and cognitive science. Stanford University Press, Stanford, pp 317–329
Smith B (2007) On place and space: the ontology of the Eruv. In: Kanzian C (ed) Cultures: conflict—analysis—dialogue. Ontos Verlag, Frankfurt, pp 403–416
Smith B (2009) Toward a realistic science of environments. Ecol psychology 21(2):121–130
Smith B (2012a) Classifying processes: an essay in applied ontology. Ratio 25(4):463–488
Smith B (2012b) On classifying material entities in Basic Formal Ontology. In: Interdisciplinary ontology. Proceedings of the Third Interdisciplinary Ontology Meeting. Keio University Press, Tokyo, pp 1–13
Smith B, Brogaard B (2003a) A unified theory of truth and reference. Log Anal 169–170:49–93
Smith B, Brogaard B (2003b) Sixteen days. J Med Philos 28:45–78
Smith B, Mark DM (2003) Do mountains exist? Towards an ontology of landforms. Environ Plan B Plan Des 30(3):411–427
Smith B, Varzi A (1997) The formal ontology of boundaries. Electron J Anal Philos 5:1–23
Smith B, Varzi AC (2000) Fiat and bona fide boundaries. Philos Phenom Research 60:401–420
Smith B, Varzi AC (2002) Surrounding space. On the ontology of organism-environment relations. Theory Biosci 120(2):139–162
Stove D (1991) The Plato Cult and other philosophical follies. Blackwell, Oxford
Stroll A (1988) Surfaces. University of Minnesota Press, Minneapolis
Talmy L (1995) Fictive motion in language and "ception". In: Bloom P, Peterson MA, Nadel L, Garrett MF (eds) Language and space. MIT Press, Cambridge, MA, pp 211–276
Talmy L (1996) The windowing of attention in language. In: Shibatani M, Thompson S (eds) Grammatical constructions: their form and meaning. Oxford University Press, Oxford, pp 235–238
Thomsen E, Smith B (2018) Ontology-based fusion of sensor data and natural language. Appl Ontol 13(4):295–333
Varzi AC (2001) Vagueness in geography. Philos Geogr 4:49–65
Varzi AC (2016) On drawing lines across the board. In: Zaibert L (ed) The theory and practice of ontology. Palgrave Macmillian, London, pp 45–78
Vinokurov E (2007) A theory of enclaves. Lexington Books, Lanham, MD
Vogt L (2018a) Levels and building blocks—toward a domain granularity framework for the life sciences. J Biomed Semant 10(1):4. https://doi.org/10.1186/s13326-019-0196-2
Vogt L (2018b) Bona fideness of material entities and their boundaries. In: Davies R (ed) Natural and artifactual objects in contemporary metaphysics: exercises in analytic ontology. Bloomsbury, London
Vogt L, Grobe P, Quast B, Bartolomaeus T (2011) Top-level categories of constitutively organized material entities. Suggestions for a formal top-level ontology. PLoS ONE 6(4):e18794
Vogt L, Grobe P, Quast B, Bartolomaeus T (2012a) Accommodating ontologies to biological reality—top-level categories of cumulative-constitutively organized material entities. PLoS ONE 7(1):e30004
Vogt L, Grobe P, Quast B, Bartolomaeus T (2012b) Fiat or bona fide boundary—a matter of granular perspective. PLoS ONE 7(12):e48603
Whyte B (2002) Waiting for the Esquimo, an historical and documentary study of the Cooch Behar enclaves of India and Bangladesh. Research paper 8, School of Anthropology, Geography and Environmental Studies, University of Melbourne

Chapter 8
Geographic Objects and the Science of Geography

Amie L. Thomasson

Abstract Human geography studies places—considered not just as spatiotemporal locations, but as places of human significance, such as nations, electoral districts, and parks. Such entities are generally thought of as depending on the beliefs and practices of the peoples who live there. The mind-dependence of such entities, however, leads some to doubt whether we can really make discoveries in human geography, and even whether the entities studied in human geography are real parts of our world. This paper examines the ways in which geographic entities may rightly be said to be mind-dependent, and what consequences this mind-dependence does and does not have regarding whether human geography may be a potential source of knowledge and discovery, and regarding whether we should accept that geographic entities exist.

Keywords Geographic objects · Ontology of human geography · Mind-dependence

The basic facts tracked by geographers involve such things as nations, electoral districts, population distributions, and industrial and agricultural zones. The importance of such things goes beyond the theorizing of the social sciences, however; as has been remarked (Smith and Varzi 2000: 404), people fight wars over such things as national boundaries and dedicate entire industries to the maintenance of political and property boundaries.

Yet it is a commonplace that such entities as those above—those studied by *human* geography rather than *physical* geography—depend in certain ways on the beliefs and customs of the people of the region studied. This mind-dependence leads some to doubt whether geography can really be considered a science involved in making discoveries about the world, and whether or not the purported facts studied by geographers should really be considered as existing at all. First, if these facts are in some way the products of our minds and social practices, it is often thought, they must be transparent to those involved in their creation and maintenance. As George Lakoff puts it:

A. L. Thomasson (✉)
Department of Philosophy, Dartmouth College, Hanover, NH, USA
e-mail: Amie.L.Thomasson@Dartmouth.edu

© Springer Nature Switzerland AG 2019

T. Tambassi (ed.), *The Philosophy of GIS*, Springer Geography,
https://doi.org/10.1007/978-3-030-16829-2_8

> In general, extending objectivism to include institutional facts gets one into trouble with the assumption that metaphysics is independent of epistemology. The reason is that institutions are products of culture and hence products of the human mind. They exist only by virtue of human minds (Lakoff 1987: 207).

But if the very existence of institutional and other facts studied by human geography somehow depends on our knowledge of them, it is difficult to see how geographers could be thought to make discoveries about the world.

Second, if we do have this epistemic privilege with regard to the facts and kinds of facts apparently studied by human geography, it is often said, we cannot be realists about them. For realism is often regarded as requiring what Crawford Elder calls the "doctrine of epistemic non-privilege", that "all constituents of the world exist, and are as they are, independently of whether anyone ever does or can form true beliefs about them" (Elder 1989: 440), so that as a result:

> Realists ... must either argue that members of a given culture could in fact hold shared beliefs about their own CGKs [culturally generated kinds] that were massively mistaken, or else maintain that CGKs are not genuine components of the world (Elder 1989: 427).

Thus, any privileged knowledge regarding these facts and kinds that arises in virtue of their mind-dependence might be thought to present obstacles to being realists about them.

The goal of this paper is to examine the way(s) in which geographic entities may rightly be said to be mind-dependent and to examine what consequences this mind-dependence does and does not have regarding whether human geography may be a potential source of knowledge and discovery, and regarding whether we should accept that geographic entities exist. I will suggest that arguments about whether they are mind-dependent, and whether mind-dependence in general entails epistemic privilege, or should lead us to deny the existence of the entities in question, are far too coarse. As I will argue, there are several distinct senses in which various geographic entities may be said to be mind-dependent. We must examine matters on a case-by-case basis to resolve what difference(s), if any, mind-dependence of various sorts makes to our epistemic relation to them and to the potential range of discovery of human geography. I will close by asking what impact these results should have on the issue of whether or not we should accept that there really are geographic entities of these kinds.

Varieties of Mind-Dependent Geographic Objects

Geography distinguishes itself from other social sciences by its focus on place. But the studies of human geography do not merely focus on places in the sense of abstract spatiotemporal locations or the slabs of land that form the continents of the global landscape. Instead of or in addition to land and space so considered, the places of concern to human geographers are often regions artificially singled out from the larger landscape and/or endowed with social significance (as nations, electoral districts, parks, industrial zones) by the beliefs and practices of the local culture. As a result, it is virtually a truism that the facts studied by human geography (that this piece of land is U.S. territory, is a national park, or is an industrial zone) depend in some ways on human mental states, more particularly on the collective intentionality of the people and cultures inhabiting those places.[1] Correlatively, the geographic objects so formed (this nation, park, or zone, *qua* nation, park, or zone) likewise can be said to depend on collective intentionality.

It is far less clear, however, what form this dependence takes and what that dependence entails. For although such geographic objects are clearly mind-dependent in some ways, they also have independent foundation in the pieces of land that have such properties as being bounded in a certain way, being an electoral district, or being an industrial zone.[2] Their foundation in independent tracts of land immediately distinguishes them from mere mental constructs or figments of the imagination. Thus, we need to begin by sorting out the different senses in which such diverse geographic objects as mountains, nations, and industrial zones are mind-dependent.

Recent work on fiat boundaries and the associated fiat objects has done much to move forward discussion of one sense in which many geographic objects are mind-dependent (Smith and Varzi 2000; Smith 2001). Whereas "bona fide " boundaries exist entirely independently of human cognitive activities, being based solely in "spatial discontinuities and … intrinsic qualitative differentiation", fiat boundaries fail to correspond to any genuine heterogeneity among or within entities in the world, and so exist "only in virtue of the different sorts of demarcations effected cognitively by human beings" (Smith 2001: §1). Thus, e.g., the coastal boundaries of Key West are bona fide , marking an actual difference between land above and below sea level, while the boundaries of states such as Wyoming are entirely fiat boundaries. Objects demarcated by boundaries that are even partially fiat boundaries (as, e.g., the state of Maryland) may be termed "fiat objects" (Smith and Varzi 2000: 403).

While fiat boundaries may arise in virtue of the conceptual or perceptual activities of individual agents or of collectives, I will focus here exclusively on those cases of fiat boundaries that are social in the sense of depending on the collective beliefs and customs of a group of people. For it is these social fiat boundaries that are at work in demarcating many of the borders of such objects of human geography as nations, states, electoral districts, parks, and pieces of real estate. In the cases mentioned thus far, this dependence is a dependence on the direct collective creation and

[1]I will limit discussion here to the issues raised by the apparent dependence of facts of human geography on the collective intentionality of the local people. Other issues arise regarding whether or to what extent, e.g., the regions of study explicitly introduced by *geographers themselves* are fiat objects depending on the boundary-drawing activities of geographers (not the collective beliefs and customs of "locals"). I will leave those issues to one side here, since they are not unique to human geography (or other social sciences), but rather involve general issues for the philosophy of natural as well as social science.

[2]Although there may be many sorts of fact referred to in geographic theories that are not land-based in this way, in this paper I will focus on those that are based in place, as these provide a particularly interesting case of mind-dependent facts central to the study of geography.

continued acceptance of such boundaries, establishing and maintaining boundaries by fiat despite a dearth of intrinsic differences in the reality parceled off.

Although the *boundaries* of fiat objects exist only in virtue of the performance of certain human cognitive acts, this does not entail that the bounded objects themselves are mind-dependent. As Barry Smith puts it:

> The admission of fiat objects into our ontology is then at least in one respect unproblematic: all fiat objects are supervenient on bona fide objects on lower levels, in the sense that the fixation of relevant traits at the lower levels suffices to fix the values of traits at higher levels. The interiors of fiat objects are in this sense autonomous portions of autonomous reality. Only the respective external boundaries are created by us; it is these which are the products of our mental and linguistic activity, and of associated conventional laws, norms and habits. The relevant underlying factual material is in every case unaffected thereby (Smith 2001: §8).

It is clearly true that the fact that an object is a fiat object does not entail that the object itself is mind-dependent, but only that some of its boundaries are (or that it *qua* bounded is). Such fiat objects as Mount Kinabalu provide excellent examples of fiat objects whose mere existence (as physical objects) is mind-independent, though the existence of certain of their boundaries depends on human cognition.

Many of the most interesting fiat objects, however, are also objects with important social status—nations, states, electoral districts, pieces of real estate, national parks, etc.—whose boundaries are at least in part drawn by fiat. Qua social objects, however, these things are not (even apart from the dependence of their boundaries) "autonomous portions of autonomous reality", although as Smith points out, the lower level physical objects they are based in may be. While the parcel of land belonging to the United States may be an autonomous portion of reality depending on human cognition only for its boundaries, its status as the national territory of the United States of America is not. For a nation, as such, can exist only through people recognizing the right of certain individuals to occupy, govern, and protect a certain parcel of land, and thus the existence of the nation itself (and the land's status as the national territory) depends on human agreements, beliefs and practices.

Thus, apart from the status of many geographic entities as fiat objects, which depend for their boundaries on human intentionality, there seems to be a separate sense in which geographic objects may be mind-dependent: they may also depend for their social status on forms of human intentionality. This difference, I would conjecture, lies behind the intuition noted by Smith (2001: §11) that not all fiat objects "belong in equal degree to the fiat realm", since there is some sense in which such apparently fiat objects as bays and mountains (but not nations and property) could exist in the absence of all linguistic and cultural habits. In fact, the issues of dependence-for-boundaries versus dependence-for-social-status are entirely orthogonal to one another. There may be fiat objects (such as mountains and bays) that do not involve any social status whatsoever. There may also be geographic objects with significant social status, such as the nation of Jamaica, that have only bona fide boundaries and thus are not fiat objects. Speaking of many geographic objects as "dependent on the cognitive states of human beings" is thus ambiguous between claiming that they depend on human intentionality for their boundaries and that they so depend for their social status.

The ambiguity may be resolved, however, by paying careful attention to what it is that is claimed to be mind-dependent. In the case of Jamaica, it is the fact that this island has the status of being a nation that is mind-dependent (the land itself being capable of independent existence); in the case of Mount Kinabalu, it is the fact that this lump of rock has these boundaries separating it from the surrounding landscape that is mind-dependent (the rock itself being capable of independent existence). In some cases, facts regarding boundaries, in others, facts regarding social or institutional status, etc., are mind-dependent. Provided we keep these differences in what is dependent clear, the mind-dependence across the various cases may be treated together.

Facts involving the social status and fiat boundaries of geographic objects, then, are apparently alike in the sense of being products of human minds. But how can human mental states create such facts (whether it is the fact that Mount Kinabalu ends about here or the fact that this island is an independent nation)? Although these may not be exhaustive, I will consider three major methods for the creation and maintenance of various kinds of geographic fact: Direct creation by token, direct creation by type, and indirect creation.

Direct Creation by Token

The most straightforward and obvious cases in which facts involving boundaries or social status may depend on human mental states are those cases in which the fact in question (that *x* has these boundaries, that *x* is a nation …) is established and maintained directly by its being collectively believed or accepted to be the case, such that that fact exists if and only if it is collectively believed to exist (accepted as existing, etc.). As John Searle puts the point for social entities: "Part of being a cocktail party is being thought to be a cocktail party; part of being a war is being thought to be a war" (Searle 1995: 34).

In the most basic cases, such facts are established informally and ad hoc, in the absence of any accepted general principles for generating facts of the kind (K) in question. In these cases, it is necessarily the case that, for all *x*, *x* is K if and only if *x* is believed to be K.[3] Thus, for example, Mount Kinabalu has fiat boundaries in a given location if and only if it is believed to, and a particular piece of land may be a village common if and only if villagers accept that it is common land. Such ad hoc facts are generally established by collective custom, rather than through formal declaration. Thus, e.g., the boundaries of Mount Kinabalu are not established through any formal declaration, but rather through the informal collective practices of people of Borneo regarding what pieces of land and rock do and do not "count

[3]"Believed" here should be taken as a placeholder for any of a number of appropriate intentional relations, including believed to be, regarded as, accepted as, etc.

as" part of Mount Kinabalu[4]. Similarly, according to the customs of a local group, a piece of land in a village may be a common if and only if it is accepted as such by the villagers, though no formal declaration of it as common land may be necessary (nor perhaps considered sufficient)[5].

In many other cases, the token creation of facts involving fiat boundaries or social status may proceed by formal declaration rather than informal practice. Thus, e.g., the fact that a piece of land has certain fiat boundaries may be established not through extended custom, but suddenly by some individual or group declaring there to be those boundaries (as in establishing the boundaries of Wyoming), and a particular piece of land (fiat or bona fide) may be directly declared a national park through an act of congress. In these cases, still, the existence of the particular fact in question (that these are the boundaries of Wyoming, that this land is a national park) requires recognition of it by someone—namely, at least that person or those people who declare it so. Such formal cases of declaration, however, depend on the prior collective acceptance either of general principles regarding the conditions under which such declarations can be made, or of certain (token) individuals or groups as authorized to draw such boundaries or declare such facts, as, e.g., we collectively accept that any piece of land approved of by congress as a national park counts as a national park.

Direct Creation by Type

The creation of facts by token is a slow and painstaking operation since that very piece of land must be considered as having those boundaries and/or that social status in order to have it. Much efficiency is gained when we move to the creation of facts by type rather than by token. Such facts may be created by type rather than by token only if we accept general principles that stipulate sufficient conditions for the creation of objects of that type. Although they are more typically created ad hoc, fiat objects may be created wholesale if, rather than requiring that someone demarcate each individual line and formally declare it a boundary, we accept some general mathematical principle to partition the globe by longitude and latitude lines, or to divide the farmland of the Midwest, accepting (say) that there are property boundaries every ten miles west and every five miles north of some starting point,

[4]I have intentionally chosen a mountain with local significance and name, since here I am concerned with the social acceptance of fiat boundaries within the local community regarding "their" Mount Kinabalu, rather than the geographer's drawing of fiat boundaries on a map of significant physical features. Clearly, fiat boundaries accepted by locals versus geographers may differ, and cases of the latter sort must be handled separately.

[5]Of course the informal collective concept of a common (or anything else) may ultimately be replaced by a more formal concept that provides conditions for the creation of common land by an act of the monarch or of parliament. This, however, is clearly a *replacement* concept of what it is to be common that may clash violently with the original informal collective concept. Different concepts attached to the same word may require different methods for creating something that falls under that concept.

thus creating those fiat boundaries without each boundary requiring separate and explicit consideration as such.

Similarly, facts regarding social status may be created wholesale if we collectively accept general principles regarding sufficient conditions for an object to have the social status in question. In such cases, it is necessarily the case that something is of kind K if and only if there is collective acceptance of a set of conditions C stipulating sufficient conditions for something to count as (a) K, and that thing meets all of those conditions[6]. Of course in some cases, as in those considered above for the creation of national parks, the general principles may require that some individual or group specifically accept the existence of the fact in question. This need not be required in all cases, however, for we may also accept that anything fulfilling certain general conditions has a certain social status while imposing no requirement that anyone has any beliefs or intentions regarding the particular case in question. The state of North Carolina, for example, protects "public trust rights" in ocean beaches by adopting the constitutive rule that any shoreline land below the highest high tide point counts as public land. In that case, any such land automatically counts as public property without the need for anyone to directly accept each and every such (token) stretch of land as public property. Similarly, Treasure Trove laws in England and Wales entail that "gold and silver objects which have been hidden (rather than lost or abandoned) in the soil or in buildings, and for which neither the original owner nor his heirs can be traced" are property of the Crown, regardless of whether or not anyone (currently living) has any beliefs regarding those gold or silver objects at all[7].

In such cases, the facts that there are fiat boundaries here or there still depend on human intentionality, for they could not exist were it not for the collective acceptance of certain principles regarding a set C of sufficient conditions for something being (a) K. But given that collective acceptance, anything satisfying all of those conditions C automatically counts as (a) K regardless of what anyone thinks about that particular case. Although such facts do not depend on anyone accepting that token fact itself, they do depend on intentional states regarding that kind of fact, and outlining sufficient conditions for facts of that kind to be created.

Indirect Creation

The cases considered thus far all require collective beliefs on the part of local people regarding the particular fact or kind of fact in question, but this is not universal among the facts studied by geographers. Consider, for example, the facts tracked

---

[6]Note that this must not be confused with merely verbal stipulations about what conditions are required for something to be *called* a "K". We do collectively accept certain conditions as sufficient for something to be called a "ewe", for example, but the kind *ewe* is not a constructed social kind, since it is not necessary, for something to *be* a ewe, that anyone accept any sufficient conditions for being a ewe.

[7]Department of National Heritage statement DNH 398/96, issued on 17 December 1996 (http://www.britarch.ac.uk/cba/portant3.html).

on typical geographical maps marking population distributions (e.g., of people of certain religious groups, income levels, political affiliation) or differential uses of land in diverse economic zones (separating the agricultural and industrial sectors, for example). These facts certainly are mind-dependent, for, e.g., there would not be a difference among densities of religious groups to map were there not religions, which in turn would not exist without the existence of a certain set of beliefs and practices of the faithful. Yet the existence of such facts and the corresponding population distributions or economic zones does not depend on anyone having any beliefs (either in this case or in general) that are *about population distributions or economic zones*[8].

The possibility that there may be mind-dependent facts that do not depend on mental states that are *of or about that fact itself*, or even about that *kind of* fact, is often overlooked. Thus, for example, Searle, having defined observer-relative features as features that "exist only relative to the attitudes of observers", in contrast with intrinsic features which "exist independently of observers" (Searle 1995: 11), proceeds to the conclusion that:

> It is a logical consequence of the account of the distinction ... that for any observer-relative feature F, *seeming to be F* is logically prior to *being* F, because—appropriately understood—seeming to be F is a necessary condition of being F (Searle 1995: 13).

But it is not a logical consequence of a kind of feature F's depending on observers that seeming to be F is a necessary condition of being F, since the kind of feature F may depend on intentional states regarding other features; nor in general does the mind-dependence of a certain feature F entail dependence on any *F-regarding* mental states. Searle similarly concludes that all social concepts (defined as facts that involve collective intentionality (Searle 1995: 26)) are self-referential in the sense that, for money and all other social concepts, "in order that the concept "money" apply to the stuff in my pocket, it has to be the sort of thing that people think is money" (Searle 1995: 32). But while this may hold for directly created social kinds, it is clearly not true for those that are the indirect products of collective beliefs and practices regarding other kinds of things entirely (see Thomasson 2003).

---

[8]A similar phenomenon may occur in certain cases of fiat boundaries. For although all fiat boundaries depend on human cognition, they need not be deliberately created and maintained. Thus, e.g., Smith discusses cases of individual perceptual (as opposed to collective geographic) fiat boundaries that may be unintentionally created:

> The term "fiat" (in the sense of human decision or delineation) is to be taken in a wide sense, as including not only deliberate choice, as when a restaurant owner designates a particular zone of his restaurant a no-smoking area, but also delineations which come about more or less automatically, as when, by looking out across the landscape, I create without further ado that special type of fiat boundary we call the *horizon* (Smith 2001: §2).

Similarly, the visible field of a perceiving subject has fiat boundaries created only in virtue of acts of perception, though those fiat boundaries do not require any perceptions or thoughts *about them* or *about boundaries of visual fields generally* in order to exist. In such cases, the fiat boundaries do depend on mental states, but not states that are themselves of or about those boundaries (instead they may be about the landscape or a parrot in the distance). It is more difficult to find cases of indirectly created collective geographic fiat objects.

Consequences for the Epistemology of Geography

As we have seen, many of the facts typically studied by human geography have rather striking ontological differences from the paradigm facts studied by natural sciences, insofar as they depend on human beliefs and concepts. It is less clear thus far, however, what difference this ontological dependence makes to our capacity to acquire knowledge of or make discoveries about these facts. Opinions in the literature vary between the extremes, as Lakoff (1987: 208) asserts that, "In the case of social and cultural reality, epistemology precedes metaphysics, since human beings have the power to create social institutions and make them real by virtue of their actions", meaning, I gather, that knowledge of (or beliefs about) these facts make them the case. Smith, however, takes the contrary view that "Even in regard to human institutions, however, in contrast to what Lakoff has to say, *our thinking does not make it so*" (Smith 2001: §5 n. 6). The stakes are high for determining who is right here, since a close epistemic connection to these facts would lead many to exclude them from an inventory of the world, and would seem to preclude our making discoveries about the relevant social-scientific facts. It would also seem to rule out the possibility that we could expose hidden facts in a critique of extant institutions—critiques such as those engaged in by Foucault (e.g., in 1995), in Haslanger's (2016) work on bringing to light structural explanations, and in the work of many other social critics and reformers.

We can only assess the situation properly by carefully distinguishing the forms of mind-dependence involved in each case and examining what epistemic consequences do and do not follow from the forms of mind-dependence in question. I will consider separately, in turn, the three cases delineated above.

Direct Creation by Token

According to the realist paradigm, in cases of genuine scientific inquiry the facts to be discovered are independent of whatever anyone accepts, beliefs, holds true, etc., regarding those facts. As a result, facts may exist and yet remain entirely unknown, with everyone in ignorance; and widespread or even universal belief in a given fact does not suffice to make it so. Yet although this epistemic picture is widely held and is at least plausible for descriptive facts regarding trees, fish, and electrons, it clearly does not hold for geographic and other facts that are created by token in the manner described above.

Consider first the case of facts regarding fiat boundaries created by token. A bona fide object such as an island or mineral deposit has boundaries that exist and are as they are completely independently of all beliefs about them, making such boundaries subject to genuine discovery and leaving all potential discoveries subject to the possibility of error. The fiat boundaries of fiat objects, however, may not remain unknown to everyone but must be transparent at least to those who establish them. In cases where fiat boundaries are (directly) established by declaration, the established boundaries are transparent to the creator(s) of those boundaries; if the boundaries exist she must know of them, and since she is involved in declaring them, she cannot get them wrong in the way that everyone, say, could be wrong about the boundaries of Key West. In those cases where fiat boundaries are (directly) established by custom, no particular person's beliefs about the location of the boundaries are protected from error, for the boundary depends only on the collective beliefs of a group regarding it[9]. Nonetheless, taken as a whole, the group cannot be entirely mistaken about where the fiat boundaries (e.g., of Mount Kinabalu) lie, since their beliefs (or what they customarily "count as" part of Mount Kinabalu) are constitutive of the location of the fiat boundaries. This again puts fiat boundaries in contrast with bona fide boundaries, regarding which everyone may be completely mistaken.

Those not themselves involved (by declaration or by custom) in the establishment and maintenance of fiat boundaries may, of course, be entirely mistaken regarding the location or even existence of fiat boundaries; to them, facts about such boundaries are a matter of discovery and may be just as opaque as facts about bona fide boundaries. Nonetheless, the close epistemic relation between fiat boundaries and the beliefs of those who create them does make a crucial difference regarding the method of discovery. To discover the boundaries of an island or mineral deposit, one goes directly to the object itself to track the discontinuities that form the bona fide boundary. By contrast, to discover the boundaries of a tribe's territory or a sacred mound, one cannot seek direct discontinuities in the landscape but must instead seek evidence of where individuals have declared, or people collectively believe or accept the boundaries to be. This captures part of the traditional wisdom that the study of the human sciences requires grasping the intentional states of others.

Much the same goes for social facts created by token: The existence of the fact in question entails that someone knows of its existence, namely, at least those individuals who are required to accept its existence in order for it to exist. In the case of social facts that are token created by custom, there must be collective acceptance of the fact by the relevant group. Thus, e.g., (where commons are created by custom), a common cannot exist in a village without anyone knowing of it, since the tract of land's status as a common exists only if it is collectively accepted as existing. Although any individual may be in ignorance of the fact, the relevant community as a whole cannot be. Similarly, the community as a whole is not subject to error if they accept that a particular piece of land is a common, since (in that context) their collective acceptance of a certain piece of land as a common makes it so. In the case of social objects created by declaration, most members of the sustaining community may be ignorant or in error regarding the particular token fact, but still not everyone may be. Nothing can be a national park unless someone (e.g., at least the members of Congress involved in establishing it as a national park) believes it to be or accepts it as a national park, although it is possible that most members of the local community involved in giving Congress that right remain ignorant of the

[9] Assuming, as seems reasonable, that a group may have a collective intention that P without every member of the group having the intention that P.

particular fact that Congress has so declared it, and thus that this piece of land is a national park.

Direct Creation by Type

Different but equally significant, epistemic consequences arise in those cases where the facts of human geography are directly created by type rather than token. In the case of facts created by type, if a particular piece of land fulfills the conditions accepted as sufficient for the existence of a fact of that kind, that fact exists regardless of whether anyone has any thoughts about that piece of land itself. Thus, here particular facts (e.g., that a piece of land is government property) may in principle remain unknown to everyone, and everyone may be mistaken regarding them (unless the general conditions accepted happen to require token recognition). Here, the interesting epistemic consequences arise at the level of type rather than token.

According to direct reference theories and many scientific realists, natural kinds have a nature that is entirely independent of beliefs, leaving everyone's beliefs about what it takes to be of the kind subject to ignorance and error. But for kinds of social facts directly created by type, the group involved in maintaining the facts has a certain privileged knowledge regarding the nature of the kind. First, any conditions that group collectively accepts as sufficient for something to be a member of the kind genuinely are sufficient for kind membership, for anything that has all of the features accepted as sufficient for kind membership automatically "counts as" being of that kind. Thus, members of the group maintaining such facts are protected from error in at least some of their beliefs regarding what it takes to be of that kind. Second, if there is anything of that kind, there cannot be complete ignorance regarding the nature of the kind. For if there is something of that kind, there must be certain principles accepted regarding sufficient conditions for kind membership, and as we have seen, these must be true.

Thus, we have a much closer epistemic relation to the kinds of human geography that are created by type than we seem to have regarding the natural kinds studied by the physical sciences, leaving less room for the discovery of the nature of the kind here than in the case of gold or tigers. This does not mean, however, that we are all totally immune from ignorance and error regarding the nature of such kinds. Far from it: here again, it is only the group taken as a whole whose beliefs are protected from certain forms of ignorance and error; any particular individual may go wrong or remain in ignorance. Second, the protection from error only applies to conditions accepted as sufficient for kind membership; it does not follow that any other beliefs the group might happen to hold (or necessary conditions they might accept) regarding the nature of the kind do in fact hold. Finally, these forms of epistemic privilege apply only to the group involved in establishing and maintaining the institutional kind in question; outsiders may, of course, be fully ignorant of conditions relevant to membership in the kind. Their discovery of the nature of the kind, however, is not through direct tests on kind members, but must again involve discerning what principles the people in the relevant group accept as sufficient for kind membership.

Indirect Creation

In the case of indirectly created geographic facts, the answer to the question "What difference does their mind-dependence make?" is very different from the case of facts directly created by token or by type. In both of those latter cases, it is the direct dependence of the fact in question on beliefs (or principles accepted) *about that fact or kind of fact* that leads to the epistemic privilege of certain groups. Indirectly mind-dependent facts, however, such as those regarding population distributions and de facto urban zoning by religion, income, or function, may exist without anyone having any knowledge of their existence, and facts of those kinds may exist without anyone accepting any beliefs about what it takes to establish a fact of that kind. Geographers, of course, may label such regions or even artificially sharpen their boundaries in drawing lines on maps marking various population or functional differences. But the *existence* of the differences in population, culture, or functional use tracked by these labels is independent of the geographer's own concepts and demarcations, as well as independent of any of the locals' beliefs about facts of this kind. Thus, we have reason to resist the conclusion that such formal and functional regions demarcated by human geographers, as one introductory human geography text puts it, "exist only on our maps and in our minds" (deBlij 1977: 7).[10]

As mentioned above, Searle holds the view that one feature of all social reality is that social facts are "self-referential" in the sense that in general, something is of any social kind K only if it—or things like it—is used as, regarded as, or believed to be (a) K. But although this may hold of the institutional kinds that are Searle's focus, it is by no means true of all social kinds whatsoever (see Thomasson 2003). Although the forms of mind-dependence at issue with facts directly created by token or type do entail certain forms of epistemic privilege, it should not be inferred that mind-dependence always provides a closer epistemic relation to the dependent entities; it does not in the case of indirectly created social facts. Facts of such kinds we may call "opaque" since their existence does not imply any knowledge regarding the existence of the particular facts or what it takes for there to be facts of this kind. Even in these cases, however, the method of discovery must include investigation into the beliefs and intentions of the local people involved, for those beliefs or intentions are still at

[10]The regions so marked out by geographers are clearly not pure fiat objects since (if well drawn) they will correspond to certain qualitative differences in the areas; the boundaries may be considered to be fiat boundaries only insofar as geographers' ways of demarcating such regions may impose artificially sharp fiat boundaries on what are in fact merely graded distinctions. This artificial (fiat) sharpening of boundaries, however, occurs not only in human geography but also in physical geography and other scientific representations where graded differences in data are grouped into sharply bounded categories. I shall reserve the analysis of such phenomena and their consequences for another occasion.

least a necessary condition for the existence of facts of the kind in question—though in this case they are not beliefs about that very social fact or kind of social fact.

Summary

As we have seen, the issue of whether or not "epistemology precedes metaphysics" in the case of the social facts studied by human geography is too complicated to be accommodated by either Lakoff's or Smith's general answer. In some cases, namely those of direct creation by token, knowledge (or acceptance) of the fact does make it the case. In other cases (those involving direct creation by type), token facts of the relevant sort may remain unknown or may be falsely believed to exist. But here we still have a closer epistemic relation to the *kind of* fact than we do in the case of natural kinds, since no facts of that kind can exist without certain conditions relevant to the nature of the kind being accepted, and principles collectively accepted regarding sufficient conditions for kind membership must hold, leaving the creating and sustaining group with a privileged knowledge regarding the nature of these social kinds that everyone lacks in the case of natural kinds. These results limit to a certain extent the range of possible discovery open to human geography, and necessitate the use of humanistic methods of discovery for many of these facts and for the nature of these kinds.

Yet these limitations do not unduly constrain the possibilities of geographic inquiry, much less entail that genuine discoveries in human geography are impossible. As I argued in the section on "Indirect Creation", there are many mind-dependent geographic facts the existence of which may be completely opaque to everyone, and so it can be a genuine matter of discovery, e.g., that (or how) cities are typically de facto zoned into economic, religious, or functional sectors. (Of course, formal zoning involves institutional facts that must be transparent at least to those who create and maintain them.) In these cases, Smith's position that our thinking does not make it so is apt.

Even regarding directly created social facts there is much that awaits discovery by social scientists. First, such facts as are token created and maintained by others will remain opaque and require discovery by geographers. Similarly, geographers themselves will have no privileged knowledge regarding the nature of geographic kinds, where facts of that kind are type-created by others. Second, even certain facts involving social status or fiat boundaries within one's own culture may remain opaque and in need of discovery. Indeed the most central issues pursued by geographers involve not discovering the boundaries of fiat objects such as Wyoming, nor the conditions relevant to belonging to a social kind like being a national park, but rather causal relations involving geographic entities, to answer questions such as "Why and how do states evolve and decline?", "What determines the location and spacing of cities and towns?" and the like (deBlij 1977: 3). Such facts remain as much in need of discovery as any and cannot be revealed simply by inspecting the beliefs or principles accepted by ourselves or anyone else.

Consequences for the Ontology of Geography

As we have seen, we do have some epistemic privilege with respect to some of the facts and kinds studied by human geography that we apparently lack with respect to the facts and kinds studied by the natural sciences. According to some formulations of realism, however, any epistemic privilege with regard to a certain (purported) fact or kind precludes it from being admitted to a realist's ontology. Lakoff takes it to be a central feature of objectivism that "No true fact can depend upon people's believing it, on their knowledge of it, on their conceptualization of it, or on any other aspect of cognition. Existence cannot depend in any way on human cognition" (Lakoff 1987: 164). Elder defends such a view when he writes, "I shall myself construe realism as a denial of epistemic privilege" (Elder 1989: 440), namely that:

> … for any component of the world and any set of beliefs about that component, the mere facts that those beliefs are (i) about that component and (ii) are held by the particular believers, by whom they are held, never by themselves entail that that set of beliefs is free from massive error (Elder 1989: 441).

Many of the facts and kinds of human geography, however, would fail such a test: In the case of facts created by token, the fact that those creating the fact have certain beliefs about those facts entails that those beliefs are free from massive error; while in the case of facts created by type, the mere fact that the creating and sustaining community believes certain conditions to be sufficient for membership in a certain social kind entails that those beliefs are protected from error. Thus, all such facts and kinds would be excluded from ontologies following criteria such as Elder's.

But despite the ontological dependence on human intentional states that characterizes many of the facts of human geography, and despite the epistemic privilege that results in at least some cases, we have several strong reasons to resist the conclusion that geographic facts and kinds are not genuine components of the world. The first reason follows from the section entitled "Summary" above. For as we have seen, even in those cases where we do have some epistemic privilege regarding geographic facts and kinds, there remains much about these facts and kinds that is opaque to everyone and in need of social-scientific discovery. Thus, they fail to meet the paradigm of mere creations of our minds—they are certainly not imaginary objects possessing all and only those characteristics we ascribe to them.

The source of the problem here, as so often, arises from assuming a simple dichotomy between the independent entities of nature and imaginary objects that "exist only in our minds"[11]. The objects of human geography—like most of the commonplace entities in the everyday social and cultural world—lie between these extremes. In the case of geography, we are typically concerned with entities that involve independent tracts of land, as well as boundaries or social status that depend on collective intentionality. This in-between position is also reflected in the fact that we have at best some partial epistemic privilege with regard to them, but not the

[11] For further discussion of the problem with this dichotomy and a finer-grained set of categories to deal with the in-between cases, see Chapter 8 of Thomasson 1999.

full transparency expected of total inventions of the mind. This should give pause to those inclined to accept any epistemic privilege as sufficient grounds for assimilating purported entities to the status of the merely imaginary.

As the study of geographic entities makes vividly apparent, questions about whether or not certain kinds of entities are "mind-dependent", and questions about whether mind-independence (in general) entails certain forms of epistemic privilege, are far too coarse-grained to do useful work for us. As we have seen, although all of the facts of human geography that we have considered do depend on collective human intentional states, there are many such (indirectly created) facts and kinds of facts with respect to which we lack any epistemic privilege whatsoever.

This also brings into question the validity of epistemic privilege as an ontologically relevant criterion for rejecting entities as candidates for being genuine components of the world. It would be odd, to say the least, to accept as "genuine components of the world" de facto zones of cities based merely in differential income, religious beliefs, or occupation, while rejecting directly created zones (the products of local zoning decisions) on the basis of the latter's epistemic transparency to the creators.

It might be argued that that simply shows that the "no epistemic privilege" criterion is only a necessary, not a sufficient, condition for being accepted as a genuine component of the world. Why should it be necessary, however? Presumably, the thought is that epistemic privilege is always a symptom of mind-dependence, and that realists should reject any mind-dependent entities (whether or not we have any epistemic privilege regarding them). On this score, then, we would have as much reason to reject indirectly created opaque social facts as we do to reject social facts directly created by type or by token, simply in virtue of the fact that all depend for their existence on certain forms of intentionality.

But does the realist need to reject all mind-dependent entities? This, it seems to me, is a misunderstanding of realism. To distinguish realism from various forms of anti-realism and idealism, the realist clearly needs to accept that *there are some* things that exist and are as they are independently of all human intentional states. But there seems to be no reason to think that the realist cannot accept that, in addition to, say, the independent entities of the physical world, there are also mind-dependent entities in the social world studied by human geography and other social sciences. Both Searle and Michael Devitt, while defending general realist views, are happy to allow that there may (also) be mind-dependent social entities without this interfering with a general realist thesis. Thus, Devitt notes that "The world that the Realist is primarily interested in defending is independent of us except in one uninteresting respect. Tools and social entities are dependent on us ..." (Devitt 1991: 249), while Searle describes realism as the view that if there had never been any representations, "Except for the little corner of the world that is constituted or affected by our representations, the world would still have existed and would have been exactly the same as it is now" (Searle 1995: 153).

Now it might be said that the realist can, perhaps, accept that there are facts of human geography (e.g., that this land is a national park) and objects of the geographic kinds involved (e.g., national parks), but that in virtue of their mind-dependence the realist must deny that they are part of the "furniture of the world". If this is taken to mean that they are not among the mind-independent components of nature, this is fairly unobjectionable. For as we have seen, these things do involve forms of dependence on collective intentionality that (the realist must assume) those of nature lack. If, however, it is taken to mean that there aren't *really* such things in the world to be studied, it is quite objectionable and misleading. There are at least two senses in which one might be said to be a realist with regard to things of a particular kind: (1) The sense of accepting that things of that kind exist independently of all mental states; (2) The sense of accepting that there are things of that kind.

Accepting, then, that these things are mind-dependent in the various ways described in the section entitled "Varieties of Mind-Dependent Geographic Objects" above, should we accept that there nonetheless *are* such things, that they are genuine components of the world (albeit the human, not natural, world) regarding which we may acquire genuine knowledge?

As I have argued at length elsewhere (2015), the general question "Are there any entities of kind K" is best answered by determining what it would take for there to be such entities (what the actual application conditions are for the relevant sortal term "K"), and then evaluating whether anything meets those conditions. For purported entities of some kinds, mind-dependence might be a problem. If (according to the associated application conditions for "unicorn"), for there to be unicorns there would have to be instances of a mind-independent biological kind, and it turns out that the best we can say is that unicorns were creations of human myth, then we have grounds for denying that unicorns (using the term as a term for a mind-independent biological kind) exist.

By contrast, consider what it takes for there to be national parks. For there to be national parks, it is not required that there be some independently existing natural kind with an essence opaque to everyone. Instead, it is merely required that there be pieces of land designated as national parks by Congress and protected as such by government agencies. We might naturally suspect that those who understand the meaning and use of the term "national park" in the United States, but deny that there are such things, buy into a massive conspiracy theory according to which the supposed declarations and acts of government agencies are all illusions. Of course, philosophers who deny their existence will deny that they are subscribing to conspiracy theories, so what can we say of them? It seems they are either confusing the question of whether there are such things with the question of whether they form an independent natural kind (or are independent natural objects), or are tacitly recommending that (at least for "scientific purposes") we drop terms for such mind-dependent objects from our vocabulary. But the conditions ordinarily required for there to be national parks certainly seem to be fulfilled, and so we have reason to say that there are national parks in the only sense we should have ever expected there to be. Much the same goes for nations, electoral districts, commons, and other social and fiat entities apparently referred to in the theories of human geography. These are as genuine components of the world as one should expect of instances of human kinds, and as genuine as one needs to make the study and discoveries of human geography possible.

The conclusions here are of broader significance. First (as I argue in Thomasson 2003) the conclusions about epistemic opacity generalize to other social entities—in ways that are important for understanding the social sciences and the possibilities of discovery and critique they bring. If we held that all social entities are epistemically transparent to us, then there would seem to be little room for discovery or critique in the social sciences. However, as I argue elsewhere (Thomasson 2003, 2009), acknowledging the different forms mind-dependence can take, and the different paths via which social entities of all types (not just geographic entities) may be created enables us to get a much more healthy and accurate picture of the prospects for discovery, and the potential role for critique in the social sciences at large.

Second, I have argued here that we should reject such criteria as mind-independence and lack of epistemic privilege as (across-the-board) criteria for existence. Instead, I have argued, if we want to know whether there are entities of a kind K, we should simply ask what it would take for there to be Ks—or what the application conditions for the term "K" are—and then evaluate whether anything meets those conditions. This is an approach I have argued for and developed more extensively elsewhere (Thomasson 2015), arguing that the method we should use in addressing existence conditions is to determine what application conditions are actually associated with the relevant term, and whether they are fulfilled. Once we take that path, we can also get a general argument that all substantive "criteria of existence"—including mind-independence and epistemic opacity—should be rejected (Thomasson 2015: Chap. 2). Instead, we must look at what it would take for there to be something of the kind, and whether those conditions are fulfilled. Failures of mind-independence may be a problem for some (purported) kinds of entities, but not others—where it is built into the very idea of a national park, dollar bill, or even fictional character that it depends in certain ways on human intentionality. I have also argued (Thomasson 2015) that we often can get "easy" arguments for the existence of disputed kinds of entity, from uncontroversial premises. We can get the same form of argument for geographic entities: I drive across the border between Vermont and New Hampshire every day to get to work, Vermont and New Hampshire are states, therefore there is a state boundary that I cross, therefore there are geographic entities.

Finally, as I have argued recently (Thomasson 2016), often metaphysical debates that appear worldly can be better analyzed as cases in which the disputants are implicitly engaged in what David Plunkett and Tim Sundell (2013) call "metalinguistic negotiation". Metalinguistic negotiations occur when speakers apparently use terms (rather than explicitly mentioning them) but do so not in order to share information about the world, but rather to press for changes in (or maintenance of) ways in which the relevant term is *to be used*. Accordingly, we can see some who argue that there are no geographic objects as implicitly suggesting that we remove terms for mind-dependent or epistemically nonopaque entities from our vocabulary—at least for purposes of doing serious philosophy or science. Yet once we put things in these terms, we can also see why this is a proposal we should reject. For these terms play an indispensable role in organizing our social and political lives together and establishing public norms for use of certain spaces. They also play an important part in our social-scientific theories. Rejecting them in favor of a linguistic and conceptual framework that only included terms of mind-independent entities or natural kinds would be a huge mistake.

The right response to examining the case of geographic entities and other social and cultural objects is not to reject them, or terms for them, nor to dismiss them all for failing to meet up to some criteria thought to be suitable for the entities studied by the natural sciences. Instead, we need to respect and appreciate the subtleties and variations among the entities we live with and study, and take a case-by-case look at the different ways in which entities may depend on human intentionality, and the diverse consequences this may have for our ways of knowing them.

Notes

This paper is an updated version of a paper of the same name, originally published in Topoi Vol. 20, No. 2 (September 2001): 149–159.

References

deBlij H (1977) Human geography: culture, society, and space. Wiley, New York
Devitt M (1991) Realism and truth, 2nd edn. Princeton University Press, Princeton (NJ)
Elder C (1989) Realism, naturalism and culturally generated kinds. Philos Q 39:425–444
Foucault M (1995) *Discipline and punish*, (Second Edition), Translated by Alan Sheridan. Vintage Books, New York
Haslanger S (2016) What is a (social) structural explanation? Philos Stud 173(1):113–130
Lakoff G (1987) Women, fire and dangerous things. University of Chicago Press, Chicago
Plunkett D, Sundell T (2013) Disagreement and the semantics of normative and evaluative terms. Philos Impr 13:1–37
Searle J (1995) The construction of social reality. The Free Press, New York
Smith B (2001) Fiat objects. Topoi 20:131–148
Smith B, Varzi AC (2000) Fiat and bona fide boundaries. Philos Phenomenol Res 60:401–420
Thomasson AL (1999) Fiction and metaphysics. Cambridge University Press, Cambridge
Thomasson AL (2003) Foundations for a social ontology. In: Protosociology, understanding the social II: philosophy of sociality, vol 18–19, pp. 269–290
Thomasson AL (2009) Social entities. In: R. le Poidevin et al. (ed) Routledge companion to metaphysics. Routledge, London
Thomasson AL (2015) Ontology made easy. Oxford University Press, New York
Thomasson AL (2016) Metaphysical disputes and metalinguistic negotiation. Anal Philos 57(4):1–28

Chapter 9
What a Geographical Entity Could Be

Timothy Tambassi

Abstract The main task of this article is providing a sketch of possible approaches, response attempts, conundrums and issues arising from the question: 'What is a geographical entity?'. It is shown how trying to answer this question is made particularly difficult by a multiplicity of aspects that might be summarized as follows: (1) There exist multiple conceptualizations of the geographical world. (2) Different languages and cultures may slice such a world in different ways. (3) The geographical world has changed and will change over time. (4) Also geography (as a discipline) has changed and will change over time, modifying its perspective, tools, domains of investigation and aims. Consequently, what had, has been, will be considered as non-geographic could be considered as geographic and vice versa. (5) There were, are and will be different kinds of geographies as well as different geographical branches, each of them had, have and might have different tools, aims, points of view and vocabularies. (6) The introduction of new scholarly fields and new technologies, the birth of intellectual movements or paradigm shifts and developments on other disciplinary contexts can/will influence geography as a discipline.

Keywords Geographical entity · Ontology of geography · Ontology of GIS · Definitions · Laundry-lists · Boundaries · Maps · Granularity · Ontological perspectivalism · Hierarchical structures · Geographical conceptualizations · Cultural diversities · Linguistic differences · Vagueness · Historical entities · Geographical kinds

A Chaotic List that Cries Out for Explanation

Providing a complete list of geographical entities would be a very long and (potentially) extravagant task, given the innumerable functions and purposes that geographical entities might have and the variety of (disciplinary) contexts from which they emerge.

T. Tambassi (✉)
Department of Humanities, University of Eastern Piedmont, Vercelli, Italy
e-mail: timothy.tambassi@gmail.com

ICUB, University of Bucharest, Bucharest, Romania

© Springer Nature Switzerland AG 2019

T. Tambassi (ed.), *The Philosophy of GIS*, Springer Geography,
https://doi.org/10.1007/978-3-030-16829-2_9

They might arise from the physical world such as, for example, mountains, seas, oceans, rivers, islands, deserts, and so forth (Inkpen and Wilson 2013). They can emerge from the combination of environmental features (of the physical world) and demarcations introduced by human cognition and action (i.e. bays, promontories and so on). They might also be the result of political and administrative subdivisions, law decrees, land ownerships such as nations, regions, postal districts and so on, involving social conventions on a number of different levels, generally marked by differences in the ways different societies structure the world (Smith and Mark 1998). In addition, an inventory of geographical entities can also include human-made objects such as streets, buildings and so forth (Laurini 2017).

Obviously, we could go on and on, listing kinds and sub-kinds of geographical entities or emphasizing that they may be zero-, one-, two-, or three-dimensional such as, respectively, the South Pole, the Tropic of Cancer, Canada (a two-dimensional object with a curvature in three-dimensional space), and the North Sea—that, according to the context, can refer either to the three-dimensional body of water, or to its two-dimensional surface (Smith and Mark 1998).

Moreover, geographical entities can be disconnected like countries with several islands or like France with Martinique, Guyana, New Caledonia, etc. Sometimes, they have 'holes' such as South Africa has with Lesotho. Some have sharp borders, others indeterminate boundaries. They also can be simple, made up by other geographical entities or share mereological or topological relations with other geographical entities or/with their parts (Varzi 2007).

Furthermore, some geographical entities also have some sorts of relations both with the (surface of the) Earth and with the space they occupy (Casati and Varzi 1999). Generally speaking, a relational theorist of space would say that entities are cognitively and metaphysically prior to space (there is no way to identify a region of space except by reference to what is or could be located or take place at that region). In contrast, an absolutist theorist would say that space exists as an independently subsistent individual (a sort of container) over and above its inhabitants (objects, events and spatial relations between objects and events, or without all these entities).

This chaotic list cries out for order and explanation. Is there really something such as a geographical entity? What, if anything, do geographical entities have in common? What sorts of entities are they, how are they individuated, and what are their identity conditions (Bishr 2007)? How to distinguish between what is a geographical entity from what is not? What is the difference, if any, between geographical and spatial entities? Are there geographical entities that are not spatial and/or spatial entities that are not geographical? What are the sorts of factors that might influence our inventory of geographical entities? What is the relationship between geographical entities and their representation in maps (Casti 2015)? Should we think of geographical entities in general, or is it more appropriate to assume that every geographical sub-area has a proper list/account of geographical entities?

Avoiding Univocal and Incomplete Accounts

In approaching those questions, one of the main issues will be to avoid univocal and incomplete accounts, which could be suitable for some theoretical tasks but not rich enough to grasp the complexity of our ways of representing the geographical world. Geography, indeed, has had hundreds of years to elaborate on different sorts of geographical entities for innumerable purposes. Moreover, geography (as a discipline) has changed over time, modifying its perspective, sometimes its aims, subdividing itself into different branches and weaving together with different scientific, social, and technical disciplines (Pattinson 1963; Couclelis 1998; Bonnett 2008; Sala 2009). In this sense, what can be considered as geographic has changed according to the geographical perspective we endorse. Furthermore, we should also observe that different languages and cultures have created different vocabularies and ways of slicing the words, producing (potentially) different kinds of geographical entities. The fact that geographical reality/realities can be studied, sliced and represented in different ways does not surely exclude that such alternative descriptions of the geographical world can be compared, overlapped and/or integrated with one another in order to get hold of improved accounts of reality itself.[1] Hence, by paraphrasing the words of Epstein (2017), the proposed analysis will be multifaceted and will fight the prevailing philosophical trend of simplifying the endless diversity and variation among different geographical perspectives (Elden 2009; Tanca 2012; Günzel 2001).

In light of these considerations, the aim of the next pages is to present a series of possible issues, conundrums and approaches for analyzing and explaining the nature of geographical entities. Surely, such a series should not be conceived as exhaustive, nor the approaches as isolated one from another. Instead, there can be mixed cases that might be seen as a combination of different approaches—for example, between laundry-lists, attempts of definition and others. The same can be said both for issues and conundrums, which rarely appear alone and sometimes persist also across the different approaches we discuss. Finally, we should underline that the choice of the term ‘entity’ is not neutral in this context and can be considered as problematic. Indeed, as Smith and Mark (2001) have remarked, philosophical ontologists have long been aware of the controversial character of ontological terminology. In this sense, the term that should be used for the ontological supercategory (things, entities, items, existents, realities, objects, somethings, tokens, instances, particulars, individuals) within which everything belongs is not exempt from possible criticisms. Each alternative has its adherents, yet each also brings problems and sometimes different inventories. In this case, the choice of ‘entity’ is given by the needs of generality and exhaustiveness (that is common to other terms), which is specified by the possibility of being inclusive, on the one hand for items such as relations, kinds and so forth, on the other hand for things that might be abstract, universals or non-instantiated. Accordingly, such a choice means to not exclude, a priori, the possible existence of these sorts of things, which could be easily compromised, for

[1]Therefore, my theoretical point of view may be seen as closed to *ontological perspectivalism* (Bateman and Farrar 2004; Grenon and Smith 2007; Elford 2012).

example, by the use of terms such as existents, particulars, instances and so on. With this, I am not saying that using terms other than entity is not appropriate, nor that it cannot bring (in principle) similar results. Rather, I would like to remark that the purpose of these pages is to show some different approaches, issues and conundrums which emerge from the debate on geographical entities, with the aim of drawing a boundary (at least, a theoretical boundary) for distinguishing what is geographical from what is not.

Laundry-Lists

Within the philosophical debate, when asked to provide a definition of 'ontological category', a possible answer consists in giving not some particular examples of ontological categories, but a full list of all of them, without further specifications. Surely, it is useful to know what has been regarded as an ontological category in the history of philosophy or what a particular author regards as such. But no matter how much interesting a list might be in itself, it is certainly no substitute for a definition. Rather, the list sets the stage by indicating which kinds of things our definition should incorporate (Westerhoff 2005, pp. 23–4).

Now, if the narrow number of ontological categories should guarantee an almost exhaustive list of ontological categories,[2] a list of geographical entities can difficulty strive for such an exhaustiveness. Therefore, a laundry-list position in geography will give only few (and maybe paradigmatic) examples of geographical entities, at the expense of an excessive long and tedious catalogue of all of them. But how to provide such examples? According to Varzi (2001), normally, we know how to use geographic terms without being able to provide a precise explanation of the grounds for this competence. Presumably, the model of family resemblance shows how, in ordinary circumstances, a word can be used successfully regardless of whether or not it meets the Fregean ideal of precision. We say that something is a geographic entity because it resembles (in some relevant respect) several things that have hitherto been said to be a geographic entity, even if the exact nature of this resemblance may give rise to borderline cases.

Nowadays, the laundry-list position constitutes a recurring integration of many attempts at definition of geographical entity.[3] A non-exhaustive explanation can be traced in the ambiguous epistemological status of geography (that ranges, among others, from physical and human approaches to spatial analysis), for which a laundry-list, even before a definition, seems to guarantee a continuity among different geographical sub-areas. However, some difficulties may arise in deciding whether some

[2]In theory, such a list should also be open to the inclusion of new empirical and theoretical evidences, which might modify and/or extend it.

[3]See for instance: Casati et al. (1998), Smith and Mark (2001) and the link https://definedterm.com/geographic_entity. In the geographical debate, other examples of laundry-lists (that integrate some attempts of definition of 'geographical entity') can be found in some of the more general classes of geo-ontologies. For a list (not a laundry-list) of the main geo-ontologies see Tambassi (2017a).

particular entities in the lists are (or not) geographical entities. In that case, what criteria should we use for selecting the geographical entities from the realm of entities? What does unify a nation, a mountain, a latitude and makes us classify them as geographical entities? In other words, what, if anything, do geographical entities have in common?

Attempts of Definition

A possible answer to the questions above might deal with the *definition of geographical entity*—that is, to specify what a geographical entity is by exhibiting its conditions of existence, individuation and persistence, and its criteria of identity (synchronic and diachronic). According to Bishr (2007), identity criteria provide sufficient conditions for determining the identity of concepts defined in the domain we have to describe. For the purpose of these pages, providing identity criteria might be useful for the following:

- classifying an entity as an instance of the class *geographical entity* [GE];
- individuating an entity as a countably distinct instance of GE;
- identifying two entities at a given time (synchronic identity);
- reidentifying an instance of GE across time (persistence, or diachronic identity).

Once we fix the identity criteria for geographical entities, it is essential to determine what (geographic) entities (objects, relations, kinds, facts, events, spatial regions and so forth) have to be included as fundamental. Moreover, we should also establish whether our list of geographical entities will comprehend only entities such as mountains, rivers, deserts, etc., traditionally linked to the physical geography and/or also artifacts studied by human geography (entities like socioeconomic units, nations, cities and so on). In this regard, Casati et al. (1998) have distinguished three main different positions on the existence of geographic entities, which are given as follows:

- strong methodological individualism—there are "only people and the tables and chairs they interact with on the mesoscopic level, and no units on the geographic scale at all";
- geographic realism—"geographic entities exist over and above the individuals that they appear to be related to and have the same ontological standing as these";
- weak methodological individualism—if geographic units exist, "then they depend upon or are supervenient upon individuals. One form of this position would accept both individuals and the behavioural settings in which individuals act. Larger-scale socioeconomic units would then be accounted for in terms of various kinds of connections between behavioural settings" (Casati et al. 1998, p. 79).

However, despite these clarifications, some issues remain unaffected. In particular, what entities should we classify as instances of the class 'geographical entities'? How to distinguish between what is a geographical entity from what is not? What is the definition of geographical entity? How could we possibly distinguish, among the physical entities, those that are specifically geographic? Where is the exact boundary between physical and human (geographical) entities? And between spatial and geographical entities?

On Being Portrayed on Maps

> It seems that [...] it is being portrayable on a map which comes closest to capturing the meaning of 'geographic' as this term is employed in scientific contexts. Geographers, it seems, are not studying geographical things as such things are conceptualized by naïve subjects. Rather, they are studying the domain of what can be portrayed on maps (Smith and Mark 2001, p. 609).

Smith and Mark have made the point clear: if geographers study the domain of what can be portrayed on maps, then being portrayed on maps can say something about the notion of geographical entity. Now, let's suppose that the notions of *being portrayed on maps* and *geographical entity* correspond—in other word, that:

(1) an entity is geographical if and only if it can be portrayed on maps
and
(2) something can be portrayed on maps if and only if it is a geographical entity.

Obviously, such an identity relation could easily solve some problems concerning the identification of geographical entities. However, it might also raise a number of ontological conundrums that have not yet been addressed.

The first one is that the question of the definition seems to be simply shifted from the notion of geographical entity to the notion of map. This might also lead to subordinate the notion of entity to its representation[4] and maybe, more generally, the geography to the cartography.[5]

The second conundrum concerns what to do in the face of

- nonspatial geographical entities, which can be difficulty located in a map—for example, Poland during the Era of Partition, that is, the era in which the entity in question did not have any territory to call as its own land;
- (and/or) unusual maps, for instance treasure maps, maps that also include imaginary entities (such as Atlantis), maps using GPS coordinates (such as Google Maps) and so forth.

In all these cases, would we be willing to include the treasure, Atlantis and/or maybe ourselves in the inventory of geographical entities, too?

Finally, the third conundrum is strictly related with the first one and concerns the relationship between the notions of map and geographical entity. If, on the one

[4]In which case, we should, perhaps, also ask whether it is more appropriate to talk about cartographic entities rather than geographical entities.

[5]About the non-correspondence between geography and cartography, and more in general, on the critique of 'cartographic reason', see: Farinelli (2003, 2009).

side, we can have some difficulties in imagining geographical entities that cannot be portrayed on map; one the other side, some issues might arise with extremely detailed maps that represent not only entities such as seas, nations, streets, etc. but also trees, sidewalks, lampposts, for which we would probably have more trouble in classifying them as properly geographic. But then, what would we be willing to include among the geographical entities?

Maps, Granularity of Interest and Multiple Levels of Details

A possible way of answering the previous question might be to include, among the geographical entities, also entities like trees, sidewalks and lampposts. But in the face of more detailed maps, the risk is that our list of geographical entities also comprehends the leaves of those trees, the columns of those lampposts, a blade of grass of a garden and so forth. An alternative can be to consider (only) maps which are not so detailed, that is, maps, containing only geographical things. But, how to build such maps?

One of the issues in this matter might be the concept of granularity of interest, according to which geographic objects can mutate in the following two different ways:

- as the scale diminishes, an area will mutate into a point and then will disappear;
- as the scale enlarges, something might appear as a point and then mutate into an area.

Of course, a conceptualization of geographic space may have several levels of granularity, each of which has a specific inventory of geographical entities at different levels of detail. However, nothing excludes that once the scale is enlarged, the granularity of interest might also contain manipulable objects (see section "On *What* and *Where*") that, according to Egenhofer and Mark (1995), should not be properly included in the geographic space (and within geographical entities). Rather, the two authors maintain that geographic space shall include entities such as 'hotel with its many rooms, hallways, floors, etc.', 'Vienna, with its streets, buildings, parks, and people', 'Europe with mountains, lakes and rivers, transportation systems, political subdivisions, cultural variations, and so on'. In other words, geographical space represents the space in which we move around and that may be conceptualized from multiple views, which are put together (mentally) like a jigsaw puzzle. To put it differently, it is the level of granularity that coincides with the mesoscopic stratum of spatial reality. (The other stratum is the microphysical one that may be conceived as a complex edifice of molecules). The mesoscopic stratum is the real-world counterpart of our nonscientific cognition and action in space, and has three different types of components:

1. objects of a physical sort (such as rivers, forests, seas) that are studied also by physics but which, within the mesoscopic stratum, have different sorts of

properties—this is in virtue of the fact that our naive cognition endows its objects with qualitative rather than quantitative features and with a social and cultural significance that is absent from the microphysical realm;
2. objects like bays and promontories, which are also in a sense parts of the physical world but exist only in virtue of demarcations induced by human cognition and action;
3. geopolitical objects such as nations and neighbourhoods that are more than physical, and which exist only as the hybrid spatial products of human cognition and action (Smith and Mark 1998, p. 313).

However, also with these clarifications, some issues remain unsolved. In particular, which maps should we properly refer to? What is the minimum level of granularity for a map that represents exclusively geographic entities? What is the difference between geographic and manipulable objects? Such questions seem to reveal some limits of the correspondence between the notions of 'being portrayed on maps' and 'geographical entity', highlighting a sort of primacy of the latter notion (or, at least, of the evaluation of what is properly geographic) over the notion of 'being portrayed on maps'. But then, how to distinguish between what is a geographical entity from what is not?

On *What* and *Where*

The theory of spatial location investigates the relation between geographical entities and the regions of space they occupy or in which they are located. According to Casati et al. (1998), specifying such a relation also means choosing, in term of representation, between classical and non-classical geographies.[6]

Classical geography assumes that every single geographic entity is located at some unique spatial region and that every spatial region has a unique geographic entity located at it. Consequently, it defines the relation between geographical entities and the regions they occupy in terms of identity. As stated by Bishr (2007), such an identity relation also constitutes a fundamental premise of GIS and geo-ontologies, according to which a (geographical) object must have some location, even if the location can be arbitrary. In contrast, non-classical geographies consider that the relation in question is not one of identity. That means the possibility of geographical entities that are not located somewhere, of spatial regions with two or more geographical entities located at them and/or without entities on them. In other words, it licenses:

- on the one hand, nonspatial geographical entities, entities with multiple location or duplicates of the same geographical entity;

[6]Among the most significant works that investigate the notion of 'classical geography' in a more geographical sense and analyze its relations with the concepts of spatial location and representation in a totally different perspective than what is being discussed here, see: Lukermann (1961), Geus and Thiering (2014).

– on the other hand, maps with regions that are assigned no entity, or two or more competing units.

By discussing the relation between what and where, the theory of spatial locations also allows not to consider geographical objects as larger versions of the everyday objects and kinds studied in cognitive science. Indeed, according to Smith and Mark (1998), geographic objects are not merely located in space, as are the manipulable objects of table-top space or roughly human scale such as birds, pets, toys and other similar phenomena. For such entities, the 'what' and the 'where' are almost independent. On the contrary, in the geographic world, the 'what' and the 'where' are intimately intertwined. To be more precise, geographical object are tied intrinsically to space, in a manner that implies that they inherit from space many of its structural (mereological, topological, geometrical) properties. Obviously, that is not the only difference. According to the authors, geographic reality comprehends mesoscopic entities, many of which are best viewed as shadows cast onto the spatial plane by human reasoning and language (and by the associated activities). Because of this, geographic categories are much more likely to show cultural differences in category definitions than are the manipulable objects of table-top space. Furthermore,

> In the geographic world, categorization is also very often size- or scale-dependent […]. In the geographic world, to a much greater extent than in the world of table-top space, the realization that a thing exists at all may have individual or cultural variability. In the geographic world, too, the boundaries of the objects with which we have to deal are themselves salient phenomena for purposes of categorization. […] Moreover, the identification of what a thing is may influence the location and structure of the boundary (Smith and Mark 1998, p. 309).

Drawing the Contour

Another strategy for the identification and the individuation of (autonomous) geographical entities starts with the specification of their boundaries, in terms of location and typology. To be more precise, the strategy consists in sketching a taxonomy of boundaries, from which it may derive a corresponding categorization of the different sorts of geographical entities that boundaries determine and/or demarcate.[7] The basic idea is that an analysis on (and a classification of) geographical boundaries might be functional for determining what kinds of geographic entities exist and have to be included as fundamental.

Smith (1995) and Galton (2003) have provided the two most cited examples of comprehensive classifications of geographical boundaries in the geo-ontological domain. Both the classifications take the form of a hierarchical tree structure with a top-level distinction, which is considered, by the authors, as absolute, exhaustive and mutually exclusive. Galton distinguishes between institutional and physical boundaries. Such a distinction is the result of the different distribution of matter and energy

[7] Cfr. Smith (1995), Smith and Varzi (1997), Casati et al. (1998), Smith and Mark (1998), Smith and Varzi (2000), Galton (2003).

in space and time, from which the existence of boundaries depends. For institutional boundaries, the dependence of the boundary on the material facts is mediated by individual or collective human intentionality. For physical boundaries, there is not such a meditation. Conversely, Smith's main distinction is between *bona fide* and *fiat* boundaries. Bona fide boundaries exist even in the absence of all delineating or conceptualizing activity on our part, independently of all human cognitive acts and demarcations. On the contrary, the existence of the *fiat* boundaries depends on our delineating or conceptualizing activity. Despite the two authors do not share the same terminology, the examples they use for the entities belonging to such categories seem to indicate an overlap between the distinctions above. Indeed, both the authors include entities such as coasts, river banks, seaboards, among the prototypical examples of bona fine/physical boundaries. In contrast, entities like political and administrative boundaries, state and provincial borders, property lines and borders of postal districts provide examples of fiat/institutional boundaries.

Now, the main issue arising from this strategy concerns whether we can really affirm that the notion of boundary is, in some way, prior to the notion of geographical entity. If no, we might, in principle, assume the existence of geographical entities without boundaries. Consequently, such a position can hardly be considered as exhaustive in providing a complete inventory of geographical entities. If yes, we should analyze the ontological status of (geographical) boundaries and (maybe) choose whether:

- to consider them as higher order entities as some eliminativist theories do (cfr. Section "Boundaries");
- or, conversely, to include them within the list of geographical entities (as mountains, rivers, cities are).

The latter option requires to explain how a class (or a sub-class) of geographical entities can play a normative role in the definition of such entities, avoiding a *petitio principii*.

Another issue might arise from the claim of exhaustiveness of the taxonomies above, which should not appear as a restriction for the existence of other kinds of geographical boundaries. On one side, we should consider a certain degree of arbitrariness regarding both what is categorized and how it can be categorized. In this sense, also the functions of boundaries that we want to categorize might assume a significant role. On the other side, we could also change the classification system (or propose a new one) and then some of our boundaries might move, some of them disappear, new ones might have to be created. Moreover, it is important to remember that the natural language (and its evolution over time) and, more generally, cultural diversities in addition to human beliefs have contributed (and still contribute) to the categorization and the generation of (new kinds of) boundaries.

Finally, paraphrasing the words of Galton, even the distinctions purposed can be not entirely clear-cut and some cases can be classified in different ways depending on how they are interpreted. On the one hand, we may find intermediate cases, which seem to occupy a middle ground between two positions in the classification (Galton 2003, p. 152). On the other hand, there can be several cases in which a boundary of one type can evolve into or give rise to a boundary of another type and vice versa (Galton 2003, p. 159).

Cultural (Geographical) Entities

As Smith and Mark (1998) have remarked, we should also consider that geographic entities (and, more in general, geographic subdivisions) might involve a degree of human-contributed arbitrariness that is generally marked by differences in the ways different languages (and their evolutions over time), beliefs and, in particular, cultures structure or slice our world. According to the authors, such (cultural) differences might act differently depending on the entities we want to categorize:

- bona fide entities (seas, mountains, lakes, deserts) are more likely to be objects of categorizations that enjoy a high degree of cross-cultural invariance;
- fiat entities (nations, provinces, postal districts), in contrast, as far as they are inculcated into the world by cognition, are more likely to show cultural dependence.

Accepting that some geographical entities (included in our categorizations) might be, in some way, culturally influenced may leave the door open to the introduction of:

- cultural (geographical) entities in our classifications;
- (as well as) categorizations of geographical entities which (in turn) may have an influence on cultural diversities, human beliefs and individual or collective behaviours.

With regard to the first point, we should also consider that the modalities through which cultural differences might influence the classification of the geographical entities (and vice versa) operate, at least, at three different levels that should not be (improperly) equate. To be more specific, by using the notion of geographical boundaries, we can specify that cultural dependency[8] can occur, at least, at the following level:

1. the *definition of geographical boundaries* that determines what should be included in (the full list of entities belonging to) our classification;
2. the identification of (some) *different kinds of boundaries*, which determines the classes of our taxonomy—for example, the inclusion of the 'property boundaries' in our taxonomy is determined by the acceptance of the notion of property. In contrast, such boundaries will disappear in a society that does not know/accept the concept of property;

[8]Obviously, the study of the (mutual) relation between geographical boundaries and cultural elements is not unique to ontological analysis, but it is, for example, one of the main assumptions of border studies—according to which boundaries are generally understood as social constructs rather than being naturally given entities. In this respect, see for example: Kolossov (2005), Newman (2006), Agnew (2008), Newman (2010), Kolossov and Scott (2013), Paasi (2013a, b), Yachin (2015).

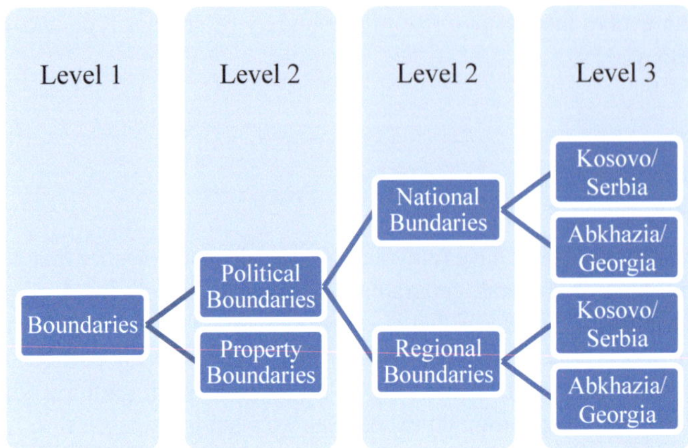

Fig. 9.1 Levels of cultural dependency

3. the *categorization of a specific boundary* within the different classes of boundaries previously identified/accepted, i.e. the boundary between Abkhazia and Georgia that, without taking into account other possible alternatives, can be regarded as a national or a regional one, according to our culture and/or beliefs[9] (Fig. 9.1).

GIS, Knowledge Engineering and Geographic Objects

If most of the approaches and considerations above generally adopt a speculative viewpoint, the perspective of Laurini (2017) is just to describe the notion of geographical object within the (applied) domain of GIS and knowledge engineering. In this context, the author maintains that any geographic object should have:

- an ID named GeoID, which will be an identifier only used for storing;
- a geographic type;
- a geometric shape (the most accurate possible)—and when necessary other less accurate representations will be derived quickly by using generalization algorithms;
- zero, one or many different toponyms;

[9]A similar example is provided by the recognition, among others, of Kosovo and Pridnestrovian Moldavian Republic, which is supported only by some (or none) of the members of the United Nations. In other words, some members of the UN might consider Kosovo and Pridnestrovian Moldavian Republic as proper nations, while other members do not. Consequently, the categorization of these entities changes according to the member of UN that classifies them. Of course, the concept of recognition is neither a prerogative of the United Nations nor of the notion of nation. On the contrary, it may be applied to, in principle, other geographical notions and/or institutions. See, for example, Italy with Lunezia (section "On Nonexistent and Abstract Geographical Entities").

- links with other geographical objects by spatial or geographic relations or even by structures.

Such a list extends and specifies the three different facets that, according to the author, characterize the peculiarity of geographical objects within such a domain: geometry, identification and semantic.

By their geometric facet, Laurini distinguishes two main categories of geographical objects:

- crisp objects that have well-defined boundaries such as administrative objects (countries, regions, natural parks, etc.) and manmade objects (streets, buildings, and so forth);
- (and) fuzzy objects which have undetermined boundaries (mountains, marshes, deserts, etc.).

The first category of geographical objects might be represented by using conventional geometry (that should also take in account issues coming from the curvature of the Earth), whereas the second one requires models deriving from fuzzy sets.

From the point of view of identification, Laurini maintains that geographic objects can have names, sometimes several names and that the same name might also be assigned to various different entities. The introduction of gazetteers and computer identifies (IDs) allows us to solve some ambiguities arising from toponymy, even though in different databases the same features can have different identifiers.

Finally, due to their semantics, geographical objects might be considered as conventional objects. However, some issues can emerge from the fact that different languages might:

- confer different names to the same geographical entity (for example, Mount Everest is known in Nepali as Sagarmāthā and in Tibetan as Chomolungma);
- (and in particular) use different categories for the geographic kind within which a specific entity is classified (for example, the geographic kind 'river' has two translations in French: 'fleuve' when a river flows to the sea, 'rivière' in any other circumstance).

Rivers, *Fleuves* and *Revières*

Considering conventional objects, we have already said that geographic categorizations can be marked by differences in the ways different languages slice the world (section "Cultural (Geographical) Entities"). "Terms like 'strait' and 'river' represent, in this sense, arbitrary partitions of the world of water bodies. The English language might have evolved with just one term, or three terms, comprehending the range of phenomena stretching between strait and river or, in French, between 'detroit' and 'fleuve'" (Smith and Mark 1998, p. 317).

Different languages might also contain different categories for the classification of the geographical entities. Taking the example of the previous paragraph, the geographical kind 'river' has indeed two translations in French: 'fleuve' when a river flows to the sea, 'rivière' for all the other rivers. "Notice that there is a topological relation between 'fleuve' and sea, and between 'fleuve' and 'rivière', whereas 'river' does not bear this kind of relation" (Laurini 2017, p. 62). Therefore, the Tiber might belong to two different categories, namely 'river' and 'fleuve', according to whether the native language of the person who categorizes such an entity is, respectively, English or French. By using the same example, another issue may emerge if a French speaker sees a natural flowing watercourse not knowing its topological relations: in this case, is he seeing a fleuve or a rivière? Maybe, we should point out that any categorization, in general, seems to require a good knowledge of the domain we want to categorize.

Moreover, in light of these considerations, we need to specify whether different languages require different classifications—since concepts can be different or differently organized. If they do, the challenge concerns how to match such different classifications of the same (geographical) domain. On the contrary, if they do not, we should select a language for our classification, consider the possibility of integrating translations in different languages, and try not to lose the conceptual richness emerging from different languages. For example, a classification of water bodies in English will lose the topological relations expressed by the (French) dichotomy between fleuve and rivière.

Danube, *Donau* and *Дунав*

If the considerations expressed in the previous paragraph are generally focused on common names, we should now consider that some geographical entities have proper ones. In the realm of physical geography, only few points have proper names, such as the North and the South Poles and some mountain summits, few lines, such as the Equator, the Tropic of Cancer, the Tropic of Capricorn, the Greenwich Meridian, the Polar Circle, and some solids such as lakes, seas, oceans and so forth. Conversely, within the human geography the list of proper names is so long that we may look at the discipline as the realm of geographical proper names. Such names might give rise to a number of conundrums on toponymy, which can also be interesting for the debate on geographical entities. According to Laurini (2017), the conundrums might regard:

- homonymy—the fact that a proper name can be the name of two (or more) different geographical entities (i.e. 'Mississippi' is the name of a river and of a state);
- endonym/toponym—the former is the local name in the official language of the country or in a well-established language occurring in that area where the feature is located (i.e. Venezia in Italian). However, potentially every geographical entity may also have different names (several toponyms) in countries with different official languages (i.e. Brussel in Flemish, Bruxelles in French);

- exonym, which is a name in languages other than the official one (i.e. Venice in English or Venice in French);
- archeonym that is a name that existed in the past (i.e. Byzantium and Constantinople for Istanbul);
- abbreviations (L.A. for Los Angeles) and nicknames (Big Apple for New York);
- place with multiple names (i.e. in New York City, the 'Sixth Avenue' is also known as 'Avenue of the Americas');
- variations about the way to write some names (i.e. 3rd Street, Third Street, Third St);
- transcriptions, for example, Peking became Beijing after a change of transcription to the Roman alphabet, but the capital of China has not modified its name in Chinese.

If the Laurini's list (and examples) is still not enough, we might add, for example, the case of the river 'Danube' that assumes different names in the different countries it crosses: 'Donau' in Germany and Austria, 'Dunaj' in Slovakia, 'Duna' in Hungary, 'Dunav' in Croatia and Serbia, 'Dunav' and 'Дунав' in Bulgaria, 'Dunărea' in Romania and in Moldova and 'Dunaj' and 'Дунай' in the Ukraine. So, what is the proper name of the river? May we assign a name to that river as a whole? If no, is the name of such a river composed by the sum of the names given by the ten nations that it flow through? Or again, which is the name of the river when it separates two different nations? Should we maybe think that the river is composed of different parts, each of which has a different proper name? More in general, might the case of Danube (as well as the previous examples) involve some sort of vagueness in the (linguistic) referent and/or in the entity/ies in question?

Vagueness

Considering the vagueness not only as a pervasive phenomenon of human thought but also of the geographical world is up for discussion in the current geo-ontological debate.[10] According to Varzi (2001), virtually every geographic word and concept suffers from it, and questions such as 'How small can a town be?', 'Where does a hill begin?', 'How long must a river be?' and 'How many islands does it take to have an archipelago?' are perfectly legitimate. Moreover, vagueness is not exclusive to common name: the name 'Everest', for example, is just as vague as mountains, hills, towns and so forth can be, giving rise to its own kind of soritical paradox.

In the same article, the author distinguishes two main different kinds of vagueness: *de re* and *de dicto*.

In line with the former, the vagueness exhibited by geographic names and descriptions should be conceived as ontological, and not as purely epistemic. Accordingly "a vague term is one that refers to a vague object, an object the spatial or temporal

[10] See Mandelbrot (1967), Sarjakoski (1996), McGee (1997), Bennett (2001), Varzi (2001).

boundaries of which are genuinely 'fuzzy'". Therefore, the name 'Everest' is vague insofar as the entity Everest is vague:

> there is no objective, determinate fact of the matter about whether the borderline hunks are inside or outside the mountain called 'Everest'. The same applies to deserts, lakes, islands, rivers, forests, bays, streets, neighborhoods, and many other sorts of geographic entities. On the *de re* reading, these entities have vague names because they are genuinely vague denizens of reality (Varzi 2001, p. 52).

Conversely, the *de dicto* (semantic) reading maintains that geographic vagueness "lies in the representation system (our language, our conceptual apparatus) and not in the represented entity". In other words, "to say that the referent of a geographic term is not sharply demarcated is to say that the term vaguely designates an object, not that it designates a vague object". Accordingly, there is no such thing as a vague mountain. Rather, there are many things where we conceive a mountain to be, each with its precise boundary, and when we say 'Everest' we are just being vague as to which thing we are referring to. That is to say that there are several different "ways of tracing the geographic limits of Mount Everest, all perfectly compatible with the way the name is used in ordinary circumstances". In the end, each one of a large variety of slightly distinct aggregates of molecules has an equal claim to being the referent of the vague name 'Everest'. And each such thing is precisely determinate (Varzi 2001, p. 54–55).

(Geographical) Kinds and Properties

In these pages, the term 'entity' has been generally used as synonym of 'object' for indicating, in the realm of geography, something like regions, parcels of land, water bodies, roads, buildings, bridges, and so on, as well as parts and aggregates of all these things. However, the association between entity and object risks to be too restrictive for the description of the geographic domain insofar as such a domain may also comprehend other sorts of entities like kinds, properties, relations, boundaries, events, processes, qualities and so forth.

Geographical kinds, for example, tell us under which category an object falls, in other words: what an object is. For instance, if we consider the following three sentences:

1. Nile is a river
2. Bucharest is a city
3. Everest is a mountain

the terms 'river', 'city' and 'mountain' represent three (possible) examples of geographical kinds that have objects, respectively, Nile, Bucharest, and Everest, as their instances. Generally speaking, Rosch has proposed that (natural) kinds are seen as possessing a radial structure, having prototypes of more central or typical members surrounded by a penumbra of less central or less typical instances (Rosch 1973,

1978). In the geographical domain, Casati et al. (1998) have also emphasized that the entities to which geographers refer are of a different kind and can be distinguished in two main categories, corresponding to the traditional distinction between physical and human geography. On the one hand, there are mountains, rivers, deserts. On the other hand, there are socioeconomic units: nations, cities, real-estate subdivisions—the spatial shadows cast by different sorts of systematically organized human activity. The correspondence between (these two) branches of geography (human and physical) and different sorts of geographical kinds seems to support the idea that, in principle, each different branch (and subbranch) of geography might be characterized by specific sorts of geographical kinds.

Finally, we should consider that geographical kinds and objects might also be characterized by (geographical) properties: that are entities which can be predicated of objects and kinds or attributed to them (Orilia and Swoyer 2017). Examples of geographical properties may be 'has a population of', 'has a catchment area of' and so forth. In addition to expressing what things are said to bear, possess or exemplified, the examples help us in the categorization of different geographical kinds and objects. For instance, the property 'elevation' (as well as 'volume', 'relief', etc.) might help us in categorizing a landforms as a mountain, a hill and so forth.

Relations, Fields and Time

In addition to kinds and properties, our list of geographical entities can also comprehend items such as relations, which, in turn, might be divided into the following:

- Mereological,[11] Topological,[12] spatial[13] relations;
- (as well as) different sorts of mixed cases of relations among geographical objects.[14]

But, how may we consider a relation as properly geographic? Is there a difference, for example, between spatial and geographic relations? According to Laurini, despite such a difference might be not so clear-cut, "we can say that spatial relations are seen more abstract whereas geographic relations are grounded in the Earth" that is, link two or more objects located in the Earth (Laurini 2017, p. 83). Obviously, this does not mean that also spatial relations are not commonly used in the geographical

[11]It is the relation 'is a part of' that can also include some temporal parameters which help to specify the criteria of identity for the entities and their constitutive parts. See Simons (1987), Smith and Mark (1998), Casati and Varzi (1999), Mark et al. (1999).

[12]Examples of topological relations are connection, overlapping, containment, distance, separation, discontinuity and so on. See Smith (1994, 1995, 1996), Varzi (2007).

[13]In general, spatial relations might be conceived as relations between objects and the regions of space they occupy or in which they are located. See Casati et al. (1998).

[14]For example, the mereotopology that is the connection between mereology and topology (Smith 1995; Breysse and De Glas 2007), or again the relation between the notions of topology and border (Casati et al. 1998; Smith and Varzi 2000; Varzi 2007).

domain, even if they can be used also in other domains such as robotics, medical imagery, etc. Examples of geographical relations can be the relation 'is a north/south of', as well as specific connections between geographical objects such as rivers ('is tributary of'), roads ('crosses'), city and country ('is the capital of') and so forth.

(Geographical) continuous fields represent another ontological conundrum in the domain of geographical entities. Indeed, on the one hand, we have the position of Smith and Mark (1998), according to which an adequate ontology of geographic kinds should embrace not only categories of discreta but also *categories* that arise in the realm of continuous phenomena. On the other hand, Laurini (2017) says that the introduction of a theory of continuous fields might help us especially in *representing* continuous phenomena such as temperature, pressure, wind, elevation or air pollution, which can be matters of geographical interest. This also means to underline that, particularly with geographic information systems (GIS), "there is also conceptual interaction with geographical entities that is mediated through mathematical models and through computer representations" (Smith and Mark 1998, p. 312). Now, if the point concerns whether to consider (continuous) fields as properly geographical *entities* or *tools* involved in the representation of geographical entities, such a doubt seems to involve the relation between the geographical reality and the tools that help to describe it. In other words, should we include also such tools or, more in general, entities coming from the domain of geographic representation in our list of geographical entities?

Finally, we should also spend a few words on the dimension of time, thus avoiding to consider the geographic reality in a static perspective. This means not only contemplating the diachronic and synchronic identity of geographic entities but also, according to Egenhofer and Mark (1995), regarding geographic space and time as tightly coupled. For instance:

> Many cultures have pre-metric units of area that are based on effort over time (Kula 1983). The English *acre* (Jones 1963; Zupko 1968, 1977), the German *morgen* (Kennelly 1928), and the French *arpent* (Zupko 1978) all are based on the amount of land that a person with a yoke of oxen or a horse can plow in one day or one morning. There have been similar measures for distance, such as how far a person can walk in an hour, or how far an army can march in a day (Egenhofer and Mark 1995, p. 7).

Boundaries

As stated in Section "Drawing the Contour", one of the many approaches to identify geographical entities starts with the specification of their boundaries. But what are geographical boundaries? What is their relation to the entities they demarcate? Is it mereological? Might boundaries exist also without the entities they separate? Should we include them in our list of geographical entities?

Without claiming to be exhaustive, we can say that the geo-ontological debate[15] has generally distinguished two main sorts of theories on (geographical) boundaries: realist and eliminativist theories (Varzi 2015). Realist theories consider boundaries as lower dimensional entities: boundaries are ontological parasites, which cannot be separated and exist in isolation from the entities they bound. Realist theories may differ significantly, however, with regard to how such dependent, lower dimensional entities relate to the extended entities they bound. With reference to the boundary between Maryland and Pennsylvania, Varzi has distinguished four main views of such theories:

1. the first view maintains that the boundary may belong neither to Maryland nor to Pennsylvania;
2. according to the second one, the boundary must belong either to Maryland or to Pennsylvania, though it may be indeterminate to which of the two states it belongs;
3. the third says that the boundary may belong both to Maryland and to Pennsylvania, "but the relevant overlap is *sui generis* precisely insofar as it involves lower dimensional parts. Boundaries do not *take up* space and so, on this theory, it is not implausible to say that (for example) the Mason–Dixon line belongs to both Maryland and Pennsylvania";
4. the last one maintains that there really may be two boundaries, one belonging to Maryland and one belonging to Pennsylvania, "and these two boundaries would be co-located—that is, they would coincide spatially without overlapping mereologically" (Varzi 2015).

Conversely, eliminativist theories move from the idea that talking of boundaries involves some sort of abstraction. Among such theories, substantivalists about space–time "see the abstraction as stemming from the relationship between a particular and its spatiotemporal receptacle, relying on the topology of space-time to account for our boundary talk when it comes to specific cases". If one is not a substantivalist about space and/or time, one can describe the abstraction as invoking the idea of ever thinner layers of the bounded entity. On this account, "boundary elements are not included among the primary entities, which only comprise extended bodies, but they are nonetheless retrieved as higher order entities, viz. as equivalence classes of convergent series of nested bodies" (Varzi 2015).

On Nonexistent and Abstract Geographical Entities

On Tuesday 5 July 1955, the Australian newspaper *The Age* wrote that the Philippine Air Force was searching the South China Sea for a mysterious island settlement

[15] See for example Mark and Csillag (1989), Smith (1995), Burrough and Frank (1996), Zimmerman (1996), Smith and Varzi (1997, 2000), Casati et al. (1998), Smith and Mark (1998), Casati and Varzi (1999), Varzi (2007, 2016), Russell (2008).

called the *Kingdom of Humanity*. The reason for this mission was that the Philippines President wanted to know whether such a place actually existed. If that had been, the Philippines President wanted to determine whether it was a legitimate settlement within the territorial Philippines (Middleton 2015). But, if that was not the case, could we have been included the *Kingdom of Humanity* within the list of geographical entities? In other words, does the notion of existence determine what we can legitimately consider as a geographical entity?

Another example may be represented by Lunezia, a geographical region that is meant to include the Italian provinces of La Spezia, Massa-Carrara, Parma, Piacenza, Reggio nell'Emilia, Mantua and part of the territories of Cremona and Lucca. As from 1946, the debate on the possible constitution of Lunezia has not (yet) led to the institution of such a geographical region. But, does the (ongoing) debate legitimize the inclusion of Lunezia within the geographical entities? Or does the fact that Lunezia never had a spatiotemporal existence on the Earth exclude such an entity from the realm of geographical entities?

If all those conundrums are not enough, let's go back to the Era of Partition, when Poland did not have a spatial location—or rather, when Poland did not have any territory to call as its own land. Now, should we include Poland among the geographical entities also in that era? More precisely, if we wanted to carry out an inventory of geographical entities of that period, does the fact the Poland did not have a territory (or a spatiotemporal existence during such an era) allow us to exclude Poland from that inventory? If no, we could include, within the list of geographical entities, also entities that has not (and maybe that will no longer have) a spatial location, such as the Holy Roman Empire or the Maritime Republics,—and (maybe) entities such as Kosovo, Timor Est and South Sudan, which had not (yet) had a spatiotemporal existence during that period of time. If yes, we should perhaps justify how, for example, a non-geographical entity can give the right to (re)claim a territory as its own land, such as Poland after the Era of Partition or nowadays with Kurdistan.

Historical Entities

> Until McCarthy completed his work, Siamese provinces were not geographically well-described. A province existed in a particular place but the place did not define it. The land itself was almost coincidental. What mattered were the people. And where a boundary did exist, it was seldom a continuous line. It wasn't even a zone. In fact it only occurred where it was needed, such as along a track or pass used by travellers. In other places, where people seldom set foot, there was no point in deciding a boundary. Further, borders between adjacent kingdoms did not necessarily touch, often leaving large unclaimed regions of forest, jungle or mountains. And in practice it was quite possible for towns to have multiple hierarchical relations of authority with more than one ruler and hence – disturbingly for Mr McCarthy – to be part of more than one state (Middleton 2015).

In Section "Cultural (Geographical) Entities", it has been said that different cultural frameworks (as well as different languages and beliefs) may describe the same geographical reality in diverse ways, in terms of categorizations, entities, boundaries, and so forth. That means that cultural environment plays a fundamental role in determining our list of geographical entities. However, we should also remark that such a cultural framework does not change only on the basis of the geographical context. Indeed, also the advancements of geography as a discipline and the historical context can have a strong influence on it.

About the influence of the historical context—besides the case of Siamese provinces provided by Middleton—we can, for example, consider if there is a difference between contemporary (military) encampments and Roman Castra (or Hiberna)—regardless of whether or not Castra had become cities. So, should we include such entities in our geographical inventory? Do contemporary military encampments and Roman Castra represent the same geographical entity? Another issue might arise from territories occupied by nomadic populations, which could change according to seasons, food resources and so forth. In this case, we could ask is there a geographical entity defined by the territory occupied by a population in a specific period of time, even if that population had no ongoing territory to call as its own land? If yes, may it be an entity that describes the ancient world but not the contemporary one? More generally, do we use the same geographical concepts that, for example, Greek and/or Roman used? Had the notion of boundaries the same meaning that it has today? Did, for example, the term Gaul denote a crisp region with clear-cut boundaries or rather the territory occupied by Celtic tribes with de re vague boundaries?

To conclude, we should consider mythological places such as Atlantis, Biringan City, Cloud cuckoo land, Paititi and Mu. Are they geographical entities, at least for some cultures in certain period of time? If yes, should we include them in our list of geographical entities? Just to add further hurdles, we might also consider the puzzling case of Thule and the several theories about its possible location, which include, among others, the coastline of Norway, Iceland, Greenland, Orkney, Shetland, Faroe Islands and Saaremaa. Obviously, if we imagine a map that shows all these locations, then we would be hardly inclined to consider Thule as the mereological sum of all the locations ascribed to it. At the same time, it would be unlikely to consider the various Thule represented on the map (with different conditions of identity) as duplicates of the same geographical entity. Perhaps, we could take the various points that locate Thule on that map as indicating different geographical entities, to which different authors have attributed the same connotation. Perhaps, we could also consider the possibility of geographical entities with multiple locations.

Complex Geographical Entities

Generally, geographic objects are complex entities: that is, they have proper parts and/or components. Moreover, geographic objects can be connected or contiguous, but they can also be scattered or separated. Sometimes they are closed (e.g. lakes), and some others are open (e.g. bays). Note that the above concepts of contiguity and closure are topological notions, and thus an adequate ontology of geographic objects must contain also a topology, a theory of boundaries and interiors, of connectedness and separation, that is integrated with a theory of parts and wholes, or mereology (Smith 1996).

To say that some geographical entities may be complex means that such entities are made up by other geographical entities: for example, a nation can be divided into regions, provinces and so forth, a city can contain geographical entities such as buildings, streets and so on. They can all be seen either from a mereological approach (part/whole relations) or from a topological point of view (contain relation). However, we should remark that a geographical entity might also have components which are not strictly geographical. To put it clear, if a geographical entity such as a forest might be defined a large area covered chiefly with trees and undergrowth, may we consider these trees, their leafs, roots and atoms as geographical entities? Moreover, a geographical entity such a forest might also have (arbitrary) spatial parts: for example, the north side of the forest and the south one. But then, should we include such spatial parts within our list of geographical entities?

Another point to mention is that the hierarchical structure exhibited by, for example, the relations between a nation with its regions, administrative subdivisions and so forth is not the only possible structure that geographical entities might show. Indeed, according to Laurini, as there are different kinds of roads, turnpikes, streets, etc. seldom a sort of hierarchy can be defined. Moreover, some geographical entities can contain specific parts of other geographical entities. In other words, a geographical entity may, in principle, belong to two or more different geographical entities, which makes it difficult to think about a hierarchical structure. For example, Via Emilia (SS 9) crosses different Italian provinces such as, among others, Rimini, Bologna, Reggio nell'Emilia, Parma. Furthermore, a geographical entity may also belong to two or more different hierarchies that, for instance, describe different branches of geography (as a discipline). In this sense, also hierarchies can presuppose overlaps. For example, Lake Iseo can be seen as an instance of the class Lakes that, in turn, is a subclass of the class Water Bodies (physical geography). At the same time, Lake Iseo can be considered as belonging to the region Lombardy that in turn is a proper part of the nation Italy and so forth (political geography). However, we should not forget that the presence of a hierarchy does not exclude eventual relations among classes at the same level or belonging to different branches of the same hierarchy (Bittner and Smith 2008).

Hierarchical Structures

To talk about (geographical) hierarchies, it may be useful to introduce the meaning of two different terms that I use in this paragraph: *hyperonym* and *hyponym*. The two terms are the (opposite) names of places with a hierarchy: for instance, Europe is a hyperonym of Italy, whereas Italy is a hyponym of Europe (Laurini 2017). In contrast, a *meronym* may be considered as a name of a part of a place without a hierarchy: for instance, the Adriatic Sea is a meronym of the Mediterranean Sea.

Now, could we benefit from thinking in term of hierarchy in distinguishing between what is geographical and what is not? Perhaps, we should first consider whether or not the hierarchy can be inclusive for all the geographical entities in our list. Accordingly, the point might be to circumscribe such a hierarchy, starting from the top hyperonym and lowest hyponyms.

About the top hyperonym, a fundamental question might be: is there something geographic to which anything uncontentiously belong? Semantically speaking—- given that the term geography comes from the Greek words *gê* ('Earth') and *graphein* ('to write, draw') and thus it means 'to write and draw about the Earth'—a possible answer can be the Earth: every geographical entity belongs to the all-inclusive geographical entity *Earth*. Now, if such an answer may have some supporters, we should, however, pay attention to, at least, two different issues. The first one is to keep geography from collapsing into its cartographical dimension (or better, to do not reduce geography to cartography). The second issue, strictly related with the first one, concerns the fact that geography is also devoted to the study of human activities, cultures, economies, interaction with the environment and relations with and across space and place. Of course, such human dynamics can have effects on the Earth, by producing something that can be analyzed through a study of the Earth. However, we can also assume that, despite the fact that human dynamics might have an impact on the Earth, they are something more. Accordingly, the Earth does not complete the entire domain of geography.

Now, what about lowest hyponyms? An idea might be to consider only those geographical entities that are not complex. Consequently, a lowest geographical hyponym (LGH) is a geographical entity that does not contain (or that is not composed by) other geographical entities. (Obviously, that does not mean that a LGH cannot contain other entities, which, in turn, should not be geographic.) However, without a definition of geographical entities a clear-cut identification of a LGH might be difficult. For example, if considering a street as a geographical entity seems to be uncontentious, might we say the same also for shoulders, (emergency, cycle) lanes, roadways of that street? What is/are the LGH(s) in this context? And what if we consider the relation between ponds and lakes? Are ponds hyponym of lakes or are lakes and ponds both categories at the basic level, mainly distinguished by size? What is/are the LGH(s) among a forest and its north and south sides? Finally, we should also consider that LGHs can change according to the different branches of geography we investigate—and consequently every branch of geography might have, in principle, a proper list of geographical entities. For instance, shoulders, roadways and so forth can be seen as potential examples of LGHs for transportation geography but not for health geography; entities such as airports and tracks may be considered as geographic for some branches of human geography but not for classical geography (Luckermann 1961) and so on.

Three Thin Red Lines

The aim of this paper has been to provide a (non-exhaustive) sketch of possible approaches, response attempts, conundrums and issues arising from the question: 'What is a geographical entity?'. Trying to answer this question is made particularly difficult by the multiplicity of aspects that might influence our answer and defies a clear-cut systematization. Without claiming completeness, we might summarize such aspects as follows.

The first one emerges from the fact that we can use (many) different conceptualizations for describing geographic space. On the one hand, according to Egenhofer and Mark (1995), such conceptualizations of geographic space may:

- reflect the differences between perceptual and cognitive space (Couclelis and Gale 1986);
- be based on different geometrical properties, such as continuous versus discrete (Egenhofer and Herring 1991; Frank and Mark 1991);
- depend on scale or difference in the types of operations we would typically employ in everyday life and/or in scientific reasoning (Zubin 1989).

On the other hand, as I have often remarked in these pages, different conceptualizations of geographical space can also emerge from the ways in which different languages and cultures—as well as the various geographical branches and perspectives—structure and systematize the world itself (Oakes and Price 2008). In this sense, as Smith and Mark (2001) suggest, work involving formal comparisons of geospatial and cartographic data standards and dictionary definitions in a variety of languages might also provide an important starting point for combining quantitative, i.e. measurable geographic phenomena described by different scientific disciplines, with qualitative geographical descriptions of reality also emerging from areas of human-geographical reasoning.

The second factor is that sometimes we may have some difficulties in distinguishing the domain of the real world from the domain of computational and mathematical representations, and both of them from the cognitive domain of reasoning, language, and human action (Smith and Mark 1998). Of course, it might sometimes be difficult to provide a clear-cut distinction between the real world and the tools that we can use to describe it (Laurini 2017). For example, should we consider a compass as a tool capable of describing parts of the geographical world or also as a proper geographical entity? And what about items such as GPS coordinates, longitude, latitude and so forth (Crampton 2010)? Do mathematical entities exist in the geographical world? And geometric ones? And geographical entities which derive from technology? Can GIS enrich our geographical inventory with new kinds of geographical entities?

The third factor concerns geography itself and specifically its development (and/or advancement), which does not only affect geography as a discipline but also the world that it describes. Take for example modifications of boundaries, the formulation of (the notion of) nation-states, the presence of airports on our current maps or, again, the possibilities given by augmented reality for geography. Take also the introduction of new scholarly fields in geography such as night studies (Gwiazdzinski and Chausson 2015) and border studies (Newman 2006; Kolossov and Scott 2013), or the birth of intellectual movements or paradigm shifts such as the spatial turn (Warf and Arias 2009). Take finally examples that represent potential changes in some parts of contemporary geography (Gomez and Jones 2010) if compared, for instance, to classical geography (Lukermann 1961; Bianchetti 2008), in terms of assumptions, tools, methods of investigation and domain to describe. Now, may we assume that (at least) some of these changes, developments and/or advancements, which introduced new ways of slicing, shaping and interpreting the geographical world, could/can/will create new perspectives for distinguishing between what is geographical and what is not?

From Multiple (Ways of Doing) Geographies to Multiple (Kinds of) Geographical Entities

If yes, as I presume, providing an exhaustive definition of geographical entity (as well as a full list of them) would be made even more difficult, to the point of running the risk of being too restrictive for what a geographical entity could be in the past and will be in the future. For that, although not offering a definition can hardly seem very precise in distinguishing what is geographical from what is not, I think that the possible imprecision of such a definition would be even worse. The issue, in this case, would be to hinder the process of theory-construction, especially for what concerns how best to interpret new (possible) geographical evidence. In other words, the idea is that since geographers (as well as GIS scientists and geo-ontologists) approach the task of theory-construction under the guidance of some ontological assumptions, the greatest contributions of analyzing the notion of geographical entity would be essentially two. The first one is simply to chart the possibilities of existence (Lowe 1989, 2006). The second contribution is providing us with the conceptual tools wherewith to categorize the world's contents in view of the heterogeneity of the geographical debate, trying to keep open minds as to how we might interpret new geographical aims, perspectives and points of view.

Accordingly, the idea behind these pages is that of subordinating every normative claim on the notion of geographical entity to (as well as of enriching our descriptive approaches with) the factors we underlined, which can be summarized as follows. (1) There exist multiple conceptualizations of the geographical world. (2) Different languages and cultures may slice such a world in different ways. (3) The geographical world has changed and will change over time. (4) Also geography (as a discipline) has changed and will change over time, modifying its perspective, tools, domains of investigation and aims. Consequently, what had, has been, will be considered as non-geographic could be considered as geographic, and vice versa. (5) There were, are and will be different kinds of geographies as well as different geographical branches, each of them had, have and might have different tools, aims, points of view and vocabularies. (6) The introduction of new scholarly fields and new technologies, the birth of intellectual movements or paradigm shifts and developments on other disciplinary contexts (such as geometry, topology and so forth) can/will influence geography as a discipline.

This means that there are multiple, alternative and overlapping views on geographical reality, and the same reality can be represented and sliced in different ways. Accordingly, the aim of the ontology of geography (and of an investigation on the notion of geographical entity) should be to provide some platforms for integrating of such alternative views. Its task is thus practical in nature, and is subject to the same practical constraints experienced in all scientific activity. Consequently, even a geo-ontological framework will always be a partial and imperfect edifice subject to correction and enhancement, so as to meet new scientific needs (Smith and Klagges 2008).

Acknowledgements Parts of this article draw on previous materials. In particular, Sections "Laundry-Lists"–"Drawing the Contour" and "Boundaries" have some overlaps with Tambassi (2017b); Section "Attempts of Definition" has some overlaps with Tambassi (2017a); Section "Cultural (Geographical) Entities" has some overlaps with Tambassi (2018). Thanks are due to Sorin Cheval, Matthew R. X. Dentith, Mihnea Dobre, Maurizio Lana, Giulia Lasagni, Cristina Meini, Iulia Nitescu and Achille Varzi for providing comments and feedback, and for their invaluable support.

References

Agnew J (2008) Borders on the mind: re-framing border thinking. Ethics Glob Polit 1(4):175–191

Bateman J, Farrar S (2004) Towards a generic foundation for spatial ontology. In: Varzi AC, Vieu L (eds) Formal ontology in information systems. IOS Press, Amsterdam, pp 237–248

Bennett B (2001) What is a forest? On the vagueness of certain geographic concept. Topoi 20(2):189–201

Bianchetti S (2008) Geografia storica del mondo antico. Monduzzi, Bologna

Bishr Y (2007) Overcoming the semantic and other barriers to GIS interoperability: seven years on. In: Fischer P (ed) Classics from IJGIS. Twenty years of the international journal of geographical information science and systems. CRC Press, Boca-Raton-London-New York

Bittner T, Smith B (2008) A theory of granular partitions. In: Munn K, Smith B (eds) Applied ontology. An introduction. Ontos-Verlag, Berlin

Bonnett A (2008) What is geography?. Sage, London

Breysse O, De Glas M (2007) A new approach to the concepts of boundary and contact: toward an alternative to mereotopology. Fundam Inform 78:217–238

Burrough PA, Frank AU (eds) (1996) Geographic objects with indeterminate boundaries. Taylor & Francis, London

Casati R, Smith B, Varzi AC (1998) Ontological tools for geographic representation. In: Guarino N (ed) Formal ontology in information systems. IOS Press, Amsterdam, pp 77–85

Casati R, Varzi AC (1999) Parts and places. MIT Press, Cambridge (MA)

Casti E (2015) Reflexive cartography. A new perspective on mapping. Elsevier, Amsterdam, Oxford, Waltham (MA)

Crampton JW (2010) Mapping. A critical introduction to cartography and GIS. Wiley-Blackwell, Oxford, New York

Couclelis H (1998) Space, time, geography. Geogr Inf Syst 1:29–38

Couclelis H, Gale N (1986) Space and spaces. Geogr Ann 68(B):1–12

Egenhofer M, Herring J (1991) High-level spatial data structures for GIS. In: Maguire D, Goodchild M, Rhind D (eds) Geographical information systems, vol 1. Principles. Longman, London, pp 147–163

Egenhofer M, Mark DM (1995) Naive geography. In: Frank AU, Kuhn W (eds) Spatial information theory: a theoretical basis for GIS. In: Proceedings of the second international conference. Springer, Berlin, Heidelberg, pp 1–15

Elden S (2009) Philosophy and human geography. In: Kitchen R, Thrift N (eds) International encyclopaedia of human geography. Elsevier, Oxford, pp 145–150

Elford W (2012) A multi-ontology view of ergonomics: applying the cynefin framework to improve theory and practice. Work 41(1):812–817

Epstein B (2017) What are social groups? Their metaphysics and how to classify them. Synthese. https://doi.org/10.1007/s11229-017-1387-y

Farinelli F (2003) Geografia. Un'introduzione ai modelli del mondo. Einaudi, Torino

Farinelli F (2009) La crisi della ragione cartografica. Einaudi, Torino

Frank A, Mark DM (1991) Language issues for GIS. In Maguire DJ, Goodchild MF, Rhind DW (eds) Geographical information systems. Principles and applications, vol 1. Longman, London, pp 147–163

Galton A (2003) On the ontological status of geographical boundaries. In: Duckham M, Goodchild MG, Worboys MF (eds) Foundation of geographic information science. Taylor & Francis, London, New York, pp 151–171

Geus K, Thiering M (2014) Common sense geography and mental modelling: setting the stage. In: Geus K, Thiering M (eds) Features of common sense geography. Implicit knowledge structures in ancient geographical texts. LIT, Wien

Gomez B, Jones JP III (eds) (2010) Research methods in geography: a critical introduction. Wiley-Blackwell, West Sussex, UK

Grenon P, Smith B (2007) Persistence and ontological pluralism. In: Kanzian C (ed) Persistence. Springer, New York, pp 33–48

Günzel S (2001) Geophilosophie. Nietzsches philosophische Geographie. Akademie, Berlin

Gwiazdzinski L, Chausson N (eds) (2015) Urban nights. J Urban Res 11 (Special issue). https://journals.openedition.org/articulo/2595

Inkpen R, Wilson G (2013) Science, philosophy and physical geography. Routledge, London

Jones S (1963) Weights and measures: an informal guide. Public Affairs Press, Washington, D.C

Kennelly A (1928) Vestiges of pre-metric weights and measures persisting in metric-system Europe, 1926–1927. The Macmillan Company, New York

Kolossov V (2005) Border studies: changing perspectives and theoretical approaches. Geopolitics 10(4):606–632

Kolossov V, Scott J (2013) Selected conceptual issues in border studies. Belgeo 1. http://belgeo.revues.org/10532

Kula W (1983) Les Mesures et Les Hommes. Paris: Maison des Sciences de L'Homme [Translated from Polish by J Ritt; Polish edition 1970.]

Laurini R (2017) Geographic knowledge infrastructure: applications to territorial intelligence and smart cities. ISTE-Elsevier, London

Lowe EJ (1989) Kinds of being: a study of individuation, identity and the logic of sortal terms. Basil Blackwell, Oxford, New York

Lowe EJ (2006) The four-category ontology: a metaphysical foundation for natural science. Clarendon Press, Oxford

Luckermann F (1961) The concept of location in classical geography. Ann Assoc Am Geogr 51(2):194–210

Mandelbrot B (1967) How long is the coast of Britain? statistical self-similarity and fractional dimension. Science 156:636–638

Mark DM, Csillag F (1989) The nature of boundaries on 'area-class' maps. Cartographica 26:65–78

Mark DM, Smith B, Tversky B (1999, August 25–29) Ontology and geographic objects: an empirical study of cognitive categorization. In Freksa C, Mark DM (eds) Proceedings of spatial information

theory. Cognitive and computational foundations of geographic information science. International conference COSIT'99 Stade, Germany. Springer, Berlin, Heidelberg, pp 283–298
McGee V (1997) Kilimanjaro. Can J Philos 23:141–195
Middleton N (2015) An atlas of countries that don't exist. A compendium of fifty unrecognized and largely unnoticed states. MacMillan, London
Newman D (2006) The lines that continue to separate us: borders in our 'borderless' world. Prog Hum Geogr 30(2):143–161
Newman D (2010) Territory, compartments and borders: avoiding the trap of the territorial trap. Geopolitics 15:773–778
Oakes TS, Price PL (eds) (2008) The cultural geographer reader. Routledge, New York
Orilia F, Swoyer C (2017) Properties. In: Zalta EN (ed) The Stanford encyclopedia of philosophy. https://plato.stanford.edu/archives/win2017/entries/properties/
Paasi A (2013a) Borders and border crossings. In: Johnson N, Schein R, Winders J (eds) A new companion to cultural geography. Wiley-Blackwell, London, pp 478–493
Paasi A (2013b) Borders. In: Dodds K, Kuus M, Sharp J (eds) The Ashgate research companion to critical geopolitics. Ashgate, London, pp 213–229
Pattinson WD (1963) The four traditions of geography. J Geogr 63(5):211–216
Rosch E (1973) On the internal structure of perceptual and semantic categories. In: Moore TE (ed) Cognitive development and the acquisition of language. Academic Press, New York, pp 111–144
Rosch E (1978) Principles of categorization. In: Rosch E, Lloyd BB (eds) Cognition and categorization. Erlbaum, Hillsdale, NJ, pp 28–49
Russell JT (2008) The structure of gunk: adventures in the ontology of space. Oxf Stud Metaphys 4:248–274
Sala M (2009) Geography. In: Sala M (ed) Geography. Encyclopedia of life support systems. EOLSS Publisher, Oxford, pp 1–56
Sarjakoski T (1996) How many lakes, islands and rivers are there in Finland? A case study of fuzziness in the extent and identity of geographic objects. In: Burrough PA, Frank AU (eds) Geographic objects with—indeterminate—boundaries. Taylor & Francis, London, pp 299–312
Simons P (1987) Parts: an essay in ontology. Clarendon Press, Oxford
Smith B (1994) Fiat objects. In: Guarino N, Pribbenow S, Vieu L (eds) Parts and wholes: conceptual part-whole relations and formal mereology. In: Proceedings of the ECAI94 workshop. Amsterdam, ECCAI, pp 15–23
Smith B (1995, September 21–23) On drawing lines on a map. In: Frank A, Kuhn W (eds) Spatial information theory—a theoretical basis for GIS. In: Proceedings, international conference Cosit'95, Semmering, Austria. Lecture notes in computer science, vol 988. Springer, Berlin, pp 475–484
Smith B (1996). Mereotopology: a theory of parts and boundaries. Data Knowl Eng 20:287–303
Smith B, Klagges B (2008) Bioinformatics and philosophy. In: Munn K, Smith B (eds) Applied ontology. An introduction. Ontos-Verlag, Berlin
Smith B, Mark DM (1998) Ontology and geographic kinds. In Poiker TK, Chrisman N (eds) Proceedings of the eighth international symposium on spatial data handling (Burnaby, British Columbia, International Geographical Union), pp. 308–320
Smith B, Mark DM (2001) Geographical categories: an ontological investigation. Int J Geogr Inf Sci 15(7):591–612. http://idwebhost-202-147.ethz.ch/Courses/geog231/SmithMark_GeographicalCategories_ IJGIS2001@2005-10-19T07%3B30%3B52.pdf
Smith B, Varzi AC (1997) Fiat and bona fide boundaries: towards an ontology of spatially extended objects. In: Hirtle SC, Frank AU (eds) Spatial information theory: a theoretical basis for GIS. Lecture notes in computer science, vol 1329. Springer, Berlin, pp 103–119
Smith B, Varzi AC (2000) Fiat and bona fide boundaries. Res 60(2):401–420
Tambassi T (2017a) The philosophy of geo-ontologies. Springer, Cham
Tambassi T (2017b) Sulla nozione di entità geografica: mappando il dibattito geo-ontologico. Geostorie 2–3(2017):93–115

Tambassi T (2018) From geographical lines to cultural boundaries. Mapping the ontological debate. Riv di Estet 67:150–164

Tanca M (2012) Geografia e filosofia. Franco Angeli, Milan

Varzi AC (2001) Vagueness in geography. Philos Geogr 4(1):49–65

Varzi AC (2007) Spatial reasoning and ontology: parts, wholes and location. In: Aiello M, Pratt-Hartmann I, van Benthem J (eds) Handbook of spatial logics. Springer, Berlin, pp 945–1038

Varzi AC (2015) Boundary. In: Zalta EN (ed) The Stanford encyclopedia of philosophy. https://plato.stanford.edu/archives/win2015/entries/boundary/

Varzi AC (2016) On drawing lines across the board. In: Zaibert L (ed) The theory and practice of ontology. Palgrave Macmillian, London, pp 45–78

Warf B, Arias S (eds) (2009) The spatial turn interdisciplinary perspectives. Routledge, London, New York

Westerhoff J (2005) Ontological categories. Clarendon Press, Oxford

Yachin SE (2015) Boundary as an ontological and anthropological category. In: Sevastianov SV, Laine JP, Kireev AA (eds) Introduction to border studies. Dalnauka, Vladivostok, pp 62–79

Zimmerman DW (1996) Could extended objects be made out of simple parts? Res 56:1–29

Zubin D (1989) Untitled. In: Mark D, Frank A, Egenhofer M, Freundschuh S, McGranaghan M, White RM (eds) Languages of spatial relations: initiative two specialist meeting report. Technical paper 89-2, National Center for Geographic Information and Analysis, Santa Barbara, CA, pp 13–17

Zupko R (1968) A dictionary of english weights and measures. The University of Wisconsin Press, Madison, WI

Zupko R (1977) British weights and measures: a history from antiquity to the seventeenth century. The University of Wisconsin Press, Madison, WI

Zupko R (1978) French weights and measures before the revolution: a dictionary of provincial and local units. Indiana University Press, Bloomington, IN

Part IV
More Theory, Practice and Applications

Chapter 10
Is There a Quantum Geography?

Thomas Bittner

Abstract In this paper, I argue that a quantum theory may be the appropriate tool for describing phenomena with indeterminate boundaries in the context of the classification and delineation of geographic regions. A motivation for this claim stems from the observation that fundamental aspects of information about the physical world that follow from the success of quantum mechanics also apply to information about certain classes of geographic phenomena. Those aspects include (i) information about the physical [geographic] world is fundamentally relational, (ii) information of the physical [geographic] world is fundamentally granular, and (iii) information about the physical [geographic] reflects the fundamentally indeterminate nature of certain aspects of the world at the respective scales. (The words in the brackets were added by this author). (Rovelli, Int J Theor Phys 35:1637–1678, 1996; Rovelli and Vidotto, Covariant loop quantum gravity, Cambridge University Press, Cambridge, 2015). More rigorous support for the above claim comes from recent work in theoretical physics. This work has identified three information-theoretic conditions that, when satisfied for a class of phenomena, call for a quantum theory as the appropriate theoretical framework for that class. In this paper, I show that there are geographic phenomena which satisfy two of the three conditions that call for a quantum theory. I then argue that the third criterium can be validated or refuted in the geographic context by developing a quantum theory for geographic phenomena with indeterminate boundaries with classification and delineation operations as means to obtain information about those phenomena. Such a theory then can produce predictions that will either be verified by observations on the ground and thereby confirm the need for a quantum theory, or rule it out as a viable option.

Keywords Quantum geography · Quantum information · Indeterminacy · Geographic classification and delineation · Information theory

T. Bittner (✉)
Departments of Philosophy and Geography, State University of New York at Buffalo, Buffalo, US
e-mail: bittner3@buffalo.edu

© Springer Nature Switzerland AG 2019

T. Tambassi (ed.), *The Philosophy of GIS*, Springer Geography,
https://doi.org/10.1007/978-3-030-16829-2_10

Introduction

Is there a quantum geography? This seems to be a question with an obvious answer: No! Things in the quantum world are too strange to inhabit the geographic world. We do not see geographic objects in different places at the same time. Moreover, geographic phenomena do not seem to have contradicting properties at the same time. It is the aim of this paper to suggest that things are not such obvious and that there may be a quantum geography after all.

First of all, the question "Is there a quantum geography?" is ill-posed (but catchy). Quantum mechanics is a formalism that has been used successfully to describe the behavior of entities at the subatomic scale. Therefore, a better way of expressing the above question may be "Are there geographic phenomena that can be described best by a quantum theory?". This question seems to be answerable by addressing the following (sub)-questions: (a) What is a quantum theory?, (b) Are there necessary and sufficient conditions that call for a quantum theory as an adequate description?, and (c) If there are such conditions, are there geographic phenomena for which such conditions are satisfied.

In what follows I will address question (a) at an abstract level without going into the details of the quantum formalism. The aim of the paper is not to actually develop a quantum theory of (certain) geographic phenomena but to argue for the need to develop one. The main focus of the paper is on questions (b) and (c)—necessary and sufficient conditions that, when fulfilled, rule out classical logic, and therefore classical geography which, like classical physics, is based on classical logic. Those conditions have been identified in the realm of theoretical physics in an attempt to provide an interpretation of the formalism of quantum mechanics. I will review the arguments for why theories of phenomena that satisfy the following three postulates are necessarily nonclassical but quantum mechanical (Rovelli 1996):

Postulate 1 (Limited information) *There is a maximum amount of* relevant *information that can be extracted from a system.*

Postulate 2 (Unlimited information) *It is always possible to acquire new information about a system.*

Jointly, Postulates 1 and 2 entail (as will be discussed below) that information about the systems that satisfy those postulates is subject to indeterminacy. A quantum theory then assumes the following:

Postulate 3 (Indeterminacy as probability) *The indeterminacy that arises from Postulates 1 and 2 manifests itself in the probabilistic nature of the processes in which information can be obtained by measurement/observation.*

After discussing Postulates 1–3, I am going to argue that geographic phenomena with indeterminate boundaries (Burrough and Frank 1995), such as ecoregions (Bailey 1983; Omernik and Griffith 2014) and regions that are characterized by certain types of land cover (Andereson et al. 1976), fall in the class of phenomena that are subject to those criteria. Conceptually, the information-theoretic Postulates 1 and 2 are the most interesting and relevant in the context of this paper. In particular, I will discuss both in the context of the classification and delineation of geographic regions, an area of geography in which boundary indeterminacy has been studied extensively (Bailey 1983; Omernik and Griffith 2014). Whether or not the indeterminacy that arises from Postulates 1 and 2 manifests itself probabilistically is an empirical question and can at least in principle be tested by experiment. I will sketch a toy theory that is consistent with Postulates 1 and 2 and that, when developed fully, will make predictions which are probabilistic in nature and that can be tested empirically.

There is one formalism of (nonrelativistic) quantum mechanics which was developed by Dirac (1930) and John von Neumann (1932). This formalism produces predictions that have been verified over and over since that time. By contrast, there are many interpretations of this formalism (Omnès 1994; Wikipedia 2019). Interpretations are attempts to describe the world which brings about the phenomena that have the properties that are predicted by the formalism. According to many interpretations, the world described by the formalism of quantum mechanics is essentially *nonlocal*. That is, according to many interpretations of the formalism, the phenomena described by it can interact instantaneously across arbitrary distances (Einstein et al. 1935; Maudlin 2002; Musser 2015; Redhead 1997; Romero 2012). Physicists have found ways to make this consistent with the theory of Special Relativity (Einstein 1905; Kennedy 2003) which postulates that information cannot travel faster than the speed of light. Nevertheless, the non-locality entailed by many interpretations of quantum mechanics contradicts Tobler's First Law of Geography which postulates that in geographic space "everything is related to everything else, but near things are more related than distant things." (Tobler 1970) One of the few interpretations of quantum mechanics that preserves locality is the relational interpretation of quantum mechanics (RQM) (Rovelli 1996). The fact that, on the relational interpretation, the formalism of quantum mechanics is consistent with Tobler's law in conjunction with the strong information-theoretic focus of this interpretation are the reasons for adopting it as the foundation of this paper.

The reminder of this paper is structured as follows: First, the basic ideas of the relational interpretation of quantum mechanics are discussed. In the context of this framework, Postulates 1 and 2 arise which entail that the underlying logic is nonclassical. I briefly discuss the commitments and intuitions that underly the probabilistic understanding of the indeterminacy that arises from Postulates 1 and 2. For closure, I also briefly introduce some aspects of the formalism of quantum mechanics (QM) itself. The second part of the paper argues that phenomena with indeterminate boundaries particularly in the context of the classification and delineation are subject to Postulates 1 and 2. Those who believe that Postulates 1 and 2 are true for the considered class of phenomena are then committed to believe that, if one can verify Postulate 3 experimentally, there is a quantum geography in the qualified way described above. I will close by sketching a toy example that illustrates how a quantum theory that produces such probabilistic predictions that can be either confirmed or refuted by observations on the ground in principle looks like.

Relational Quantum Mechanics

According to quantum mechanics (QM), any measurement/observation is fundamentally a physical interaction between a system S being measured and some observing system O. In relational quantum mechanics (RQM) such a physical interaction is the establishment of a *correlation* between the observed system and the observing system. This form of correlation corresponds to the notion of *information* in Shannon's information theory (Shannon 1948). Therefore, at the core of the relational interpretation of quantum mechanics (RQM) is the recognition that measurement/observation are bidirectional, information-theoretic processes (Yang 2018).

The *amount* of information that one system has about another system can be quantified as the number of the elements of a set of alternatives out of which a configuration is chosen (Shannon 1948). In this context, the set of alternatives are possible ways in which the observed and observing systems can be correlated. Information is a *discrete* quantity: there is a minimum amount of information exchangeable: a single bit or the information that distinguishes between just two alternatives. Therefore, the process of acquisition of information (a measurement/an observation) can be framed as a question that an observing system asks an observed system (Wheeler 1989). Since information is discrete, any process of acquisition of information can be decomposed into acquisitions of elementary bits of information. An elementary question that collects a single bit of information is a yes/no question. In what follows, such yes/no questions are labeled as $Q_1, Q_2, \ldots$

In RQM any system S, viewed as an observed system, is characterized by the family of yes/no questions that can be asked to it. Following Rovelli (1996), the set of yes/no questions is written as $W(S) \equiv \{Q_i \mid i \in I\}$, for some index set I. The result of a sequence of questions $(Q_1, Q_2, Q_3, \ldots)$ to S, from an observing system O, can be represented by a binary string $(e_1, e_2, e_3, \ldots)$, where each e_i is either 0 or 1 (no or yes) and represents the response of the system S to the question Q_i.

The First Postulate of RQM

The first postulate of RQM, Postulate 1, can be spelled out more precisely in Wheeler's (1989) information-theoretic framework (Rovelli 1996): For all $Q_i \in W(S)$: if Q_i can be inferred from (is determined by) an infinite string of answers $(e_1, e_2, e_3, \ldots)$, then Q_i can also be inferred from (is determined by) a finite string $[e_1, \ldots, e_N]$ of answers. Any system S has a maximal information capacity N, where N is an amount of information that is expressed in bits. N bits of information exhaust everything one can say about S.

Combinatorially, there are 2^N different binary strings of length N (left of Fig. 10.1). Since 2^N possible answers $s^{(1)}, s^{(2)}, \ldots, s^{(2^N)}$ to the N yes/no questions are (by construction) mutually exclusive, one can identify 2^N questions $Q^{(1)}, \ldots, Q^{(2^N)}$ such yes answers to the question $Q_c^{(i)}$ correspond to the string of

i	$s_c^{(i)}$ Q_1	Q_2	$Q_c^{(i)}$
0	0	0	$\neg Q_1 \wedge \neg Q_2$
1	0	1	$\neg Q_1 \wedge Q_2$
2	1	0	$Q_1 \wedge \neg Q_2$
3	1	1	$Q_1 \wedge Q_2$

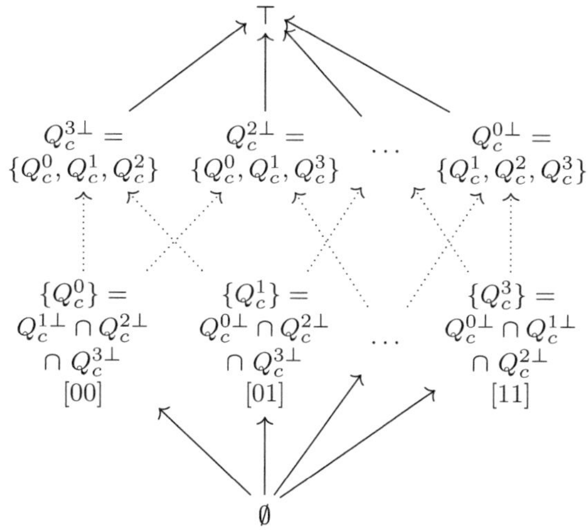

Fig. 10.1 Set $Q_c = \{Q_c^{(i)} \mid 0 \leq i < 4\}$ of 2^4 combinatorially possible complete questions $Q_c^{(i)}$ formed by two yes/no questions Q_1, Q_2 (left) and the distributive orthomodular lattice (a Boolean algebra) formed by the complete questions $Q_c^{(i)}$ (right). See also Hughes (1981) for details and illustrations

answers $s^{(i)}$. This is illustrated in the left part of Fig. 10.1 for the specific case of two yes/no questions Q_1 and Q_2 which give rise to the set $Q_c = \{Q_c^{(i)} \mid 1 \leq i \leq 2\}$ of $2^2 = 4$ combinatorially possible *complete* questions (Beltrametti et al. 1984; Hughes 1981; Rovelli 1996).

The set-theoretic unions of sets of complete questions $Q_c^{(i)}$ (of the same family c), give rise to a Boolean algebra (see Appendix "Boolean Algebra and Orthomodular Lattices") that has singleton sets of the form $\{Q_c^{(i)}\}$ as atoms (right of Fig. 10.1). Intuitively, the atoms of the Boolean algebra are the 2^N different states of S that can be distinguished given N bits of information provided by answers to the yes/no questions $Q_1, \ldots, Q_N$. The nonatomic nodes of the Boolean algebra describe disjunctions of the form $Q_c^i \vee Q_c^j$ in which there is less than N bits of information available. The maximal element of the Boolean algebra has minimal information. By contrast, the atoms have maximal information. In the right of Fig. 10.1, the Boolean algebra that arises from the set Q_c of $2^2 = 4$ combinatorially possible complete questions.

The fact that there are 2^2 distinct pattern of answers to two yes/no questions logically/combinatorially possible does not guaranty that all the logical possibilities are also physically possible. For example, of the set Q_c of 2^4 logically possible complete questions only Q_S^1 and Q_S^2 are assumed to be physically possible and Q_S is the set of physically possible complete questions, i.e., $Q_S = \{Q_S^1, Q_S^2\}$. This results in the table in the left of Fig. 10.2. The corresponding Boolean algebra is depicted in the right of the figure.

	$s_S^{(i)}$		
i	Q_1	Q_2	$Q_S^{(i)}$
1	0	1	$\neg Q_1 \wedge Q_2$
2	1	0	$Q_1 \wedge \neg Q_2$

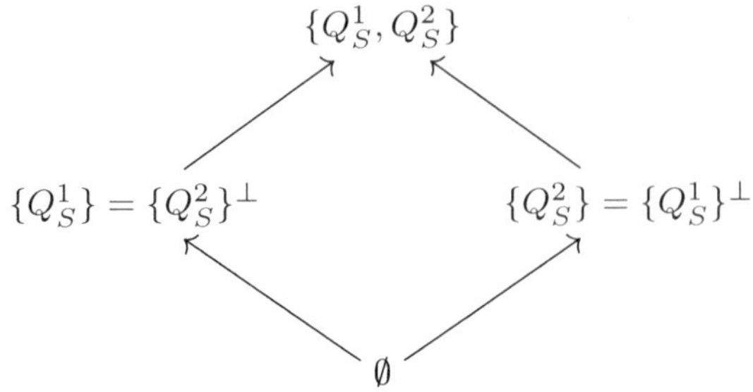

Fig. 10.2 (left) Of the set Q_c of 2^4 logically possible complete questions only Q_S^1 and Q_S^2 are physically possible; (right) the distributive orthomodular lattice (a Boolean algebra) formed by the physically possible complete questions Q_S^1 and Q_S^2

The Second Postulate

In the previous section, a single family c of complete questions $Q_c^{(i)}$ was considered by an observer O to gather N bits of information about the observed system S. Alternatively, O could use a different family b of N complete questions $Q_b^{(i)}$ to gather N bits of information about S. The answers to $Q_b^{(i)}$ will still have a maximal amount of information about S formed by an N-bit string. Again, unions of sets of complete questions $Q_b^{(i)}$ (of the same family b) give rise to a Boolean algebra that has the logically/combinatorially possible $\{Q_b^{(i)}\}$ as atoms. In the context of the example illustrated in Figs. 10.1 and 10.2, this means that there may be a second set of two yes/no questions $\{R_1, R_2\}$ which give rise to a set $Q_R = \{Q_R^{(i)} \mid 1 \leq i \leq 2\}$ of physically possible complete questions $Q_R^{(i)}$. The logical and algebraic structures of the questions in Q_R mirror those of the questions in Q_S as they are displayed in Fig. 10.2.

Postulate 2 of RQM captures what happens if, after having asked the N questions such that the maximal information about S has been gathered, the system O asks a further question Q_{N+1}. According to RQM, there are two extreme possibilities. First, the answer to the question Q_{N+1} is fully determined by the answers $[e_1, \ldots, e_N]$ to the previous questions and no new information is gained. The second possibility is captured in the Postulate 2 demanding that it is always possible to obtain new information.

Jointly, Postulates 1 and 2 can be understood as follows Rovelli (1996): Since the amount of information that O can have about S is limited by Postulate 1, it follows that, if O has a maximal amount of information about S, then, when new information about S is acquired by O, O must loose information. In particular, if a new question Q_{N+1} (not determined by the previous information gathered) is asked, then O looses (at least) one bit of the previous information. So that, after asking the question Q_{N+1}, new information is available, but the total amount of information about the system does not exceed N bits. For more details on the bidirectional nature of measurement/observation see (Yang 2018).

Table 10.1 Two sets of complete questions Q_S and Q_R for 2 bits of information (adapted from Calude et al. (2014))

	$s_S^{(i)}$			$s_R^{(i)}$			
$Q_S^{(i)}$	Q_1	Q_2	$\bigwedge_i Q_i$	R_1	R_2	$\bigwedge_i R_i$	$Q_R^{(i)}$
Q_S^1	1	0	$Q_1 \wedge \neg Q_2$	1	0	$R_1 \wedge \neg R_2$	Q_R^1
Q_S^1	1	0	$Q_1 \wedge \neg Q_2$	0	1	$\neg R_1 \wedge R_2$	Q_R^2
Q_S^2	0	1	$\neg Q_1 \wedge Q_2$	1	0	$R_1 \wedge \neg R_2$	Q_R^1
Q_S^2	0	1	$\neg Q_1 \wedge Q_2$	0	1	$\neg R_1 \wedge R_2$	Q_R^2

At the logic/algebraic level this is captured by the fact that Postulates 1 and 2 imply that the set $W(S)$ as a whole—with the families of complete questions $Q_b^{(i)}$, $Q_c^{(i)}$, …, which classically form Boolean algebras—has the structure of an orthomodular lattice (Grinbaum 2005). The nonclassical nature of systems that adhere to Postulates 1 and 2 manifests itself algebraically in the fact that, unlike Boolean algebras, orthomodular lattices may lack the property of distributivity.

Consider the families $Q_S = \{Q_S^1, Q_S^2\}$ and $Q_R = \{Q_R^1, Q_R^2\}$ of complete questions and the associated two bits of information as displayed in Table 10.1. An orthomodular lattice that satisfies Postulates 1 and 2 is displayed in Fig. 10.4. To see how this structure comes about, consider the diagram in Fig. 10.3. The diagram displays the lattices that arise from the ordering of the subsets of Q_S (left) and Q_R (right) as discussed above (Fig. 10.2). The bottom element of both lattices is the empty set of questions and represents the possibility of "no answer". The nodes of the intermediate level represent the yes answers to exactly one question which each yield two bits of information (maximal amount of information). The top elements of the lattices represent the situation of a positive answer to at least one question in the respective set of questions. Logically, this corresponds to a yes answer to the question $Q_S^1 \vee Q_S^2$ from which neither a yes answer to the question Q_S^1 nor a yes answer to the question Q_S^2 can be inferred. As discussed above, the top node of the Boolean algebra asso-

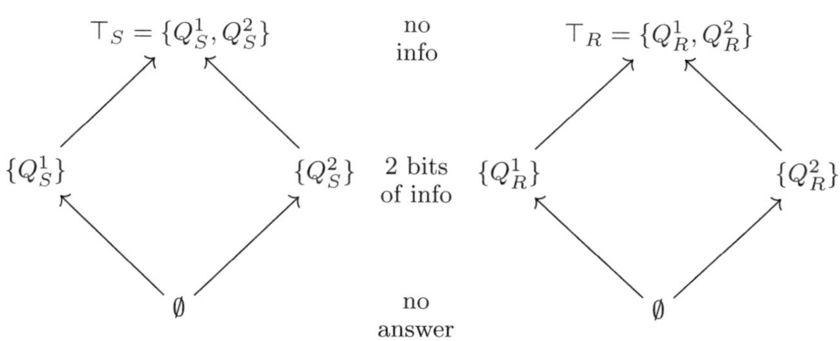

Fig. 10.3 Information content associated with sets of complete questions

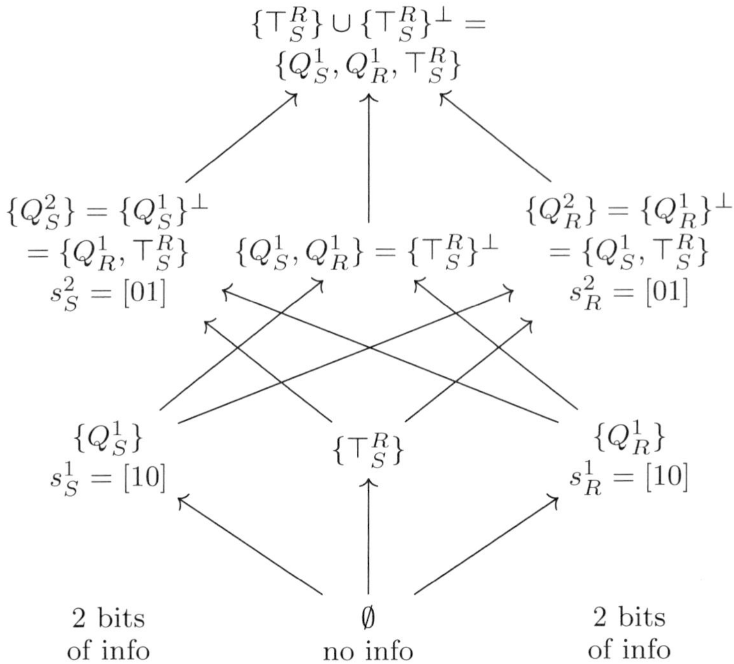

Fig. 10.4 An orthomodular lattice of the two sets of complete questions Q_S and Q_R where the arrows indicate subset relations between subsets of Q_S and Q_R and unions thereof. (Calude et al. 2014)

ciated with minimal information. Similarly, for the top element of the lattice formed by the complete questions in Q_R and the information gained from a yes answer to the question $Q^1_R \vee Q^2_R$.

The orthomodular lattice which is consistent with Postulates 1 and 2 can be constructed from the two lattices in Fig. 10.3 as follows: (i) the bottom nodes in both lattices which do not yield an answer are identified and form the bottom element of the combined lattice and (ii) a new atomic node, $\{\top^R_S\}$ is introduced in the combined lattice, which identifies the top nodes $\top_S$ and $\top_R$ of the lattices in Fig. 10.3. Both, $\top_S$ and $\top_R$, stand for a yes answer to a disjunctive question in which minimal information is gained. The intuition is that in the same sense in which there is only one empty set which represents the absence of an answer to any question, there is only one atomic node that represents the lack of information; (iii) the other two atoms of the combined lattice are $\{Q^1_S\}$ and $\{Q^1_R\}$, each of which is associated with two bits of information; (iv) the nodes $\{Q^2_S\}$ and $\{Q^2_R\}$, respectively, arise as complements of $\{Q^1_S\}$ and $\{Q^1_R\}$ and as such yield two bits of information each; (v) the node $\{Q^1_S, Q^1_R\}$ is identical to the complement of the node ${\top^R_S}^\perp$. Since the latter represents indeterminacy, the former needs to represent indeterminacy. This is consistent with (a) the disjunctive reading of $\{Q_S^1, Q_R^1\}$ and (b) with the fact that yes answers to both Q_S^1 and Q_R^1 will result in identical bits of information and thus can not distinguished.[1]

The lattice in Fig. 10.4 is an algebraic realization of the fact that, due to the finite amount of possible information, distinct sets of complete questions (Q_S and Q_R in this case) are *incompatible* in the sense that asking a question of the form $Q_S^i \vee Q_R^i$ will fail to yield determinate information as discussed above. Algebraically, this nonclassical nature is reflected by the fact that the distributive law does not hold in this structure:

$$(Q_R^1 \wedge Q_S^1) \vee (Q_R^1 \wedge \top_S^R) = \emptyset \neq Q_R^1 \wedge (Q_S^1 \vee \top_S^R) = Q_R^1$$

Probabilities

As discussed above, Postulates 1 and 2 imply that the information that can be obtained in a setting that satisfies both postulates cannot be fully deterministic. The formalism of quantum mechanics models this indeterminacy probabilistically. That is, the formalism of quantum mechanics provides means to quantify indeterminacy by predicting the probability for sequences of responses that can be obtained from observing a system. This specific understanding of indeterminacy is independent of the Postulates 1 and 2 and needs to be captured by additional postulates (Rovelli 1996; Trassinelli 2018).

In first approximation, the reading of indeterminacy as probability can be captured in constraints on a family of functions of the form $p : Q_b \times Q_c \to \Re$. Those functions take the members of two sets of complete questions, Q_b and Q_c, to real numbers in a way that gives rise to a $N \times N$ matrix p^{ij} via the assignment $p^{ij} = p(Q_b^{(i)}, Q_c^{(j)})$. The aim is to constrain the functions of the form $p(Q_b^{(i)}, Q_c^{(j)})$ in such a way that their outcome can be interpreted as follows Rovelli (1996):

> $p^{ij} = p(Q_b^{(i)}, Q_c^{(j)})$ is the probability that a yes answer to the question $Q_b^{(i)}$ of the b-family of complete questions will follow the strings$^{(j)}$ of information that results from a yes answer to the question $Q_c^{(j)}$ of the c-family of complete questions. (10.1)

For the outcome of $p^{ij} = p(Q_b^{(i)}, Q_c^{(j)})$ to be interpretable as a probability in this sense, functions of this form need to satisfy some basic properties of probability functions (Rovelli 1996): (i) $0 \leq p^{ij} \leq 1$; (ii) $\sum_i p^{ij} = 1$ and (iii) $\sum_j p^{ij} = 1$. That is, the function p gives rise to a $N \times N$ matrix of functions that yield real numbers between 0 and 1. All columns and rows of this matrix sum up to 1 for any families of complete questions.

---

[1] Usually, the lattice in Fig. 10.4 is constructed starting from the standard two-dimensional Hilbert space (e.g., Calude et al. 2014).

Table 10.2 The outcome of $p^{ij} = p(Q_S^{(i)}, Q_S^{(j)})$, $p^{ij} = p(Q_S^{(i)}, Q_R^{(j)})$, and $p^{ij} = p(Q_R^{(i)}, Q_R^{(j)})$ for $1 \leq i \leq 2$

p^{ij}	Q_S^1	Q_S^2	p^{ij}	Q_R^1	Q_R^2	p^{ij}	Q_R^1	Q_R^2
Q_S^1	1	0	Q_S^1	0.5	0.5	Q_R^1	1	0
Q_S^2	0	1	Q_S^2	0.5	0.5	Q_R^2	0	1

The probability functions for the sets Q_S and Q_R of complete questions are displayed in Table 10.2. The table has three sub-tables which are interpreted as follows. Questions that belong to the same family are compatible and yield determinate predictions. This is displayed in the left and right subtables of Table 10.2. The probability that a yes answer to the question $Q_S^{(1)}$ of the S-family of complete questions will follow the string $s_S^{(1)}$ of information that results from a yes answer to the question $Q_S^{(1)}$ of the S-family of complete questions is one. That is, a sequence of identical questions always yields the same information. By contrast, the probability that a yes answer to the question $Q_S^{(2)}$ of the S-family of complete questions will follow the string $s_S^{(1)}$ of information that results from a yes answer to the question $Q_S^{(1)}$ of the S-family of complete questions is zero. That is, questions in the same family of complete questions are mutually exclusive and asking a sequence of questions of the same family does not introduce indeterminacy. In both case, the information that comes from the answer to the second question is already contained in the information that is provided by a yes answer to the first question.

The sub-table in the middle of Table 10.2 illustrates that the situation is very different if questions from distinct families of complete questions are asked. Both questions yield a maximum amount of information and thus new information obtained from the observed system S must overwrite existing information in the observer O. This introduces indeterminacy which probabilistically expresses itself as follows. The probability that a yes answer to the question $Q_S^{(i)}$ of the S-family of complete questions will follow the string $s_R^{(j)}$ of information that results from a yes answer to the question $Q_R^{(j)}$ of the R-family of complete questions is completely random, i.e., all possibilities are equally likely.

As pointed out above, the set $W(S)$ consists not only of sets of complete questions $Q_b^{(i)}, Q_c^{(i)}, \ldots$, but for each set of complete questions Q_c, $W(S)$ also contains the questions that correspond to non-singleton subsets of Q_c. Thus, for each family Q_c of complete questions one needs to consider all the questions in the lattice generated by the subsets of Q_c as illustrated in Figs. 10.1, 10.3, and 10.4. As above in section "The Second Postulate", the notation $Q_c^{(j)} \vee Q_c^{(k)}$ is used to represent the question corresponding to the set $\{Q_c^{(j)}\} \cup \{Q_c^{(k)}\}$ in the respective lattices. Again, it is important to note that there is the answer yes to the question $Q_c^{(j)} \vee Q_c^{(k)}$ if and only if either there is a yes answer to the question $Q_c^{(j)}$ or there is a yes answer to the question $Q_c^{(k)}$. The amount of information that is associated with a yes answer to the question $Q_c^{(j)} \vee Q_c^{(k)}$ is less than the maximal amount of N bits of information that is associated with a answer yes to the questions $Q_c^{(j)}$ and $Q_c^{(k)}$ when asked separately.

The questions $Q_c^{(j)}$ and $Q_c^{(k)}$ are complete questions. By contrast, if $j \neq k$ then $Q_c^{(j)} \vee Q_c^{(k)}$ is not a complete question. Since functions of the form $p : Q_b \times Q_c \rightarrow \Re$ are restricted to complete questions, expressions such as $p(Q_b^{(j)} \vee Q_b^{(k)}, Q_c^{(i)})$ are not defined. What is definable are *conditional* probability functions of the form $\overline{p} : \mathcal{P}(Q_b) \times Q_c \rightarrow \Re$, where $\overline{p}(\{Q_b^{(j)}, Q_b^{(k)}\}, Q_c^{(i)}) \equiv \overline{p}(Q_b^{(j)} \vee Q_b^{(k)}, Q_c^{(i)})$ is interpreted as the probability that a yes answer to $Q_b^{(j)} \vee Q_b^{(k)}$ which is associated with less than N bits of information will follow a yes answer to the question $Q_c^{(i)}$ which is associated with N bits of information. Postulates for $\overline{p}$ are as follows:

Postulate 4 (Conditional probability (Trassinelli 2018))

$$(a)\ \overline{p}(\{Q_b^{(j)}\}, Q_c^{(i)}) \geq 0$$
$$(b)\ \overline{p}(\{Q_b^{(j)} \mid 1 \leq j \leq N\}, Q_c^{(i)}) \equiv \overline{p}(\textstyle\bigvee_{j=1}^{N} Q_b^{(j)}, Q_c^{(i)}) = 1$$
$$(c)\ \overline{p}(Q_b^{(j)} \vee Q_b^{(k)}, Q_c^{(i)}) = \overline{p}(\{Q_b^{(j)}\}, Q_c^{(i)}) + \overline{p}(\{Q_b^{(k)}\}, Q_c^{(i)})$$

These properties of $\overline{p}$ imply as special cases the properties of $p^{ij} = p(Q_b^{(i)}, Q_c^{(j)})$ as stated above (Trassinelli 2018). Postulate 4 is a more precise statement of Postulate 3 and thereby supersedes it. In what follows, I will refer to Postulate 4 in place of Postulate 3.

The Formalism of QM

Trassinelli (2018) and others have shown that Postulates 1, 2, and 4 are sufficient to derive the formalism of quantum mechanics within the framework of complex vector spaces (See Appendix "Hilbert Space"). For the purpose of this paper, it will be sufficient to briefly sketch some relevant aspects of it. The point here is to illustrate representational (and non-dynamic) aspects of the formalism that focus on the formalism's consistency with Postulates 1 and 2 about the nature of information. An understanding of the formalism at this level will facilitate understanding of the probabilistic reading of indeterminacy in Postulate 4 as well as its viability in the geographic context.

Algebraic Structure

The formalism of quantum mechanics which actually "implements" structures and functions with the properties postulated in section "Relational Quantum Mechanics" now arises as follows.[2] Assume that there are systems O and S both of which have a maximal information capacity of N bits with respect to one another. That is, there are families of complete questions that O can ask S and vice versa.

In Table 10.1 and Fig. 10.3, two complete sets of questions $Q_S = \{Q_S^1, Q_S^2\}$ and $Q_R = \{Q_R^1, Q_R^2\}$ for acquiring 2 bits of information were presented. In the standard Hilbert space formulation of QM (Dirac 1930; John von Neumann 1932) (see Appendix "Hilbert Space"), sets of complete questions such as Q_S and Q_R give rise to bases of a two-dimensional complex vector space $\mathcal{H}$ (a Hilbert space). The two complete questions in Q_S correspond to a system of base vectors of $\mathcal{H}$, in the sense that the question Q_S^1 corresponds to the base vector $|Q_S^1\rangle$ and Q_S^2 corresponds to the base vector $|Q_S^2\rangle$. The base of $\mathcal{H}$ that is formed by the vectors corresponding to the members of Q_S is called the Q_S-base of $\mathcal{H}$. Similarly, the two complete questions in Q_R correspond to a different system of base vectors of $\mathcal{H}$—the Q_R-base—in the sense that the question Q_R^1 corresponds to the base vector $|Q_R^1\rangle$ and so on. This is illustrated in Fig. 10.5.

Every vector $|Q_a^i\rangle \in \mathcal{H}$ ($a \in \{S, R\}$ and $i \in \{1, 2\}$) gives rise to the "line" $Q_a^i \approx \{\alpha|Q_a^i\rangle \mid \alpha \in \mathcal{C}\}$ that emerges when multiplying the vector $|Q_a^i\rangle$ by a complex number $\alpha \in \mathcal{C}$. The "line" Q_a^i is a one-dimensional subspace of $\mathcal{H}$ induced by the span of the vector $|Q_a^i\rangle$. Jointly, the vectors $|Q_a^1\rangle$ and $|Q_a^2\rangle$ form the basis of a two-dimensional subspace of $\mathcal{H}$ by spanning a plane by means of vector addition and scalar multiplication. The plane spanned by $|Q_a^1\rangle$ and $|Q_a^2\rangle$ is designated by $Q_a^1 \vee Q_a^2$ and specified as $Q_a^1 \vee Q_a^2 \approx \{\alpha|Q_a^1\rangle + \beta|Q_a^2\rangle \mid \alpha, \beta \in \mathcal{C}\}$. One is justified to use the notation $Q_a^1 \vee Q_a^2$ to designate a two-dimensional subspace of $\mathcal{H}$ because (i) the subspaces associated with complete questions of the form Q_a^i and $Q_a^1 \vee Q_a^2$ in conjunction with the the subset relation $\subseteq$ between the subspaces of $\mathcal{H}$ form a lattice $\mathcal{L}_a$ (Fig. 10.5 (middle)); (ii) in this lattice, the least upper bound of the subspaces associated with the questions Q_a^1 and Q_a^2 is the plane associated with the question $Q_a^1 \vee Q_a^2$ via the span of the vectors $|Q_a^1\rangle$ and $|Q_a^2\rangle$. Similarly, the question $Q_a^1 \wedge Q_a^2$ is associated with the greatest lower bound of the associated subspace with respect to the subset relation of the underlying lattice $\mathcal{L}_a$.

The base vectors $|Q_R^1\rangle$ and $|Q_R^2\rangle$ share the origin with $|Q_S^1\rangle$ and $|Q_S^2\rangle$ but are rotated by 45 degree (Fig. 10.5 (left)). Jointly, $|Q_R^1\rangle$ and $|Q_R^2\rangle$ span the same subspace as $|Q_S^1\rangle$ and $|Q_S^2\rangle$—the Hilbert space $\mathcal{H}$ as a whole. Since both systems of base vectors have the same origin and span the same space, $\mathcal{H}$, they form a joint lattice structure $\mathcal{L}_{S\oplus R}$ in Fig. 10.5 (right). With the join and meet operations $\vee$ and $\wedge$ defined as above, it is easy to verify that the lattice $\mathcal{L}$ is non-distributive as given below:

$$(Q_R^1 \wedge Q_S^1) \vee (Q_R^1 \wedge Q_R^2) = \emptyset \neq Q_R^1 \wedge (Q_S^1 \vee Q_R^2) = Q_R^1$$

In addition, one can verify that the lattice $\mathcal{L}_{S\oplus R}$ is also orthomodular. As pointed out above, orthomodular lattices are structures in which Postulates 1 and 2 are satisfied.

For a more intuitive argument of the why Postulates 1 and 2 are satisfied in a two-dimensional Hilbert space with the bases induced by the sets of complete questions

[2]Historically, the formalism was developed long before the insights into its interpretation.

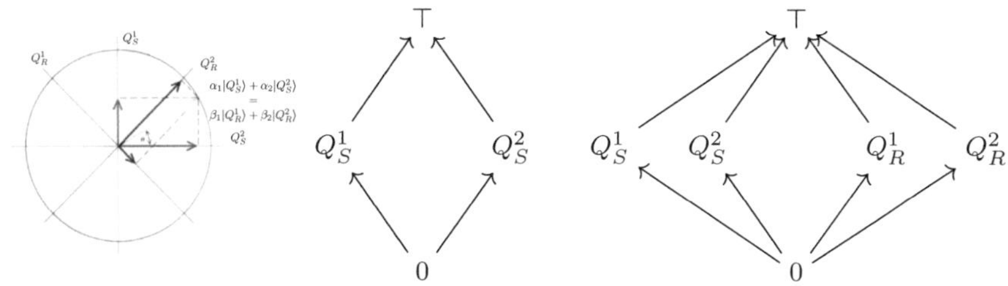

Fig. 10.5 Two complete sets of questions Q_S and Q_R for 2 bits of information in a (projection of a) two-dimensional Hilbert space (left) and the lattices $\mathcal{L}_S$ (middle) and $\mathcal{L}_{S \oplus R}$ (right) (adapted from Calude et al. (2014) and Hughes (1981))

Q_S and Q_R, consider Fig. 10.5. By construction, the amount of information provided by yes answers to any of the questions Q_R^1, Q_R^2, Q_S^1, Q_S^2 is maximal. Understanding the two sets of complete questions Q_S and Q_R as forming two bases of the same two-dimensional vector space expresses in a mathematical language that both sets of questions yield the same amount information in virtue of describing *the same* vector space in different bases. The amount of information that can be obtained is constrained by the dimension of the Hilbert space. New information can be gained by closing a different base, i.e., by switching to a different set of complete questions, and thus focusing on different aspects of the described object.

The relation between the lattice in the right of Fig. 10.5 and the lattice in Fig. 10.4 can be established via embeddings that are described, for example, in Calude et al. (2014). In general, one can prove that although both lattices are orthomodular, there does not exist an embedding that preserves all the properties associated with the lattice arising from QM into lattices that arise from classical logic and classical information theory (Kochen and Specker 1967; Calude et al. 2014). Intuitively, in QM, one can *quantify* indeterminacy. In a (semi-)classical framework, one can *qualitatively* distinguish determinate from indeterminate situations as they arise from measurement/observation interactions.

Indeterminacy as Probability

Every vector in a vector space can be represented as a superposition of a system of base vectors. That is, if $|\phi\rangle$ is a vector of $\mathcal{H}$ that is described in the Q_S-base, then there are complex numbers $\alpha, \beta \in C$ such that $|\phi\rangle = \alpha|Q_S^1\rangle + \beta|Q_S^2\rangle$. The formalism of QM requires that the square modulus $|\langle\phi|\phi\rangle|$ of the inner product $\langle\phi|\phi\rangle$ of the vector $|\phi\rangle$ is equal to one (see Appendix "Hilbert Space"). For given system of base vectors, this requirement allows for the distinction of two kinds of vectors in a Hilbert space: (a) vectors that, when expressed in that base, are such that one of the coefficients α, β is equal to one and the other coefficients are equal to zero; (b) vectors $|\phi\rangle$ that, when expressed in that base, are such that none of the coefficients is equal to one, but jointly the square of the modulus of $\langle\phi|\phi\rangle$ is equal to one.

The case (a) covers all the situations in which an observing system can obtain two bits of information about the observed system in the form of a yes answer to the question associates with the base vector of the nonzero coefficient. That is, case (a) covers all the situations where there is determinate information. By contrast, case (b) covers all the situations in which an observing system cannot obtain determinate information about the observed system. Since the coefficients range over complex numbers, there is a huge number of indeterminate situations that can be distinguished. This is very different from the classical framework. Consider the orthomodular lattice of Fig. 10.4. This lattice represents a classical understanding of the indeterminacy that arises in systems with two bits of information that satisfy Postulates 1 and 2. On a classical view of the sort that is presented in Fig. 10.4, there are exactly three cases of indeterminacy that can be distinguished as the nodes $\{\top^S_R\}$, $\{\top^S_R\}^\perp$, and $\top$ in the depicted lattice. This illustrates that QM as a formalism for indeterminacy is capable of *quantifying* indeterminacy rather than only identifying indeterminate situations qualitatively.

In the formalism of QM, indeterminacy is quantified probabilistically in terms of the likelihood that a yes answer to a specific complete question is obtained. In the base formed by vectors corresponding to the members of Q_S (the Q_S-base), a vector $|\phi\rangle = \alpha|Q^1_S\rangle + \beta|Q^2_S\rangle$ represents a state of an observed system S with respect to the observing system O. If S is in the state $|\phi\rangle$ with respect to O, then the probability that O receives a yes answer to the question Q^i_S is $|\langle\phi|Q^{(i)}_S\rangle|$. Here, the expression $|\langle\phi|Q^{(i)}_c\rangle|$ encodes in the object language of QM the conditional probability $\overline{p}$ of Postulate 4.

An Information-Theoretic View of Geographic Information

The systematic investigation of the nature of geographic information and geographic information processing from an information-theoretic perspective was pioneered by Sinton (1978). According to Sinton, geographic information has three components that are logically interrelated but need to be treated independently: information of geographic qualities; information about the spatial location and the temporal location of the geographic phenomena that have those qualities.[3] In addition, Sinton postulates that information about the three components cannot be measured/observed at once. One component has to be fixed, one component has to be controlled, and one component can be measured/observed. In the language of Wheeler's (1989) information-theoretic view of measurement/observation processes, Sinton's conception of the nature of geographic information can be expressed as follows:

[3]Since geographic phenomena are strictly nonrelativistic, it is consistent with RQM to treat spatial location and temporal location as independent.

Postulate 5 (Sinton and Wheeler) *(S1) For a string of bits of information to count as geographic information, it must be constituted by bits that result from answers to yes/no questions that fall in three broad classes: (i) yes/no questions about measurable/observable (geographic) qualities; (ii) yes/no questions about spatial location in geographic space; and (iii) yes/no question about temporal location.*

(S2) For a string of bits of information in the sense of (S1) to count as geographic information: First, one type of yes/no questions which answers determine the bit string needs to be fixed, that is, limited to the yes answer to one question of this type. Second, one type of yes/no questions needs to be controlled, that is, limited to yes answers to a fixed number of questions—control questions—that yield information about some domain that is subject to a fiat subdivision.[4] *Third, one type of yes/no questions needs to be measured, that is, every yes answer to a control question is complemented by a yes answer to a question from the class of questions that is neither fixed nor controlled—a yes/no question in Wheeler's standard understanding.*

That is, in Wheeler's (1989) information-theoretic language, Sinton's paradigm requires that: (i) there is one yes answer to k yes/no question, $Q_f^1, \ldots, Q_f^k$, pertaining to fixed information; (ii) there are n yes answers to n yes/no questions $Q_c^1 \ldots Q_c^n$ pertaining to controlled information; and (iii) there is one yes answer to yes/no questions $(Q_m)_1^1 \ldots (Q_m)_n^h$ for every bit of control information pertaining to information obtained by measurement/observation. Here h is the number of yes/no questions from the class of questions that represent possible measurement/observation outcomes. With those $k + n + (n * h)$ yes/no questions, there is associated an amount of $k + n + (n * h)$ bits of information and there are $2^{k+n+(n*h)}$ combinatorially possible bit strings of the form sketched in Eq. 10.2.

s_S	Q_f^1	$\ldots$	Q_f^k	Q_c^1	$\ldots$	Q_c^n	$(Q_m)_1^1$	$\ldots$	$(Q_m)_1^h$	$(Q_m)_2^1$	$\ldots$	$(Q_m)_2^h$	$\ldots$	$(Q_m)_n^h$
s_S^1	1	...	1	1	...	1	1	...	1	1	...	1	...	1
s_S^2	1	...	1	1	...	1	1	...	1	1	...	1	...	0
...	...	...	...	...	...	...	...	...	...	...	...	...	...	...
s_S^L	0	...	0	0	...	0	0	...	0	0	...	0	...	0

(10.2)

with $L = k + n + (n * h)$

As indicated in Eq. 10.2, there is a set, s_S, of bit strings of length $k + n + (n * h)$ which has $2^{k+n+(n*h)}$ members. The paradigm of fix/control/measure reduces these combinatorial possibilities to the set S_S of possibilities that are consistent with the paradigm (Eq. 10.3).

[4]A subdivision which boundaries are not aligned with physical discontinuities of the domain that is subdivided (Smith 1995, 2001; Smith and Varzi 2000).

$$
\mathsf{S_S} = \left\{ s_S^i \in s_S \;\middle|\; \begin{array}{ll} \Sigma_{j=1}^{k} s_S^i[j] = 1 \;\& & \text{(one yes-bit of fixed information for each } s_S^i\text{)} \\ (\Sigma_{j=k+1}^{k+n+1} s_S^i[j]) = n \;\& & (n \text{ yes-bits of control information for each } s_S^i\text{)} \\ (\bigwedge_{j=0}^{n-1} (\Sigma_{l=l_1}^{l_m} s_S^i[l] = 1)) = 1 & \text{(one yes-bit for measurement for each yes-control-bit for each } s_S^i\text{)} \\ l_1 = k + n + (j * h) + 1; & \\ l_m = k + n + (j * h) + h & \end{array} \right\} \tag{10.3}
$$

Example 1 Consider Fig. 10.6 and suppose that (a) the information that is obtained by measurement/observation is information about about the quality of elevation; (b) that the information about spatial location is controlled by projecting a fiat (Smith 1995, 2001; Smith and Varzi 2000) raster-shaped partition onto the ground as indicated in the image; (c) information about temporal location is fixed by allowing for a single time stamp. That is, a yes answer to one of the $Q_t^1, \ldots, Q_t^{10}$ picks out a particular time stamp. Yes answers to the yes/no questions $Q_c^1, \ldots, Q_c^{36}$ pick out

645	650	654	658	653	648
664	666	670	672	668	659
678	682	684	693	689	680
703	708	714	721	719	716
728	732	738	744	745	732
730	739	744	749	748	735

time	Q_t^1	Time is 11/20/2018?
	...	...
	Q_t^{10}	Time is 11/29/2018?
location	Q_l^1	Location is cell 1?
	...	...
	Q_l^{36}	Location is cell 36?
quality	Q_q^1	Quality measure is 640?
	...	...
	Q_q^{110}	Quality measure is 750?

$s_S \backslash Q_S$	Q_t^1	...	Q_t^{10}	Q_l^1	...	Q_l^{36}	$(Q_q)_1^1$	...	$(Q_q)_1^{110}$	$(Q_q)_2^1$	...	$(Q_q)_2^{110}$	...	$(Q_q)_{36}^{110}$
s_S^1	1	...	1	1	...	1	1	...	1	1	...	1	...	1
s_S^2	1	...	1	1	...	1	1	...	1	1	...	1	...	0
...	...	...	...	...	...	...	...	...	...	...	...	...	...	...
s_S^L	0	...	0	0	...	0	0	...	0	0	...	0	...	0

where $L = 1 + 36 + (36 * 110)$

$$
\mathsf{S_S} = \left\{ s_S^i \in s_S \;\middle|\; \begin{array}{ll} (\Sigma_{j=1}^{10} s_S^i[j]) = 1 = s_S^i[1] \;\& & \text{(fix time)} \\ (\Sigma_{j=11}^{46+1} s_S^i[j]) = 36 \;\& & \text{(36 yes-bits of control information)} \\ (\bigwedge_{j=0}^{36-1} (\Sigma_{k=k_1}^{k_m} s_S^i[k] = 1)) = 1 & \text{(one measurement bit is 1 for each control location)} \\ k_1 = 10 + 36 + (j * 110) + 1 & \\ k_m = 10 + 36 + (j * 110) + 110 & \end{array} \right\}
$$

$$
Q_S^{img} = Q_t^1 \wedge \neg Q_t^2 \wedge \ldots \wedge \neg Q_t^{10} \wedge Q_c^1 \wedge \ldots \wedge Q_c^{36} \wedge \neg (Q_q)_1^1 \wedge \ldots \wedge (Q_q)_1^6 \wedge \ldots \wedge \neg (Q_q)_1^{110} \wedge \neg (Q_q)_2^1 \wedge \ldots \wedge (Q_q)_2^{11} \wedge \ldots \wedge \neg (Q_q)_{36}^{110}.
$$

Fig. 10.6 Sinton's paradigm of geographic information where temporal location is fixed (11/20/2018), spatial location is controlled by fiat, and a geographic quality is measured. Q_S^{img} is the complete yes/no question the yes answer to which yields L bits of information encoded in the image in the top left. (The image in the top left is from Bolstad (2005).)

particular cells in the grid structure. For every control region picked out by a control question Q_c^i, there is a yes answer to one of the yes/no questions $Q_m^1, \ldots, Q_m^{110}$. By imposing those constraints, the paradigm of fixing time, controlling spatial location, and measuring/observing qualities of control regions, reduces these combinatorial possibilities to the set S_S of possibilities that are consistent with the paradigm (Eq. 10.3). The yes/no questions, the yes answers to which give rise to the bit strings of information in S_S are the members of the set of complete questions Q_S. Consider the yes/no question $Q_S^{img} \in Q_S$ as depicted in Fig. 10.6. A yes answer to Q_S^{img} yields L bits of information. This information is encoded in the string $s_S^{img} \in \mathsf{S}_\mathsf{S}$. The information encoded in s_S^{img} corresponds to the information encoded in the image in the top left of Fig. 10.6. □

The Nature of Geographic Information

The success of quantum mechanics in physics reveals three important aspects of information about the physical world (Rovelli 1996; Rovelli and Vidotto 2015): (i) information about the physical world is fundamentally relational (according to RQM); (ii) information of the physical world is fundamentally granular; and (iii) information about the physical world reflects the fundamentally indeterminate nature of certain aspects of the world. These properties of information manifest themselves logically in Postulates 1, 2, and 4 as discussed above. If there is a quantum theory that captures at least certain classes of geographic phenomena, then, in analogy to (i), (ii), and (iii), information of those phenomena is fundamentally relational, granular, and affected by indeterminacy and, logically, subject to Postulates 1, 2, and 4. In what follows the relational and granular nature of geographic information and the way geographic information is affected by indeterminacy is discussed within Sinton/Wheeler framework of geographic information processing.

The Relational Nature of Geographic Information

The inherently relational nature of Sinton's paradigm is revealed in the explicit focus on the aspect of control that is asserted by the observing system and targeted toward the observed system in form of a fiat subdivision (Smith 2001) of some aspect of the observed system (Postulate 5 (S2)). The assertion of control on how certain bits of information are obtained in Sinton's framework corresponds to the idea of a granular partition (Smith and Brogaard 2002; Bittner and Smith 2003; Bittner and Stell 2003). The theory of granular partitions (TGP) (Smith and Brogaard 2002) emphasizes the bidirectional nature of the interrelationship between observing and observed systems in geographic contexts such as the one sketched in Fig. 10.6. That is, control asserted by the observing system cannot be arbitrary. It has to adhere to certain features of the observed system. According to TGP, features include structural aspects such as mereology (Leonard and Goodman 1940; Simons 1987) as well as aspects of granularity and scale. In the context of a quantum theory, aspects of granularity and scale are of particular importance.[5] The ways in which the theory of granular partitions extends Sinton's paradigm can be seen as an investigation in the nature of the information transfer via correlations between observed and observing systems.

The Granular Nature of Geographic Information

In Example 1 (pg. xxx), there does not seem to be a limit to the amount of information about elevation that can be had by an observer. More information can be obtained by refining cells and asking yes/no questions about the elevation in these refined cells.[6] Similarly, more information can be obtained by allowing for more precise elevation measurements.[7] It follows:

Proposition 1 *When reading the qualities in Example 1 and Fig. 10.6 as elevation, then Example 1 constitutes a counter example to Postulate 1 and a supporting example for Postulate 2.*

Consider, again, Fig. 10.6, but now suppose that the classificatory in nature such as the quality of land cover and land use types (Andereson et al. 1976). In virtue of the classificatory nature, there is a maximal number of land cover types that can be distinguished in a given classification scheme. In addition, there *is* a limit to the degree to which the raster cells can be subdivided while still being meaningfully associated with land cover and/or land use (or other classificatory) qualities. This is because there cannot be a land cover of type forest in a region that is too small to contain a sufficient number of trees. This is captured in Postulate 6.

Postulate 6 (Granularity) *If the Sinton/Wheeler scheme is applied in contexts where classificatory qualities* $Q_m^1 \dots Q_m^h$ *of geographic regions are measured/ observed, time is fixed and space is controlled via control cells referenced by yes answers to a control questions of the form* $Q_l^1 \dots Q_l^n$*, then there is a minimal size of control cells—a finest level of resolution/granularity—for which yes/no questions of the form:*

[5]The theory of granular partitions was originally linked to Griffiths' (1984) consistent history interpretation of quantum mechanics (Smith and Brogaard 2002). However, none of the assumptions in the formalism of TGP restrict it to the consistent history interpretation.

[6]Of course, the notion of elevation ceases to be meaningful if the refinement of cells reaches the atomic scale. However, this is so far outside of the realm of geography that it can be ignored here.

[7]According to RQM, there *are* minimal units of length (Rovelli and Vidotto 2015). This can be ignored here.

$$Q_S^k = \text{“}\textit{Does the cell referenced by a yes answer to the question } Q_l^i \textit{ have the quality } Q_m^j\text{?” } (1 \leq i \leq n, 1 \leq j \leq h)$$

still have an answer. For cells of less than minimal size—below the finest level of resolution/granularity—there is no answer to a question such as Q_S^k.

Example 2 Suppose that in this example the questions $Q_q^1, \ldots, Q_q^{110}$ of Fig. 10.6 are designed to obtain information about land cover and land use qualities. In particular, suppose that the symbol 645 designates the land cover type “forrest”, the symbol 670 designates “industrial area”, and so on. In the context of Sinton's scheme, the yes/no questions

$$Q_t^1, \ldots, Q_t^{10}, Q_l^1, \ldots Q_l^{36}, (Q_q)_1^1, \ldots, (Q_q)_{36}^{110}$$

play the same roles as specified in Example 1. Again, the answers to those yes/no questions give rise to the set S_S of bit strings that emerge from yes answers to complete questions such as $Q_S^{img} \in \mathcal{Q}_S$. □

Proposition 2 *On the classificatorial interpretation of Example 2, the complete questions in* $\mathcal{Q}_S$ *satisfy Postulate 1, only if the control questions* $Q_l^1 \ldots Q_l^{36}$ *acquire information about cells of* maximal *resolution.*

Proof Every question $Q_S^i \in \mathcal{Q}_S$ is by construction *complete* (in the sense of section “The First Postulate of RQM”) and adheres to Sinton's scheme. Therefore, a yes answer to any of the complete questions in $Q_S^i \in \mathcal{Q}_S$ yields the same amount of L bits of information about land use and coverage about the target area that is picked out by the control questions $Q_l^1 \ldots Q_l^{36}$. More information about land use and land cover in the target area can be had only by further subdividing control cells, but this would render the question Q_S^i meaningless because it cannot be answered for cells that are smaller than cells of maximal of resolution. Thus, the amount of information of complete questions Q_S^i associated with control questions $Q_l^1 \ldots Q_l^{36}$ that acquire information about cells of *maximal* resolution is maximal. Hence, Postulate 1 is satisfied. □

Unlimited Amounts of Limited Information

On the classificatority interpretation of Fig. 10.6 in Example 2, the question arises, if there are other sets of complete questions (such as $\mathcal{Q}_R$ in Table 10.1) that can serve in support of Postulates 1 and 2. To address this issue, consider the yes/no questions $Q_t^1, \ldots, Q_t^k, Q_l^1, \ldots, Q_l^n$ where the fixed yes question that picks out the time stamp is Q_t^i with $1 \leq i \leq k$ and, as above, the control questions $Q_l^1, \ldots Q_l^n$ acquire information about cells of maximal resolution. Now suppose that there are yes/no questions $(Q_{i/b})_j^1$ and $(Q_{i/b})_j^2$:

- $(Q_{i/b})^1_j$: "Is the cell associated with a yes answer to the control question Q^j_l an interior part of a region of a homogeneous land cover type?"
- $(Q_{i/b})^2_j$: "Does the cell associated with a yes answer to the control question Q^j_l contain a boundary between distinct land cover types?"

Suppose that the observing system has a maximal amount of information about the observed system in form of the bit string s^i_S which stems from a yes answer to the complete question Q^i_S associated with control cells at finest level of resolution as in Example 2. By assumption, yes answers to the control questions $Q^1_l, \ldots Q^n_l$ pick out cells which have (mostly) *fiat* boundaries, i.e., boundaries that that do not correspond to discontinuities in the geographic world (Smith 1995, 2001; Smith and Varzi 2000). This means that the boundaries between control cells in general do not coincide with boundaries between regions with distinct types of land uses. This is illustrated in Example 3 and Fig. 10.7.

Example 3 Consider the image of Fig. 10.7 which displays the cells 1, 2, 7, and 8 of Fig. 10.6. On the classificatority interpretation, the numbers of the image in Fig. 10.6 are interpreted as land cover types and the respective cells had the classificatory values 645, 650, 664, and 666. As indicated in the image of Fig. 10.7, the classification of the cells in Fig. 10.6 is consistent with the land cover types of the actual regions on the ground. However, the actual boundaries that demarcate the regions on the ground lie skew to the boundaries of the raster cells that are picked out by the control questions Q^1_l, Q^2_l, Q^7_l, and Q^8_l. The tables in Fig. 10.7 illustrate, in analogy to the tables in Fig. 10.6, how the set $Q_{i/b}$ of complete questions and the strings of information $\mathsf{S}_{i/b}$ that emerge from yes answers to questions in $Q_{i/b}$ arise in a way that is consistent with Sinton's paradigm. $Q^{img}_{i/b}$ is the complete question corresponding to the image in the top of Fig. 10.7. □

Proposition 3 *The complete questions in $Q_{i/b}$ of Example 3 and Fig. 10.7 satisfy Postulate 1, only if the control questions $Q^1_l \ldots Q^{36}_l$ acquire information about cells of* maximal *resolution.*

Proof Consider complete questions of the form $Q^i_S \in Q_S$ of Fig. 10.6 and complete questions of the form $Q^i_{i/b} \in Q_{i/b}$ of Fig. 10.7. Since both Q^i_S and $Q^i_{i/b}$ contain the same control questions, it follows that if Q^i_S yields a maximal amount of (classificatory) information, which is determined by the maximal resolution of the raster cells picked out by the control questions. Therefore, the question $Q^i_{i/b}$ must yield the maximum amount of information about (in)homogeneity associated with the classification underlying Q_S and vice versa. Thus, Proposition 3 is true, if Proposition 2 is true. □

It now remains to investigate whether, jointly, the complete questions analyzed in Examples 2 and 3, satisfy Postulate 2:

Proposition 4 *Jointly, the sets of complete questions Q_S and $Q_{i/b}$ of Examples 2 and 3 satisfy Postulate 2, only if the control questions $Q^1_c \ldots Q^n_c$ acquire information about cells of* maximal *resolution.*

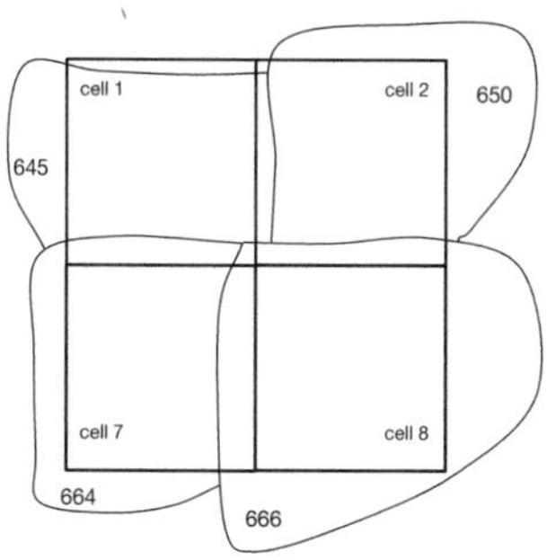

time	Q_t^1	Time is 11/20/2018?
	...	...
	Q_t^{10}	Time is 11/29/2018?
location	Q_l^1	Location is cell 1?
	...	...
	Q_l^{36}	Location is cell 36?
quality	Q_q^1	Cell is land-cover homogeneous ?
	Q_q^2	Cell is land-cover inhomogeneous?

$s_{i/b}$	Q_t^1	...	Q_t^k	Q_c^1	...	Q_c^{36}	$(Q_{i/b})_1^1$	$(Q_{i/b})_1^2$	$(Q_{i/b})_2^1$	$(Q_{i/b})_2^2$	...	$(Q_{i/b})_{36}^2$
$s_{i/b}^1$	1	...	1	1	...	1	1	1	1	1	...	1
$s_{i/b}^2$	1	...	1	1	...	1	1	1	1	1	...	0
...	...	...	...	...	...	...	...	...	...	...	...	...
$s_{i/b}^{L_{i/b}}$	0	...	0	0	...	0	0	0	0	0	...	0

where $L_{i/b} = 10 + 36 + (36 * 2)$

$$\mathsf{S}_{i/b} = \left\{ s_{i/b}^i \in s_{i/b} \,\middle|\, \begin{array}{ll} (\Sigma_{j=1}^{10} s_{i/b}^i[j]) = 1 = s_{i/b}^i[1]\ \& & \text{(fix time)} \\ (\Sigma_{j=2}^{36+1}\ s_{i/b}^i[j]) = 36\ \& & \text{(36 yes bits of control information)} \\ (\Sigma_{j=0}^{36-1}\ (\Sigma_k^{k_m}\ s_{i/b}^i[k] = 1) = 36) & \text{(one measurement bit is 1 for each control location)} \\ k = 1 + 36 + (j * 2) + 1 & \\ k_m = 1 + 36 + (j * 2) + 2 & \end{array} \right\}$$

$$Q_{i/b}^{img} = Q_t^1 \wedge \neg Q_t^2 \wedge \ldots \wedge \neg Q_t^{10} \wedge Q_c^1 \wedge \neg Q_c^2 \wedge \neg Q_c^7 \wedge \neg Q_c^8 \wedge \neg (Q_q)_1^1 \wedge (Q_q)_1^2 \wedge \neg (Q_q)_2^1 \wedge (Q_q)_2^2 \wedge \neg (Q_q)_7^1 \wedge (Q_q)_7^2 \wedge (Q_q)_8^1 \wedge \neg (Q_q)_8^2.$$

Fig. 10.7 Sinton's paradigm of geographic information where temporal location is fixed (11/20/2018), spatial location is controlled by fiat, and the quality of (in)homogeneity of land coverage is measured/observed. (The image in the top left corresponds to the four top left cells (cells 1, 2, 7, 8) in the image of Fig. 10.6.) $Q_{i/b}^{img}$ is the complete question corresponding to the image in the top left

Proof Consider the complete questions Q_S^i and $Q_{i/b}^j$. Question Q_S^i yields $10 + 36 + (36 * 110)$ bits of information and question $Q_{i/b}^j$ yields $10 + 36 + (36 * 2)$ bits of information. Thus, if Q_S^i and $Q_{i/b}^j$ are complete questions at extract information on the finest level of granularity, then the maximal amount of information that an observer can have about the observed phenomenon is $10 + 36 + (36 * 110)$ bits. Suppose that the image in Fig. 10.6 is the graphical representation of a yes answer to question Q_S^i. A yes answer to Q_S^i yields $10 + 36 + (36 * 110)$ bits, and thereby exhausts the amount of information that can be had. Now suppose that the observer asks question $Q_{i/b}^j$. A yes answer to this question yields $(36 * 2)$ bits of *new* information. This information is new because, by assumption, the boundaries between the cells picked out by the control questions are created by fiat, and therefore may or may not coincide with discontinuities in the observed phenomenon. Thus, a yes answer to Q^i_S does not contain information discontinuities at the finest level of granularity. By contrast, a yes answer to $Q^j_{i/b}$ does yield information about homogeneity and inhomogeneity and thus *new* information about discontinuities in the observed phenomena at the finest level of granularity.

Since a yes answer to Q^i_S yields the maximal amount of information an observer can have about the observed phenomenon, the new information obtained by a yes answer to $Q^j_{i/b}$ must *overwrite* old information, which therefore is lost. Asking Q^i_S again and receiving a yes answer will yield genuinely new information, i.e., information that was erased by the information that was obtained by a yes answer to $Q^j_{i/b}$. The questions Q^i_S and $Q^j_{i/b}$ are geographic examples of what in quantum mechanics are called *complimentary* questions or complimentary qualities. As in the example of Q^i_S and $Q^j_{i/b}$, every question in a sequence of complementary questions, when answered with yes, will yield new information. □

Corollary 1 *The complete questions in Q_S and $Q_{i/b}$ satisfy both, Postulate 1 and Postulate 2 only if the control questions $Q^1_c \dots Q^n_c$ acquire information about cells of maximal resolution.*

Of course, the examples discussed in the past two sections are specific instances of the famous cluster of problems that arise in the realm of the classification and delineation of geographic regions (Bailey 1983; Omernik and Griffith 2014). The question Q^i_S is an example of the formulation of a classification problem in the language of Wheeler's (1989) information-theoretic view of measurement/observation processes. By contrast, the question $Q^j_{i/b}$ is an example of the formulation of a delineation problem in Wheeler's (1989) language. Thus, this is an information-theoretic argument in support of the view that the classification and delineation of geographic regions are complementary measurement/observation processes at the geographic scale. Similar points were made from non-information-theoretic perspectives in Bittner (2011, 2017).

Information Density

The arguments of the previous section about the complementary nature of complete questions, Q^i_c, about classification and complete question, $Q^j_{i/b}$, about delineation, depended critically on the assumption that the control questions (that are part of $Q^i_c \in Q_c$ as well as in $Q^j_{i/b} \in Q_{i/b}$) refer to cells at the finest level of granularity. Linking a maximal amount of information to a minimal unit of space, as it is evident in Postulate 6 and Propositions 2 and 3, makes explicit that in the context of the processing of information about the classification and delineation of geographic regions there is a maximal *information density*. The notion of maximal information density then opens the possibility that larger amounts of information can be had at coarser levels of granularity.

Maximal Information Density

Consider Fig. 10.8 and suppose that in the context of the classification and delineation of geographic phenomena the employment of Sinton's paradigm has led to fixed time, controlled space and the measurement/observation of land cover types and land cover (in)homogeneity. Suppose further that (a) the control results in a raster

time	Q_t	Time is 11/20/2018?
location	Q_l^1	Location is cell 1?
	Q_l^2	Location is cell 2?
quality	$Q_{i/b}^1$	Cell is land-cover homogeneous ?
$Q_{i/b}$	$Q_{i/b}^2$	Cell is land-cover inhomogeneous?
quality	Q_q^1	Cell is land-cover-type A?
Q_q	Q_q^2	Cell is land-cover-type B?

$Q_c^{1,1} \equiv Q_t \wedge Q_l^1 \wedge Q_q^1$, $Q_c^{1,2} \equiv Q_t \wedge Q_l^1 \wedge Q_q^2$, $Q_c^{2,1} \equiv Q_t \wedge Q_l^2 \wedge Q_q^1$, and $Q_c^{2,2} \equiv Q_t \wedge Q_l^2 \wedge Q_q^2$;
$Q_{i/b}^{1,1} \equiv Q_t \wedge Q_l^1 \wedge Q_{i/b}^1$, $Q_{i/b}^{1,2} \equiv Q_t \wedge Q_l^1 \wedge Q_{i/b}^2$, $Q_{i/b}^{2,1} \equiv Q_t \wedge Q_l^2 \wedge Q_{i/b}^1$, and $Q_{i/b}^{2,2} \equiv Q_t \wedge Q_l^2 \wedge Q_{i/b}^2$.

		$s_c^{(i)}$					
$\mathcal{H}_c$	$Q_c^{(i)}$	$Q_c^{1,1}$	$Q_c^{1,2}$	$Q_c^{2,1}$	$Q_c^{2,2}$		
	–	1	1	1	1	–	
	–	1	1	1	0	–	
	–	1	1	0	1	–	
	–	1	1	0	0	–	
	–	1	0	1	1	–	
$\lvert Q_c^{(1)}\rangle$	$Q_c^{(1)}$	1	0	1	0	$Q_{i/b}^{(1)}$	$\lvert Q_{i/b}^{(1)}\rangle$
$\lvert Q_c^{(2)}\rangle$	$Q_c^{(2)}$	1	0	0	1	$Q_{i/b}^{(2)}$	$\lvert Q_{i/b}^{(3)}\rangle$
	–	1	0	0	0	–	
	–	0	1	1	1	–	
$\lvert Q_c^{(3)}\rangle$	$Q_c^{(3)}$	0	1	1	0	$Q_{i/b}^{(3)}$	$\lvert Q_{i/b}^{(2)}\rangle$
$\lvert Q_c^{(4)}\rangle$	$Q_c^{(4)}$	0	1	0	1	$Q_{i/b}^{(4)}$	$\lvert Q_{i/b}^{(4)}\rangle$
	–	0	1	0	0	–	
	–	0	0	1	1	–	
	–	0	0	1	0	–	
	–	0	0	0	1	–	
	–	0	0	0	0	–	
		$Q_{i/b}^{1,1}$	$Q_{i/b}^{1,2}$	$Q_{i/b}^{2,1}$	$Q_{i/b}^{2,2}$	$Q_{i/b}^{(i)}$	$\mathcal{H}_{i/b}$
		$s_{i/b}^{(i)}$					

Fig. 10.8 Constructing a two sets, Q_c and $Q_{i/b}$ of complete questions that each capture four bits of information about the land cover type or four bits of information about the land cover homogeneity of two *minimal* raster cells

structure that has two cells and that each cell is of the minimal size (i.e., of maximal resolution) associated with land cover types (in the sense of Postulate 6); (b) the measurement/observation of land cover types distinguishes two types: type A and type B; and (c) the measurement/observation of land cover homogeneity distinguishes two types: homogeneous and inhomogeneous. The information-theoretic representation of what can be measured/observed within this example is encoded in eight yes/no questions. The questions are labeled $Q_c^{1,1}$, $Q_c^{1,2}$, $Q_c^{2,1}$, $Q_c^{2,2}$, $Q_{i/b}^{1,1}$, $Q_{i/b}^{1,2}$, $Q_{i/b}^{2,1}$, $Q_{i/b}^{2,2}$ and specified as conjunctions of yes/no questions that extract fixed, controlled, and measured information as outlined in the middle of Fig. 10.8.

Due to the complementary nature of the classification qualities and delineation qualities of minimal control cells, there are two sets of complete questions as given below:

$$Q_c = \{Q_c^1, Q_c^2, Q_c^3, Q_c^4\} \quad Q_{i/b} = \{Q_{i/b}^1, Q_{i/b}^2, Q_{i/b}^3, Q_{i/b}^4\}$$

The amount of information that can be obtained by answering each of the four questions in the two groups is four bits.[8] That is, the maximum information density is *four bits per cell of minimal size*. The 2^4 combinatorially possible pattern of yes/no answers to these questions are listed in Fig. 10.8. Within the constraints of Sinton's paradigm, only four pattern of yes/no answers are possible.

Example 4 Consider Fig. 10.9 and suppose that the image in the middle represents a specific situation on the ground—the system S that is constituted by two regions, one of land cover type A and one of land cover type B. Suppose further, that the cells marked "cell 1" and "cell 2" are fiat subdivisions imposed by the observer O on S which are of minimal size with respect to the measured/observed quality (land cover types and land cover (in)homogeneity). If O can have only four bits of information at the finest level of granularity, then the image in the middle of Fig. 10.9 is either characterized by a yes answer to the question $Q_c^{(2)}$ which yields the information $s_c^{(2)} = [1001]$, or by a yes answer to the question $Q_{i/b}^{(3)}$ which yields the information $s_{i/b}^{(2)} = [0110]$. □

Interpreted in the context of the limited amount of information that observer O can have about S at finest level of granularity, the image in the middle of Fig. 10.9 is an illusion. The information that can be had by O at any given time at the finest level of granularity is either a string of bits $s_c^{(2)}$ obtained as a yes answer to the question $Q_c^{(2)}$ or a string of bits $s_{i/b}^{(2)}$ obtained as a yes answer to the question $Q_{i/b}^{(2)}$, but not both.

The orthomodular lattice that arises from the two sets of complete questions Q_c and $Q_{i/b}$ at the finest level of granularity is displayed in Fig. 10.10. In analogy to Fig. 10.4, the lattice in Fig. 10.10 is an algebraic expression of the fact that, due to the limited amount of possible information, the distinct sets of complete questions Q_c and $Q_{i/b}$ are *incompatible* in and asking questions of the form $Q_c^i \vee Q_{i/b}^i$ will fail to yield determinate information.

[8] In the context of this example, it is ignored that there are more bits of information "hidden" in the $Q_c^{(i)}$ and $Q_{i/b}^{(j)}$.

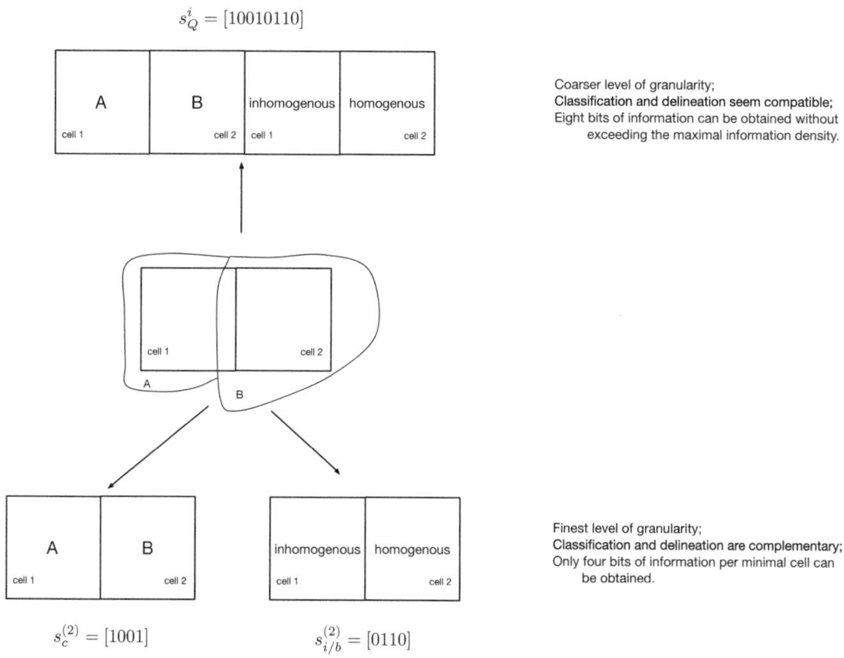

Fig. 10.9 Amount of information that can be obtained at non-finest level of granularity is 8 bits (top) and the amount of information that can be obtained at finest level of granularity is 4 bits (bottom)

Information at Coarse Levels of Granularity

To illustrate that the notion of maximal information density is consistent with larger amounts of information at coarser levels of granularity consider Example 5, which is given below:

Example 5 Consider, again, Fig. 10.9 but now suppose that the cells labeled "cell 1" and "cell 2" are of a size that is much larger than the minimal size (say 10 times larger or so) with respect to the measured/observed quality. That is, on this interpretation the figure represents a specific configuration S' at a coarse level of granularity. At this coarser level of granularity, the observing system O can have an amount of (at least) eight bits of information about the observed system S' *without* exceeding the maximal information density of four bits per cell at the finest level of granularity. The eight bits of information specify: Cell 1 is inhomogeneous *and* of type A and cell 2 is homogeneous *and* of type B. The string of eight bits of information that corresponds to the image in the middle of Fig. 10.9 described at a coarser level of granularity is $s_Q^i = [10010110]$ as indicated in the top of the figure. □

In contrast to the non-distributive orthomodular lattice that arises from the two sets of complete questions Q_c and $Q_{i/b}$ at the finest level of granularity (Fig. 10.10),

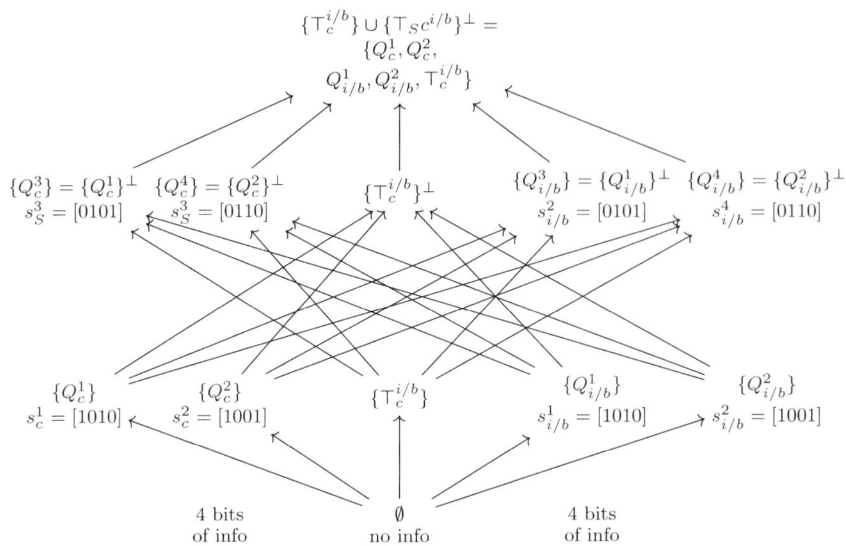

Fig. 10.10 An orthomodular lattice of the two sets of complete questions Q_c and $Q_{i/b}$ where the arrows indicate subset relations between subsets of Q_c and $Q_{i/b}$ and unions thereof. (Calude et al. 2014)

at coarser levels of granularity at which (at least) eight bits of information can be obtained per cell, the indeterminate nodes $\{\top_c^{i/b}\}$ and $\{\top_c^{i/b}\}^{\perp}$ of Fig. 10.10 disappear and the lattice becomes a Boolean algebra of the form sketched in the right of Fig. 10.1. This represents at the algebraic level that at the finest level of granularity the theory that describes the information that an observer can have about the observed system is very nonclassical, i.e., logical conjunction and disjunction are non-distributive. At coarser levels of granularity, descriptions become more classical, i.e., logical conjunction and disjunction are distributive.

Sketch of a Quantum Theory

The discussion of the previous sections supports the hypothesis that a quantum theory may be the proper theory of the information that arises in the context of the classification and delineation of geographic phenomena at finest level of granularity. This is because the analysis of the classification and delineation of geographic phenomena from an information-theoretic perspective provides examples that satisfy both, Postulates 1 and 2, which, according to the relational interpretation of QM, in conjunction with Postulate 4, demand a quantum theory. At this point, however, it is not clear whether in the the context of the classification and delineation of geographic phenomena the indeterminacy that is entailed by Postulates 1 and 2, manifests itself probabilistically. To test whether or not this is indeed the case one needs to build a quantum theory and test its probabilistic predictions. It goes beyond the scope of this paper to actually develop such a theory. Nevertheless, it will be useful to sketch a toy theory that illustrates how such a theory could look like. The examples developed in section "Information Density" will continue to serve as an illustration.

Analogous to the discussion in section "The Formalism of QM", the Hilbert space of the standard formalism of QM can be constructed from the sets Q_c and $Q_{i/b}$ of complete questions. The four complete questions in Q_c correspond to a system of base vectors—the Q_c-base—of a four-dimensional Hilbert space, $\mathcal{H}$. That is, the question Q_c^1 corresponds to the base vector $|Q_c^{(1)}\rangle$, etc. Similarly, the four complete questions in $Q_{i/b}$ correspond to a different system of base vectors of $\mathcal{H}$—the $Q_{i/b}$-base. This is illustrated in Fig. 10.8.

As in the example in the left of Fig. 10.5, the systems of base vectors corresponding to the sets of complete questions Q_c and $Q_{i/b}$ are rotated with respect to one another in the four-dimensional Hilbert space $\mathcal{H}$. The orthomodular lattice that is formed by the subspaces generated by the vectors in this four-dimensional Hilbert space is displayed in Fig. 10.11. In analogy to the lattice in the right of Fig. 10.5, there are two complementary sets of atoms in the lattice of Fig. 10.11.

Every vector in a vector space can be represented as a superposition of a system of base vectors. That is, if $|\phi\rangle$ is a vector of $\mathcal{H}$ that is described in the Q_c-base, then there are complex numbers $\alpha, \beta, \gamma, \delta \in \mathcal{C}$ such that $|\phi\rangle = \alpha|Q_c^{(1)}\rangle + \beta|Q_c^{(2)}\rangle + \gamma|Q_c^{(3)}\rangle + \delta|Q_c^{(4)}\rangle$. If the vector $|\phi\rangle$ represents the state of an observed system S with respect to an observing system O, then the probability that O receives a yes answer to the question $Q_c^{(i)}$ is $|\langle\phi|Q_c^{(i)}\rangle|$.

Base vectors are just vectors. Thus, one can express the base vectors of the base associated with the questions in $Q_{i/b}$ (the $Q_{i/b}$-base) in terms of the base vectors associated with the questions in Q_c (the Q_c-base). The fact that the complete questions in Q_c are incompatible with or complementary to the complete questions in $Q_{i/b}$ is represented in the formalism of QM as follows: If a vector $|\phi\rangle$ is determinate when expressed in the Q_c-base (case (a) in section "Indeterminacy as Probability"),

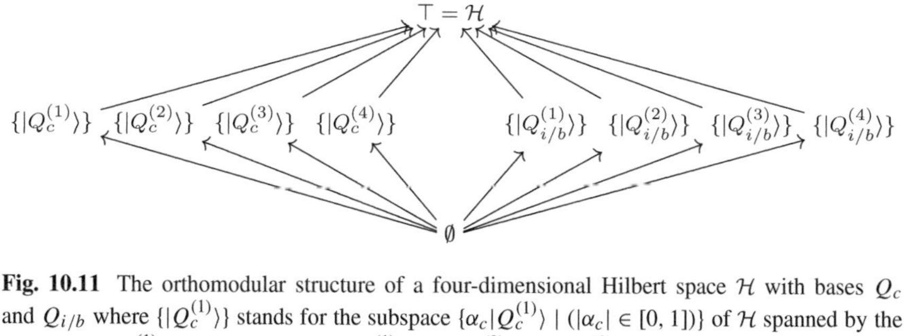

Fig. 10.11 The orthomodular structure of a four-dimensional Hilbert space $\mathcal{H}$ with bases Q_c and $Q_{i/b}$ where $\{|Q_c^{(1)}\rangle\}$ stands for the subspace $\{\alpha_c|Q_c^{(1)}\rangle \mid (|\alpha_c| \in [0,1])\}$ of $\mathcal{H}$ spanned by the base vector $|Q_c^{(1)}\rangle$ and $\top = \{|\phi\rangle = \alpha_c|Q_c^{(1)}\rangle + \beta_c|Q_c^{(2)}\rangle + \gamma_c|Q_c^{(3)}\rangle + \delta_c|Q_c^{(4)}\rangle \mid (|\langle\phi|\phi\rangle| = 1)\} = \{|\psi\rangle = \alpha_{i/b}|Q_{i/b}^{(1)}\rangle + \beta_{i/b}|Q_{i/b}^{(2)}\rangle + \gamma_{i/b}|Q_{i/b}^{(3)}\rangle + \delta_{i/b}|Q_{i/b}^{(4)}\rangle \mid (|\langle\psi|\psi\rangle| = 1)\}$

	α	β	γ	δ	$\vert\phi\rangle = \alpha\,\vert Q_c^{(1)}\rangle + \beta\,\vert Q_c^{(2)}\rangle + \gamma\,\vert Q_c^{(3)}\rangle + \delta\,\vert Q_c^{(4)}\rangle$
1	1	0	0	0	probability of $\vert\langle\phi\vert Q_c^{(1)}\rangle\vert = 1$ to obtain a yes answer to $Q_c^{(1)}$
2	0	1	0	0	probability of $\vert\langle\phi\vert Q_c^{(2)}\rangle\vert = 1$ to obtain a yes answer to $Q_c^{(2)}$
3	0	0	1	0	probability of $\vert\langle\phi\vert Q_c^{(3)}\rangle\vert = 1$ to obtain a yes answer to $Q_c^{(3)}$
4	0	0	0	1	probability of $\vert\langle\phi\vert Q_c^{(4)}\rangle\vert = 1$ to obtain a yes answer to $Q_c^{(4)}$
5	$\frac{1}{\sqrt{2}}$	$\frac{1}{\sqrt{2}}$	0	0	equal probability of $\vert\langle\phi\vert Q_c^{(1)}\rangle\vert = \vert\langle\phi\vert Q_c^{(2)}\rangle\vert = \frac{1}{2}$ to obtain a yes answer to $Q_c^{(1)}$ or $Q_c^{(2)}$ and no probability to obtain a yes answer to $Q_c^{(3)}$ or $Q_c^{(4)}$ ($\vert\langle\phi\vert Q_c^{(3)}\rangle\vert = \vert\langle\phi\vert Q_c^{(4)}\rangle\vert = 0$.
6	$\frac{1}{\sqrt{2}}$	0	$\frac{1}{\sqrt{2}}$	0	equal probability of $\vert\langle\phi\vert Q_c^{(1)}\rangle\vert = \vert\langle\phi\vert Q_c^{(3)}\rangle\vert = \frac{1}{2}$ to obtain a yes answer to $Q_c^{(1)}$ or $Q_c^{(3)}$ and no probability to obtain a yes answer to $Q_c^{(2)}$ or $Q_c^{(4)}$.
...	...	...	...	...	...
m	$\frac{1}{\sqrt{3}}$	$\frac{1}{\sqrt{3}}$	$\frac{1}{\sqrt{3}}$	0	equal probability of $\vert\langle\phi\vert Q_c^{(1)}\rangle\vert = \vert\langle\phi\vert Q_c^{(1)}\rangle\vert = \vert\langle\phi\vert Q_c^{(3)}\rangle\vert = \frac{1}{3}$ to obtain a yes answer to $Q_c^{(1)}$ or $Q_c^{(2)}$ or $Q_c^{(3)}$ and no probability to obtain a yes answer to $Q_c^{(4)}$.
...	...	...	...	...	...
n	$\frac{1}{\sqrt{4}}$	$\frac{1}{\sqrt{4}}$	$\frac{1}{\sqrt{4}}$	$\frac{1}{\sqrt{4}}$	equal probability of $\frac{1}{4}$ to obtain a yes answer to $Q_c^{(1)}$ or $Q_c^{(2)}$ or $Q_c^{(3)}$ or $Q_c^{(4)}$.

Fig. 10.12 Examples of vectors in the base corresponding to the complete questions in Q_c and their probabilistic interpretation

then the vector $|\phi\rangle$ is maximally indeterminate when expressed in the $Q_{i/b}$-base (case (b) in section "Indeterminacy as Probability"). In particular, if $|\langle\phi|Q_c^{(i)}\rangle| = 1$ for some $i \in 1 \ldots 4$, then $|\langle\phi|Q_{i/b}^{(i)}\rangle| = \frac{1}{4}$ for every $i \in 1 \ldots 4$. That is, determinacy expressed probabilistically as certainty and maximal indeterminacy expressed probabilistically as complete randomness. Similarly, if $|\langle\phi|Q_{i/b}^{(i)}\rangle| = 1$ for some $i \in 1 \ldots 4$, then $|\langle\phi|Q_c^{(i)}\rangle| = \frac{1}{4}$ for every $i \in 1 \ldots 4$. Some examples of intermediate degrees of indeterminacy and their probabilistic representations are displayed in Fig. 10.12.

Conclusion

So, is there a quantum geography? Or better: Are there geographic phenomena that can be described best by a quantum theory? The answer to this question is definitively not "No". In fact, there seem to be good reasons to believe that the answer is "Yes". In support of this answer stand regional geographic phenomena with indeterminate boundaries that typically are identified by classification and delineation processes. This was illustrated above in the context of the classification and delineation of geographic regions that are characterized by their land use and land coverage.

The argument of why a quantum theory may be a good tool to describe those phenomena has three major premises: First, there are three necessary and sufficient conditions that call for a quantum theory as an adequate description (Postulates 1–3). Second, The above class of phenomena satisfy two of those conditions (Postulates 1 and 2). Third, a quantum theory can be developed along the lines sketched above. This theory would produce predictions that, if verified empirically, indicate that Postulate 3 is satisfied. The focus of this paper was on first two premises. The truth of the third premise is yet to be determined.

Boolean Algebra and Orthomodular Lattices

A Boolean algebra is a specific form of an orthomodular lattice, which in turn is a special lattice—a partially ordered set $(X, \leq)$, with join and meet operations $(\vee, \wedge)$, with unique maximal $(\top)$ and minimal elements $(\emptyset)$. The join and meet operations $\vee$ and $\wedge$ are defined in the standard way such that $\vee$ yields the least upper bound of its arguments in the underlying partially ordered set and $\wedge$ yields the greatest lower bound (Birkhoff 1948; Grinbaum 2005; Wikipedia contributors 2018). In addition, a Boolean algebra has an *ortho-complementation*, a function that maps each element a to an orthocomplement $a^{\perp}$ such that (Beltrametti et al. 1984; Wikipedia contributors 2016): (i) $a^{\perp} \vee a = 1$ and $a^{\perp} \wedge a = 0$; (ii) $a^{\perp\perp} = a$; (iii) if $a \leq b$ then $b^{\perp} \leq a^{\perp}$. A Boolean algebra is an ortho-complemented lattice that is distributive, i.e., $a \wedge (b \vee c) = (a \wedge b) \vee (a \wedge c)$, etc. is always true. By contrast, an orthomodular lattice is a ortho-complemented lattice in which the weaker condition if $a \leq c$, then $a \vee (a^{\perp} \wedge c) = c$ is always true. Orthomodular lattices also describe the mathematical structure of Hilbert spaces that are exploited in quantum mechanics (Beltrametti et al. 1984; Hughes 1981; Rovelli 1996).

Hilbert Space

A Hilbert space $\mathcal{H}$ is a *complex* vector space with an *inner product*. In what follows Dirac's notation for vectors in Hilbert spaces (Dirac 1930) is used. The members of a Hilbert space $\mathcal{H}$ are written as ket vectors of the form $|\phi\rangle$ where ϕ is a name/label. As vector spaces Hilbert spaces are closed under vector addition and scalar multiplication. That is, if $|\phi\rangle, |\psi\rangle \in \mathcal{H}$ then $\alpha|\phi\rangle + \beta|\psi\rangle \in \mathcal{H}$, where α and β are complex numbers that modify the length of a vector via scalar multiplication and $+$ is the vector addition. The inner product $\langle\psi|\phi\rangle$ of the vectors $|\psi\rangle, |\phi\rangle \in \mathcal{H}$ (defined below) is a complex number.

A base $\mathcal{B} = |Q_1\rangle, \ldots, |Q_n\rangle$ of a n-dimensional Hilbert space $\mathcal{H}$ is a system of vectors such that every member of $\mathcal{H}$ can be expressed as a vector sum of the base vectors. A base is orthonormal if the inner product of distinct base vectors is zero and all base vectors are of unit length, i.e., $\langle Q_i|Q_j\rangle = 1$ if $i = j$ and $\langle Q_i|Q_j\rangle = 0$ otherwise. If the vector $|\phi\rangle = \alpha_1|Q_1\rangle + \ldots + \alpha_n|Q_n\rangle$, then there exists a dual vector $\langle\phi| = \overline{\alpha}_1|Q_1\rangle + \ldots + \overline{\alpha}_n|Q_n\rangle$ where $\overline{\alpha}_i$ is the complex conjugate of α_i. If $|\phi\rangle =$

$\alpha_1|Q_1\rangle + \ldots + \alpha_n|Q_n\rangle$ and $|\psi\rangle = \beta_1|Q_1\rangle + \ldots + \beta_n|Q_n\rangle$, then the inner product of $|\phi\rangle$ and $|\psi\rangle$ designated by $\langle\psi|\phi\rangle$ is the sum of the products of the components of $\langle\psi|$ and $|\phi\rangle$ computed as $\sum_i \overline{\beta_i}\alpha_i$. In what follows $|\langle\phi|\psi\rangle|$ stands for the squared modulus of the scalar product $\langle\phi|\psi\rangle$. The value of $|\langle\phi|\psi\rangle|$ is a real number between zero and one and is interpreted in Wheeler's (1989) information-theoretic framework as the probability that a yes answer to the question encoded in $|\phi\rangle$ is followed by a yes answer to the question encoded in $|\psi\rangle$. Details can be found in any text book on quantum mechanics. The classic reference is Dirac (1930).

References

Andereson JR, Hardy EE, Roach JT, Witmer RE (1976) A land use and land cover classification system for use with remote sensor data. Technical report, Geological Survey Professional Paper 964

Bailey RG (1983) Delineation of ecosystem regions. Environ Manag 7:365–373

Beltrametti EG, Cassinelli G, Carruthers PA (1984) The logic of quantum mechanics. Encyclopedia of mathematics and its applications, vol 15, Cambridge University Press, Cambridge. https://doi.org/10.1017/CBO9781107340725

Birkhoff G (1948) Lattice theory. American Mathematical Society, New York

Bittner T (2011) Vagueness and the tradeoff between the classification and delineation of geographic regions. Int J Geogr Inf Sci (IJGIS) 25(5):825–850

Bittner T (2017) Towards a quantum theory of geographic fields. In: Clementini E, Donnelly M, Yuan M, Kray C, Fogliaroni P, Ballatore A (eds) 13th international conference on spatial information theory (COSIT 2017), Schloss Dagstuhl–Leibniz-Zentrum fuer Informatik, Dagstuhl, Germany, Leibniz international proceedings in informatics (LIPIcs), vol 86, pp 5:1–5:14, https://doi.org/10.4230/LIPIcs.COSIT.2017.5, http://drops.dagstuhl.de/opus/volltexte/2017/7766

Bittner T, Smith B (2003) A theory of granular partitions. In: Duckham M, Goodchild MF, Worboys MF (eds) Foundations of geographic information science. Taylor & Francis, London, pp 117–151

Bittner T, Stell J (2003) Stratified rough sets and vagueness. In: Kuhn W, Worboys M, Timpf S (eds) Spatial Information theory. Cognitive and computational foundations of geographic information science. International conference COSIT'03, Springer, pp 286–303

Bolstad P (2005) GIS fundamentals. Eider Press, Minnesota

Burrough P, Frank AU (eds) (1995) Geographic objects with indeterminate boundaries. GISDATA series II. Taylor and Francis, London

Calude CS, Hertling PH, Svozil K (2014) Embedding quantum universes into classical ones. arXiv:1402:5199

Dirac P (1930) Principles of quantum mechanics. Clarendon Press, Oxford

Einstein A (1905) On the electrodynamics of moving bodies. Ann Phys 17:891–921

Einstein A, Podolsky B, Rosen N (1935) Can quantum-mechanical description of physical reality be considered complete? Phys Rev 47:777–780. https://doi.org/10.1103/PhysRev.47.777

Griffiths RB (1984) Consistent histories and the interpretation of quantum mechanics. J Stat Phys 36(1):219–272. https://doi.org/10.1007/BF01015734

Grinbaum A (2005) Information-theoretic princple entails orthomodularity of a lattice. Found Phys Lett 18(6):563–572. https://doi.org/10.1007/s10702-005-1129-0

Hughes RIG (1981) Quantum logic. Sci Am 245(4):202–213. http://www.jstor.org/stable/24964585

Kennedy J (2003) Space, time, and Einstein: an introduction. McGill-Queen's University Press. http://books.google.com/books?id=laalPR8p6fgC

Kochen S, Specker EP (1967) The problem of hidden variables in quantum mechanics. J Math Mech 17(1):59–87. http://www.jstor.org/stable/24902153

Leonard HS, Goodman N (1940) The calculus of individuals and its uses. J Symb Log 5(2):45–55. http://www.jstor.org/stable/2266169
Maudlin T (2002) Quantum non-locality and relativity: metaphysical intimations of modern physics. Blackwell Publishers
Musser G (2015) Where is here? Sci Am 313(5):70. http://search.ebscohost.com.gate.lib.buffalo.edu/login.aspx?direct=true&db=f5h&AN=110284477&site=ehost-live&scope=site
von Neumann J (1932) Mathematical foundations of quantum mechanics. Springer, Berlin
Omernik JM, Griffith GE (2014) Ecoregions of the conterminous united states: evolution of a hierarchical spatial framework. Environ Manag. https://doi.org/10.1007/s00267-014-0364-1
Omnès R (1994) The interpretation of quantum mechanics. Princeton University Press, Princeton
Redhead M (1997) Quantum non-locality and relativity: metaphysical intimations of modern physics by Tim Maudlin. Philos Q 47(186):118–120. https://doi.org/10.2307/2956486
Romero GE (2012) Philosophical problems of space-time theories. In: Gravitation and cosmology
Rovelli C (1996) Relational quantum mechanics. Int J Theor Phys 35:1637–1678. https://doi.org/10.1007/BF02302261, arXiv:quant-ph/9609002
Rovelli C, Vidotto F (2015) Covariant loop quantum gravity. Cambridge University Press, Cambridge
Shannon CE (1948) A mathematical theory of communication. Bell Syst Tech J 27(3):379–423. https://doi.org/10.1002/j.1538-7305.1948.tb01338.x
Simons P (1987) Parts, a study in ontology. Clarendon Press, Oxford
Sinton DF (1978) The inherent structure of information as a constraint to analysis: mapped thematic data as a case study. In: Dutton G (ed) Harvard papers on geographic information systems. Addison Wesley, pp 1–17
Smith B (1995) On drawing lines on a map. In: Frank A, Kuhn W (eds) Conference on spatial information theory, COSIT, vol 988. Springer, Semmering, Austria, pp 475–484
Smith B (2001) Fiat objects. Topoi 20(2):131–148
Smith B, Brogaard B (2002) Quantum mereotopology. Ann Math Artif Intell 35(1–2):153–175
Smith B, Varzi A (2000) Fiat and bona fide boundaries. Philos Phenomenol Res 60(2):401–420
Tobler W (1970) A computer movie simulating urban growth in the Detroit region. Econ Geogr 46(2):234–240
Trassinelli M (2018) Relational quantum mechanics and probability. Found Phys 48(9):1092–1111. https://doi.org/10.1007/s10701-018-0207-7
Wheeler JA (1989) Information, physics, quantum: the search for links. In: Proceedings III international symposium on foundations of quantum mechanics, pp 354–358
Wikipedia (2019) Interpretations of quantum mechanics—wikipedia, the free encyclopedia. URL https://en.wikipedia.org/wikiInterpretations_of_quantum_mechanics. Accessed 16 Apr 2019
Wikipedia contributors (2016) Complemented lattice—wikipedia, the free encyclopedia. https://en.wikipedia.org/w/index.php?title=Complemented_lattice&oldid=745481682. Accessed 3 Nov 2018
Wikipedia contributors (2018) Lattice (order)—wikipedia, the free encyclopedia. https://en.wikipedia.org/w/index.php?title=Lattice_(order)&oldid=866284134. Accessed 4 Nov 2018
Yang JM (2018) A relational formulation of quantum mechanics. Sci Rep 8(1):13,305. https://doi.org/10.1038/s41598-018-31481-8, https://doi.org/10.1038/s41598-018-31481-8

Chapter 11
Disorientation and GIS-Informed Wilderness Search and Rescue

Pablo Fernandez Velasco and Roberto Casati

Abstract Nowadays, Wilderness Search and Rescue (WiSAR) operations revolve around the creation of probability maps using GIS planning tools (Doherty et al, Appl Geogr 47:99–110, 2014). Although this method has proven effective, there is a missing link between WiSAR theory and advances in other fields related to disorientation (e.g. psychology and neuroscience). A unified conceptualisation of disorientation is a crucial element for understanding the mind and behaviour of disoriented subjects. The central aim of this chapter is to explore how a unified conceptualisation of disorientation can contribute to GIS-informed WiSAR theory. The paper will open with a review of the work on disorientation coming from different fields, to then introduce the conceptual work that is needed for a unified account of disorientation. We will discuss two different approaches to WiSAR theory: a ring model and a Bayesian model. We end on a discussion on how conceptual work on disorientation and GIS-informed WiSAR theory can help each other advance.

Keywords Disorientation · Wilderness search and rescue · Unified account of disorientation

Disorientation Review

> Again something dark appeared in front of him. Again he rejoiced, convinced that now it was certainly a village. But once more it was the same boundary line overgrown with wormwood, once more the same wormwood desperately tossed by the wind and carrying unreasoning terror to his heart.
>
> –Tolstoy. Master and Man

Work for this chapter has been made possible by grants Disorientation Remediation PSL-NYU 2018, ANR-17-EURE-0017 FrontCog, ANR-10-IDEX-0001-02 PSL.

P. Fernandez Velasco (✉) · R. Casati
Département d'études Cognitives, Institut Jean Nicod, ENS, EHESS, CNRS, PSL, Paris, France
e-mail: pfernandezvelasco@ens.fr

R. Casati
e-mail: roberto.casati@ens.fr

© Springer Nature Switzerland AG 2019

T. Tambassi (ed.), *The Philosophy of GIS*, Springer Geography,
https://doi.org/10.1007/978-3-030-16829-2_11

Tolstoy's story of a disoriented man in a snowstorm is not far off from actual reports of lost person behaviour (Tolstoy 2015). After realising that he's been walking in circles, Vasili Andreevich tries to keep calm, but assailed by a mounting panic, he finds himself unable to collect himself and stay put and instead starts to wander aimlessly, which is a typical lost person behaviour (Hill 1998). In 1998, Kenneth Hill conducted over a hundred interviews with subjects who had become lost in the wilderness, and constructed a list of typical disoriented behaviour that has now become a central reference in the study of lost person psychology.

Hill claimed that when lost, people followed (at least) one of the following patterns and strategies: random travelling (often caused by confusion and high emotional arousal), route travelling (i.e. following a given route in the hope of finding something familiar), direction travelling (trying to follow a given direction), route sampling (using an intersection as a base for exploring different routes), direction sampling (using a visible landmark as a base for exploring different directions), view enhancing (aiming for a high position in order to gain visibility), backtracking, using folk wisdom (e.g. follow streams to civilisation) and staying put (which according to Hill is the most effective strategy if a SAR operation is triggered).

There has been some work relating this taxonomy of lost person behaviour with different categories of lost subjects (e.g. hunters are more likely than others to use direction sampling, while hikers are more likely to use route following). Nevertheless, there isn't any work relating the taxonomy to a broader conceptualisation of disorientation (e.g. how disorientation arises and what makes subjects follow one method over another) or to the work in other fields such as neuroscience.

In an effort to establish a link between the behavioural psychology and the neuroscience of disorientation, Dudchenko holds that humans, unlike other animals, need vision to stay oriented (Dudchenko 2010). This makes sense because landmarks are used to update one's orientation and position within a cognitive map (Scholl 1987; Knierim and Hamilton 2011). The notion of cognitive maps was originally developed by Tolman as a hypothesis to explain rat behaviour in labyrinths without reducing the said behaviour to stimulus–response connections. Tolman supports the view that rats construct something similar to a field map with results from different experiments, including shows of initial latent learning that gets activated when reward is offered later (Blodgett 1929), rats sampling of the environments (Tolman 1938) and rats finding shortcuts that they had not learned through stimulus–response conditioning (Tolman 1948).

The discovery of place cells in the rodent hippocampus further supported the idea of a cognitive map (O'Keefe and Dostrovsky 1971), which has now become central in the study of spatial cognition. Place cells fire as a function of the rodent's spatial location. Other cells that may play a role in navigation are grid cells (which fire in a hexagonal grid that corresponds with the environment floor, where the grid module is correlated with the size of the animal; Hafting et al. 2005), head direction cells (which fire depending on head orientation; Ranck 1985; Taube et al. 1990) and boundary vector cells (which fire when the rodent, moving in a certain direction, gets to a specific distance from a boundary element of the environment; Barry et al. 2006).

An observation is in order. Different classes of spatial features are coded by different cell systems, as per the following table:

Cell type	Geometric feature coded	Example
Place cells	Individual locations	The entrance of the maze, the fork, etc.
Grid cells	Metric properties	Distance covered based on number and size of grid modules
Head direction cells	Angles	Azimuth of a landmark relative to body's central plane of symmetry
Boundary vector cells	Topological properties	The middle of the room, close to the wall

This suggests that the spatial representation system could work in a constraint satisfaction mode, looking for solutions that take into account whichever topological, metric, angular and location representations are available at a given moment.

Of course, an open question is if the same type of cells and the same type of cognitive map exist in humans. Although the invasive single-cell recording techniques that are used in rats have not been used in humans, there is some empirical literature to support the idea that work on rodent spatial cognition can be extrapolated to humans to a substantial degree (see Epstein et al. 2017 for a review). For instance, a much quoted fMRI study of taxi drivers in London shows that they acquire larger right posterior hippocampi as a result of the years-long training they undertake to learn the map of London streets (Maguire et al. 2000).

If the underpinnings of human spatial cognition do not differ substantially from those of rodent spatial cognition, then the brain can store different maps for different environments, and even represent different conditions of the same environment. Different place cells would then fire for different environments, following a 'global remapping' as the subject changes environments. When the same environment changes conditions, the place cells that fire stay the same, but the rate at which they fire changes (Colgin et al. 2008). It should be noted that landmarks play an important role in anchoring these cognitive maps (Yonder et al. 2011). Finally, when it comes to navigation, the grid cell network computes the direction and distance to the navigational goal (Kubie and Fenton 2012), while the hippocampus and entorhinal cortex work together to calculate the optimal path to the goal (Bryne et al. 2007).

It is beyond the scope of this paper to explore in-depth the empirical literature on spatial cognition, but in addition to the brief review above, we should refer to some of the work that is of special relevance for the study of disorientation. An example of this work is the experiment exploring what happens to the rodent's place field map and HD cells when disorientation occurs. The geometry of the environment (disregarding useful non-geometric cues) determines the realignment of the cognitive map, which determines in turn the behaviour of the animal during navigation (Keinath et al. 2017). An interesting distinction is to be made between orientation and context retrieval. Orientation refers to one's location and direction within a cognitive map of a given place, and context retrieval refers to the retrieval of the specific map that is appropriate for the place in question. In mice, context retrieval and orientation have been found to be mediated by dissociable cognitive processes, and it is not inappropriate to think that the same could be the case for humans (Julian et al. 2015).

Although there is an increasingly clear picture of human orientation coming from spatial cognition, the study of lost person behaviour is not connected to this body of research. The result is that there is no unified account of disorientation that can draw on the different fields that deal with the phenomenon. Conceptual work is needed to link spatial cognition and the psychology of lost person behaviour (see Casati and Fernandez Velasco, forthcoming). Establishing this link will advance the understanding of disorientation and will allow WiSAR to profit from the progress coming from a broad variety of disciplines.

Disorientation: Conceptual Issues

There are several impediments that keep the different fields concerned with disorientation from associating. An important one is that these fields operate at different levels of analysis. Even within neuroscience, where most explanations appertain to the neural level, there is a chasm between the single-cell study that is possible in non-human cognition and the non-invasive neuroimaging studies done in human cognition. Moreover, functional magnetic resonance imaging cannot be done while subjects travel physically through space, which restricts investigation to non-ecological, purely represented settings. Although some lost person psychology relies on first person reports (Hill 1998), for the most part it is concerned with the third-person analysis of behaviour. In the case of WiSAR, the separation from spatial cognition goes one step further, as most of WiSAR work operates at the level of statistical patterns emerging from lost person behaviour (Koester 2008).

Another important element that separates these fields is that they tend to draw from different experimental and study settings. A big part of the work in psychology takes place in non-ecological conditions (see Li, Mou and McNamara 2012 for an example). WiSAR theory is based on ecological non-experimental conditions (i.e. analysis of existing cases of lost people), and the work in cognitive science is usually simulation based (Ruddle et al. 2011). Finally, as mentioned above, neuroimaging studies are stationary, and thus based on virtual reality, spatial recollections or imagined navigation. Consequently, in order to extrapolate results from one field to another, one has not only to extrapolate from one level of analysis to another, but also from one setting to another (e.g. from simulations to the actual behaviour of people lost in the wilderness).

A symptom of this divergence among fields is a divergence among characterisations of the phenomenon of disorientation itself. Disorientation could be understood, among others: as being unable to find one's way (Dudchenko 2010); as the failure of the way-finding process (Golledge 1999); or as not knowing the directions and distances to get to a given point (Rieser 1999). These all seem to point to the same phenomenon, but it should be noted that they each characterise disorientation in a proprietary way: as an issue of ability, failure or knowledge, respectively.

Although the above characterisations each capture some of the features of disorientation, they all have certain shortcomings. A central issue here is that while being lost in an objective condition, disorientation is a strongly subjective one. Being lost can be easily characterised as being unable to find one's way, but this characterisation does not necessarily extend to disorientation. Take the following case, for instance:

> Maggie is following a path through the woods that leads to the Dorlcote Mill. The path is very hilly and sinuous, changing directions often, which makes Maggie feel disoriented. Disregarding her disorientation, she keeps going and eventually gets to the Dorlcote Mill.

In this case, Maggie is not lost (i.e. she is able to find her way), but she is disoriented; a distinction that Duchenko's characterisation fails to capture. And while Dudchenko's characterisation fails to capture some cases of disorientation, Golledge's characterisation overshoots, mistakingly identifying certain cases of being lost as disorientation. Let us look at the following example:

> Pierre is confident that he is at Place Denfert Rochereau, when in fact he just emerged at the Port Royal crossing, because he skipped one metro stop.

Here, Pierre's way-finding process failed (he failed to keep track of the stations when he was in the metro), but he doesn't feel disoriented. Golledge would characterise Pierre as being disoriented, but a more precise characterisation would identify Pierre as someone who is lost, not disoriented. Finally, below is a last case that raises some problems for Rieser's characterisation:

> John is a tourist in Paris. He is outside the Metro stop Odeon and he wants to go to walking to Place de la Concorde. He is told that if he follows Boulevard St. Germain and then crosses the Seine he will get there. Boulevard St. Germain curves north as it advances west, but this is not a problem for John. He just follows the street without knowing the exact direction of Concorde. He ends up at the Seine, crosses the river and is at Place de la Concorde.

In the case above, John doesn't know the direction or the distance of his destination, but he is neither lost nor disoriented. Overall, there are many elements related to disorientation, such as knowledge of distances and directions, the way-finding process or navigational ability. However, casting disorientation in such terms overlooks that disorientation is primarily a feeling. Characterising disorientation as a feeling captures the subjective aspects of disorientation, and it can serve to develop a conceptualisation of the phenomenon that helps us understand how elements such as a failure in the way-finding process can lead to the emergence of the feeling of disorientation.

Elsewhere, we have proposed to sharply distinguish being lost and feeling disoriented and have argued that disorientation should be better understood as the metacognitive feeling of unconfidence in the subject's online spatial representation (Casati and Fernandez Velasco, forthcoming). That is, disorientation is a feeling that evaluates the level of confidence in the cognitive subsystem responsible for spatial representation.

Bridging the gap among these different fields is not only an empirical endeavour, but a conceptual one. An important first step is a characterisation of disorientation that can be effectively applied to different fields. As for the gap between levels of analysis, it can be bridged indirectly. That is, results at one level can be used to speculate and produce hypotheses at other levels. Ultimately, one would want to have an overall theory of disorientation that draws on the progress from the variety of fields dealing with disorientation into a unified conceptualisation of the phenomenon that can serve to generate testable hypothesis for different settings and levels of analysis.

In what follows, we will look at different models that have been developed for WiSAR theory, and speculate how novel conceptualisations of disorientation can contribute to the models generating better predictions.

GIS-Based WiSAR Models

When a person becomes lost in the wilderness, the WiSAR manager faces a difficult challenge, which is to figure out where the different SAR teams are more likely to locate the lost person. The areas that need to be covered are often vast areas of difficult terrain, and the SAR team needs to coincide with the lost person not only in space, but also in time (i.e. if the SAR team and the lost person pass the same place at different times the lost person will not be found). These difficulties should not be overlooked. WiSAR operations often have limited resources, and past 2 days, or about 50 h of search, the lost person's chances of survival decrease significantly (Adams et al. 2007).

To guide the search effort, the most common strategy is to develop a probability map stemming from the Initial Planning Point (IPP), which is usually the point where the lost person was last seen or known to be (i.e. relying on cues). Geographic Information Systems (GIS) provide WiSAR the planning tools (such as the ArcGIS planning tool, see Ferguson 2008) to create and update probability maps (Doherty et al. 2014). Here, we will discuss two methods of establishing probability maps: a ring model and a Bayesian model.

The ring model, based on mathematical search theory (Koopman 1980), is the most common way of establishing probability areas. The main dependent variable of the ring model is Euclidean distance from IPP. The statistics of past WiSAR cases are analysed to produce four concentric rings stemming from IPP. These rings represent probability areas, and are established at the distances equivalent to the probability quartiles 25, 50, 75% and then at 95%. That is to say, within the first concentric ring, the probability of finding the lost person is 25%, within the second concentric ring, the probability of finding the lost person is 50% and so on.

An important finding for WiSAR is that different subject categories correspond to different probability distributions. For example, the 25% probability area of a hiker is much larger than that of a child, because the hiker can cover bigger distances and tends to follow either trails or directions rather than staying put or walking randomly. Syrotuck first analysed 242 cases from New York and Washington states and established probability distributions for eight subject categories (Syrotuck 2000). Koester has recently created ISRID, a large database unifying thousands of cases from around the world and dividing them into 41 categories based on scenario, age, medical or mental status, and activity (Koester 2008). Although other parameters (such as the type of terrain) can be added, the subject category is the main determinant of the distribution of the probability map that will guide the search efforts.

A recent development in WiSAR is Lin and Goodrich's Bayesian approach (Lin and Goodrich 2010). The main parameters of their model are the topography, vegetation coverage and local slope of the terrain. One of the innovative elements of their model is that it uses expert opinions to determine the probability of the lost subject transitioning from one terrain type to another (e.g. a WiSAR manager's estimation of the likelihood of a given lost subject moving from a plain to a hill). This opinion-based probability is expressed in the form of mean and variance (to account for uncertainty). These probabilities are used as priors to create a state transition matrix (specifying the probability of the subject transitioning from each state to all other states) from which to generate a predictive probability distribution map. As a person can only travel to adjacent cells, all of the cells that are not adjacent to the current cell have probability zero. Of course, as the person moves, new cells will become adjacent. Assuming a process in which only present states affect the probability of future states (i.e. assuming a first-order Markov process), it is then possible to predict the lost person's trajectory as time progresses (i.e. the continuous posterior beliefs of the lost person's position at different times).

While in the ring model, the terrain is divided in concentrical rings, in Lin and Goodrich's model the whole terrain gets divided into tessellated hexagons. In their hypothetical scenario, these hexagons are 24 m wide. Each state in the transition matrix is a hexagonal cell in the tessellation. Each row in the matrix is composed of the transitional probabilities of transitioning from one cell to each cell in the tessellation (and of staying in the same cell). Given a point at which the person was last seen (in which the probability of the person being in that cell at the time when she was last seen is (1), it is possible to compute the probability of the person being in different cells at different points in time, thus generating a probability distribution map to guide WiSAR efforts.

One advantage of Bayesian models is that existing observations (e.g. GPS track logs and their associated terrain) can be used to update prior beliefs and thus reduce experts' uncertainty. It is impossible to determine the continuous posterior beliefs about the lost person's position in closed form, so Lin and Goodrich employ a Markov Chain Monte Carlo (MCMC) algorithm to approximate the posterior distribution needed to generate the updated probability distribution map (following the methods of Gelman et al. 2013). The MCMC method consists of a massive iteration of different transformations from prior distributions (based on expert opinion) to observations (based on GPS track logs) to generate a series of distributions (21 in their hypothetical scenario) that approximate the posterior distributions (see Lin and Goodrich 2010 for mathematical detail). These posterior distributions are used in the place of the original prior distributions to update the probability distribution map.

To conclude this section, the table below sums up some of the distinctions between Koester's ring model and Lin and Goodrich's Bayesian model:

Model	Sources	Updating	Terrain subdivision
Ring model	Actual cases (ISRID)	Based on time	Concentric circles
Bayesian model	Expert opinion	Based on both time and observations	Hexagonal tessellation

Conceptualising Disorientation for GIS WiSAR

The traditional ring model of WiSAR seems to be 'blind' to different conceptualisations of disorientation. What we mean by this is that what is used to generate the probability map is the statistics of different subject categories, which are unaffected by different models of lost person behaviour. For example, what determines the distribution of the probability map is not whether a subject in a given category is perceived to be more likely to give preference to a given reorientation strategy, but simply how far from the IPP subjects in the same category were found in previous WiSAR cases. However, this doesn't mean that conceptualising disorientation plays no role in WiSAR operations that use the ring model. After all, WiSAR teams use their beliefs about lost person behaviour (based on a certain conceptualisation of disorientation) to prioritise the search within the sub-areas of a given probability ring.

One of the ways in which conceptual work can help improve the ring model itself is by determining which subject categories should be used for generating the probability map. From a cohesive theory of disorientation, one can hypothesise the existence of new categories, and then use GIS to see if these categories are statistically relevant. Furthermore, a good conceptualisation of disorientation is expected to generate predictions as to what parameters (e.g. age, terrain…) might be statistically relevant for determining new subject categories.

Progress in the conceptualisation of disorientation is much more promising when it comes to Lin and Goodrich's Bayesian model. In their example, expert opinion is used to generate a prior distribution regarding the transitions between three dimensions of terrain (vegetation, slope and topography). What is promising is that their method offers a way to incorporate different sources for the generation of a prior distribution. One such source could be a unified theory of disorientation or the opinion of an expert informed by such a theory.

An adequate conceptualisation of disorientation, together with Lin and Goodrich's Bayesian model offers a way in which developments coming from a broad range of fields can be exploited for improving GIS-informed WiSAR operations. With a working theory of disorientation, knowledge about the phenomenon could then be used for establishing the prior distribution that determines the initial probability map. What is more, such distribution could be updated with GIS observation, by using the observations such as GPS logs to generate a posterior distribution. Such posterior distribution could then be used to discover elements of disorientation that have been overlooked by the overall theory, thus highlighting ways in which the theory can be improved.

Lastly, conceptual work can help us improve the models themselves and choose between different models for different scenarios. For instance, Lin and Goodrich's Bayesian model assumes that the lost person's travelling resembles a first-order Markov process, in which only present states affect the probability of future states. This might not be the case for route sampling, a strategy that involves the lost subject starting on an intersection and following different routes for a while, always returning to the intersection, as a way to explore the surrounding terrain. In this case, there is a certain length that the subject is likely to follow (say 1 km), so that the likelihood of the subject turning around is less likely after she has walked 10 m than after she has walked 10 km on the same path. The travelling of such an individual would not resemble a first-order Markov process. It might well be that Lin and Goodrich's model works better for children than for hikers (as hikers are more prone to route sampling), and a good theory of disorientation would give us the conceptual apparatus to tell if this is the case.

Conclusions

We opened this paper by reviewing the progress in different fields related to disorientation, such as lost person psychology and the neuroscience of spatial cognition. We then outlined several challenges to unite these separate fields (i.e. differences in levels of analysis and differences in settings) to show that conceptual work was needed to connect these fields. We defended the idea that an adequate conceptualisation of disorientation would permit us to use the findings in one field to produce hypothesis in a different field. Afterwards, we reviewed two different methods of GIS-informed WiSAR (a ring model and a Bayesian model) to then analyse how an adequate conceptualisation of disorientation could help WiSAR profit from the developments in a broad range of research areas. We concluded that while for the ring model conceptual work can only help develop new subject categories, for the Bayesian theory conceptual work can be used to generate more accurate probability maps. Furthermore, these Bayesian-based probability maps can in turn be used as input for improving our theories of disorientation. Finally, the main claim of our contribution is that conceptual work can help us improve the various WiSAR models (by examining their underlying assumptions) and choose between different models.

References

Adams AL, Schmidt TA, Newgard CD, Federiuk CS, Christie M, Scorvo S, DeFreest M (2007) Search is a time-critical event: when search and rescue missions may become futile. Wilderness Environ Med 18(2):95–101

Barry C, Lever C, Hayman R, Hartley T, Burton S, O'Keefe J, Burgess N (2006) The boundary vector cell model of place cell firing and spatial memory. Rev Neurosci 17(1–2):71–98
Blodgett HC (1929) The effect of the introduction of reward upon the maze performance of rats. Univ Calif Publ Psychol 4:113–134
Byrne P, Becker S, Burgess N (2007) Remembering the past and imagining the future: a neural model of spatial memory and imagery. Psychol Rev 114(2):340
Casati R, Fernández Velasco P (forthcoming) Subjective disorientation as a metacognitive feeling
Colgin LL, Moser EI, Moser MB (2008) Understanding memory through hippocampal remapping. Trends Neurosci 31(9):469–477
Doherty PJ, Guo Q, Doke J, Ferguson D (2014) An analysis of probability of area techniques for missing persons in Yosemite National Park. Appl Geogr 47:99–110
Dudchenko PA (2010) Why people get lost: the psychology and neuroscience of spatial cognition. Oxford University Press, USA
Epstein RA, Patai EZ, Julian JB, Spiers HJ (2017) The cognitive map in humans: spatial navigation and beyond. Nat Neurosci 20(11):1504
Ferguson D (2008, February). GIS for wilderness search and rescue. In: Proceedings of ESRI federal user conference, Washington DC
Gelman A, Stern HS, Carlin JB, Dunson DB, Vehtari A, Rubin DB (2013) Bayesian data analysis. Chapman and Hall/CRC, Boca Raton (FL)
Golledge R (ed) (1999) Wayfinding behavior. Cognitive mapping and other spatial processes. The John Hopkins University Press, Baltimore
Hafting T, Fyhn M, Molden S, Moser MB, Moser EI (2005) Microstructure of a spatial map in the entorhinal cortex. Nature 436(7052):801
Hill KA (1998) The psychology of lost. In: Hill KA (ed) Lost person behavior. National SAR Secretariat, Ottawa
Julian JB, Keinath AT, Muzzio IA, Epstein RA (2015) Place recognition and heading retrieval are mediated by dissociable cognitive systems in mice. In: Proceedings of the National Academy of Sciences, USA, vol 112(20)
Keinath AT, Julian JB, Epstein RA, Muzzio IA (2017) Environmental geometry aligns the hippocampal map during spatial reorientation. Curr Biol 27(3):309–317
Knierim JJ, Hamilton DA (2011) Framing spatial cognition: neural representations of proximal and distal frames of reference and their roles in navigation. Physiol Rev 91(4):1245–1279
Koester, R. J. (2008). Lost person behavior: a search and rescue guide on where to look—for land, air and water. dbs Productions LLC, Charlottesville (VA)
Koopman BO (1980) Search and screening: general principles with historical applications, vol 7. Pergamon Press, New York
Kubie JL, Fenton AA (2012) Linear look-ahead in conjunctive cells: an entorhinal mechanism for vector-based navigation. Front Neural Circuits 6:20
Lin L, Goodrich MA (2010) A Bayesian approach to modeling lost person behaviors based on terrain features in wilderness search and rescue. Comput Math Organ Theory 16(3):300–323
O'Keefe J, Dostrovsky J (1971) The hippocampus as a spatial map: preliminary evidence from unit activity in the freely-moving rat. Brain Res 34:171–175
Ranck JB (1985) Head direction cells in the deep cell layer of dorsolateral pre-subiculum in freely moving rats. In: Buzáski G, Vanderwolf CH (eds) Electrical activity of the archicortex. Akadémiai Kiadó, Budapest, Rao
Rieser JJ (1999) Dynamic spatial orientation and the coupling of representation and action. In: Golledge R (ed) Wayfinding behavior. Cognitive mapping and other spatial processes. The John Hopkins University Press, Baltimore, pp 168–190
Ruddle RA, Volkova E, Bulthoff HH (2011) Walking improves your cognitive map in environments that are large-scale and large in extent. ACM Trans Comput Hum Interact 18(2). ISSN:1073–0516
Sholl MJ (1987) Cognitive maps as orienting schemata. J Exp Psychol Learn Mem Cogn 13(4):615
Syrotuck WG (2000) Analysis of lost person behavior: An aid to search planning. Barkleigh Productions, Mechanicsburg (PA)

Taube JS, Muller RU, Ranck JB (1990) Head-direction cells recorded from the postsubiculum in freely moving rats. I. Description and quantitative analysis. J Neurosci 10(2):420–435
Tolman EC (1938) The determiners of behavior at a choice point. Psychol Rev 45(1):1
Tolman EC (1948) Cognitive maps in rats and men. Psychol Rev 55(4):189
Tolstoy L (2015) Master and man. Wildside Press
Yoder RM, Clark BJ, Taube JS (2011) Origins of landmark encoding in the brain. Trends Neurosci 34(11):561–571

Chapter 12
From Education to Job Opportunities. Defining Professional Profiles for Geographers with High Competences in GIS Environment

Cristiano Pesaresi

Abstract In this paper, the importance is underlined of promoting higher education and training courses and opportunities, focused on a close connection between GIS and geography to cover topics of remarkable and daily social interest. In the first part of the contribution, some educational key points are defined in order to fuel a virtuous circle which can provide considerable methodological and applied skills in the use of GIS applications. It is considered fundamental to broaden one's mind, to provide practice to the theories, to give enthusiasm and security in one's own skills, to create a community of Geographers able to provide important added value in different research fields and workplaces. Therefore, it seems essential to build a *trait d'union* between the university and the workplace. Particular attention is attributed to well-structured university modules founded on GIS applications, to intensive courses for postgraduates and to the geocartographic laboratories. In the second part of the paper, the focus will be on some examples of professional profiles for Geographers in order to define specific professional categories. In particular, ten application fields for Geographers with high competences and abilities in GIS environment are identified and explained with relative examples. In all these cases, a rigorous approach founded on geographical contents and a wide, updated and concrete use of GIS functionalities and extensions is required.

Keywords Education · Geographers · GIS · Intensive courses · Job opportunities · Laboratories · Professional profiles · Skills

Introduction

Higher education, training courses and opportunities for a professionalising didactics, which is able to meet the demands of the working world and which gives back tangible outcomes at the level of enthusiasm and personal satisfaction, are becoming more

C. Pesaresi (✉)
Department of Letters and Modern Cultures,
Sapienza University of Rome, Rome, Italy
e-mail: cristiano.pesaresi@uniroma1.it

© Springer Nature Switzerland AG 2019

T. Tambassi (ed.), *The Philosophy of GIS*, Springer Geography,
https://doi.org/10.1007/978-3-030-16829-2_12

and more pivotal nowadays. This is important both for the professors and experts, who organise the courses and the students and people who attend them. Particularly, teachers and experts can conduct research in didactics and have the opportunity to recognise prominent figures and potential talents, while the students and people who attend the courses must feel enriched and gratified to have gained a baggage of knowledge and abilities, which makes them more self-confident and able to work concretely using applied tools.

At the same time, the importance of a didactics is evident which might break down the barriers of mere presentations, aimed only at remembering how much it is possible to do, that is to say the barriers of too theoretical and contemplative teaching, static from an operative point of view. In fact, first of all, teachers and experts must make a big effort since they must feel their role as a mission, which must be carried out with operating modalities, by updating and spending time (often a lot of time) organising a very useful educational offer.

Moreover, there is a clear need for courses and opportunities which are not simple containers of uncritical technical skills, to be replicated in different situations without the appropriate knowledge of the facts, prior selective reasoning and without that creativeness which derives from a mix of geographical, cartographical and geotechnological contents.

The need is strongly felt to experiment with ad hoc didactic modalities, which make it possible to obtain a set of highly qualifying knowledge and abilities—spendable in the world of work and highly researched since able to provide important added value in terms of innovativeness and effects of social usefulness.

In this perspective, the Geographical Information Systems create a huge range of opportunities which can gain wide visibility, consent and interest at interdisciplinary level, in the field of research, at the institutional level (public and private). In fact, 'GIS has become a [relevant] source of employment in the geography [and not only] profession' (Artvinli 2010: 1282).

But there is an important issue which must be considered, because around GIS and related educational and professional courses 'there is plenty of interest, but there are very few models of how such a course might be constructed and taught' (Goodchild 2008: 67) and it makes necessary a wide debate and a full awareness of these problems (Kemp et al. 1992; Chen 1998).

Some Educational Key Points Through Job Opportunities

As a first basic key point, the importance of well-structured university modules founded on GIS applications must be underlined. In a Geography degree course, the acquisition of special skills for working with rigour and accuracy through GIS and geotechnologies must be considered essential. Starting from a brief history of GIS and the debate regarding GIS as tool or GIS as science, continuing with the introduction of the different data which can be used (and collected in a geodatabase) and with the main reference systems, showing several examples of important functionalities, application fields, innovations and interoperability, a hypothetical GIS course should then provide detailed methodological and operative basis and principles in order to correctly: elaborate different kind of digital maps (with quantitative and qualitative data); georeference historical cartographies and satellite images; edit and calculate geometries and define buffer zones; work with DEM; share results and maps through web applications, etc. The theoretical part must be preparatory for the successively applied part, in order to create a close connection and an ordered sequential path. The applied part, which undoubtedly represents the most enthusiastic phase for students but which cannot be conducted without the first one, makes it possible to work according to a team-based and problem-based learning, in the constructivist perspective supported by a favourable working environment which also makes it possible to empower and to enrich the individual attitude, levelling skills and bringing out potential talents. A focus is then required on the web applications and the highly communicative (when opportunely thought out and elaborated) Story Maps which seem to have a considerable involvement capacity (Kerski 2013; Marta and Osso 2015; Zamperlin and Azzari 2017), also because Internet has acquired a fundamental global function and has unhinged barriers and distances. So, '[Web]GIS have also become media for constructive dialogs and interactions about social issues', giving multiple inputs, a wide field of digital cartographic elaborations, reflections, findings and creating an online mapping community who can cover topics of considerable public interest (Sui and Goodchild 2011: 1738).

For a well-structured university module founded on GIS applications, which makes it possible to connect geographical contents and knowledge and geotechnological elaborations, for meticulous analysis, hypothesis formulation, relationship and cause-effect studies, teachers can consider two possibilities in terms of software equipment. They can propose the use of open source software (as for example QGIS, https://www.qgis.org/it/site/), or tightening conventions and collaborations with ESRI they can propose the use of the ArcGIS suite (Pesaresi 2017), and according to the chosen solution they must also adopt reference books that explain the steps to follow in detail during the guided examples and exercises. In some cases and particular conditions (classroom with few students and/or with very careful and receptive students), teachers can suggest the combined use of open source software and ArcGIS, also using specific books (Cetraro 2015), contributing to forming very adaptive and competent figures.

For the exam, in addition to an oral or written test, the delivery of a project with different elaborations can be recommended. The project presented by the students must be characterised by a well-organised issue around a specific geographical theme or problem, focused on a study area, and from a practical point of view must show the acquisition of the required competencies and abilities. Sometimes, it can be shared by two students in group work, but with the specification of the different parts and the personal contribution to the reaching of the common purpose. Any project-oriented activity is very delicate but productive if students are empowered and motivated. 'Project-oriented instruction [often] does not begin with a clearly defined list of topics, but instead the list evolves as the project progresses. Instruction under these conditions can be very demanding for the student and for the instructor, and requires a clear recognition of the range of theoretical and conceptual knowledge that students should acquire during the project and some insight into the sequence in which these principles are best acquired' (Kemp et al. 1992: 187).

As far as concerns a second key point, a relevant role is covered by intensive courses for postgraduates. In a previous contribution, the main steps of a possible scheme for intensive courses have been provided (Pesaresi and Pavia 2017: 112–114), because similar occasions can be considered crucial moments for professional training. The proposed scheme represents an example of general reference guidelines, useful and adaptable to different situations, planned in five consecutive steps, which together make it possible to provide, in a short time and with effective modes, competences to work critically on a geographical theme using targeted functionalities. In short, the five steps—tested during various activities conducted in the GeoCartographic Laboratory of the Sapienza University of Rome, generally in collaboration with the Italian Association of Geography Teachers—are: review of the aims and methods; brief introduction to gradually enter in the application–laboratory part; demonstrative–explanatory phase, which constitutes the core of the course; phase of the guided exercise, i.e. the moment of maximum personal and group production (it is also the phase of verification); open discussion and confrontation. Furthermore, during these initiatives—aimed at georeferencing and territorial screening, to the remote sensing and GIS applications for the study of temporal transformations, to the spatial analysis for supporting emergencies—the importance of the binomial geography-GIS has been underlined, and sometimes the attention has been focused on the realisation of 3D and 4D models since a further dowel, widely requested in the research, planning and institutional fields, is represented by the capacity of operating in these ways.

Students and people who attend the courses have 'to surpass technical issues and to appreciate the conceptual and functional linkages between GIS and geography's intellectual core' (Sui 1995: 578). It is important to find and promote a delicate balance and a harmonic leitmotiv. For this reason, it is strategic to '*create richer, more fulfilling educational experiences that profile the power of geography's spatial [and temporal] perspective*' and to '*create opportunities for geographers and GIScientists to consider new and innovative ways in which maps are being used*' (Sinton 2009: S7). Instead, GIS modules tend often to underestimate or ignore the crucial position of the geographical problems or questions which must be faced, 'and just cover the technical aspects of how to use GIS software' (Bearman et al. 2016: 394).

Moreover, as third key point, the geocartographic laboratories can have a strategic role, because they make it possible to carry out many activities offered in university courses, as for example activities with information technology address, traineeships and internships, postgraduate intensive courses, etc. The geocartographic laboratories represent the places of excellence where to conduct activities of applied research related to the funding of projects recognised as having a high level of technological innovation and which requires the guided use of GIS and geotechnology. In this way, it is possible to concur to the formation of a kind of Geographers–GIScientists, able to use GIS and geotechnologies to bring concrete added value to the research activities and for which these tools become the vehicles whereby to translate the disciplinary contents into practice and to open the mind in the direction of the modelling of a rigorous spatial and temporal analysis and 3D and 4D representations.

'One important component of GIS education should be a training strategy aimed at exposing students to academic research, by involving them in relevant research projects' (Bertazzon 2013: 72).

Through similar experiences in the geocartographic laboratories, young recently graduated people have the possibility to finalise their technical skills to a specific project, learning and refining methodological aspects, getting a taste of applied research, collaborating in groups (in horizontal and vertical senses), in a virtuous process of giving (freshness, ability to find tutorials, georeferenced data, blogs, using various functionalities) and receiving.

The geocartographic laboratories can be also the places where specific extensions and modelling software application can be used in a purposeful and stimulating environment where it is possible to experiment and share ideas, project proposals, refined elaborations, staying updated and joining technical skills, rigorous methodologies and innovative thinking. For example, for structures using ESRI products, periodicals demonstrative courses, able to open new perspectives and to promote profitable discussions, could be organised in the geocartographic laboratories, in collaboration with ESRI (and other partners) expertise, around the use and the last improvements of ArcGIS Pro, ArcGIS Online, Drone2Map, CityEngine, ArcGIS Indoors, ArcGIS for AutoCAD, and mobile devices. It also permits a panel of particularly worthy post-graduate and Ph.D. students to absorb a set of information, knowledge and abilities useful in different fields of the working world, where Geographers can play a crucial role. It also makes it possible to create a dynamic and profitable system between the university and IT companies, characterised by a direct link between the research and advanced tertiary sector and very important also in securing synergies useful for the proposal of important projects which provide application effects on society.

In so doing, it is possible to start 'creating a GIS professional community of practice' founded both on geographical content and way of thinking, and high GIS competences and abilities, in order to develop a network of Geographers-GIScientists and to build a 'bridge from higher education to the workplace' (Tate and Jarvis 2017: 327). After all, Geographers and 'GIS job applicants need to possess high levels of GIS analysis and modeling skills […] and personal and social skills […] to [hope to] obtain employment' (Hong 2016: 147) and to these they have to add a rigorous methodology and apposite contents.

Examples of Professional Profiles for Geographers

A high-level use of the Geographical Information Systems, for their notable operational usefulness and their capacity to put researchers and experts of different scientific disciplinary sectors in fruitful contact, making multiple competences interact and enriching them, are an important business card for Geographers. After all, 'GIS is at the heart of all modern spatial decision making' (Wiegand 2001: 68). Thus, if

Geographers also possessed some basic programming competences (for example for Python), following appropriate courses or creating a network for exchanging abilities with colleagues of other sectors, they could overcome some obstacles that sometimes require the use of programming languages.

Preparing 'today's students for tomorrow's workforce' is a crucial challenge for geography researchers and educators (Solem et al. 2008: 357). But merely being experts in geotechnologies is not sufficient; Geographers must be able to use and to mix specific GIS skills, disciplinary contents, tools, experience, field work and direct surveys, data collection, web applications to make a truly positive difference in the working world and at the service of society (Longley et al. 2011: 501; Favier and van der Schee 2009). In this way, Geographers can aspire to interesting job positions (Ronza 2012).

All this with the geography always in the centre, and with GIS and geotechnologies to enlighten aspects and relations, phenomena and events, transformations and possible evolutions, providing an enormous chance for digital representations and models, where it is possible to converge algorithms, statistical functions, quantitative and qualitative data, historical cartographies and satellite images, information obtained by questionnaires and participatory investigations online, textual information, photos taken on the field, for numerical and cognitive analysis and the successive sharing and diffusion of the results through web applications.

In fact, it has been affirmed that: 'Conventional GIS development has predominantly concentrated on cartographic, quantitative, computing and spatial database oriented with the goal for efficiency, which tends to be closely associated with the left-side/slow thinking capabilities of the human brain. In contrast, the emerging GIS themes focus more on narrative, qualitative, storytelling and synthesis oriented with the goal of equity, which tend to be more closely associated with the right-side/fast thinking capabilities of the human brain' (Sui 2015: 11).

As far as concerns the definition of special professional profiles, in addition to the academic profiles and the figures required in the cartographic and GIS offices or in the social–demographic offices of public institutions, ten application fields for Geographers with high competences and abilities in GIS environment can be defined.

1. *Geographers who use GIS in the civil protection field* (seismic, volcanic, hydrogeological, fire, etc.): to elaborate and analyse hazard, vulnerability, risk and risk perception maps, which can be updated contemporarily with the evolution of the geophysics and anthropic components (also considering geological–meteorological and cultural–historical aspects, and infrastructure and road network) and with the possible verification of particular phenomena; to provide added value both in the pre and post event (considering a wide series of appropriate data coming from different sources, thanks to Geographers' capacity for synthesis), contributing to elaborate scenarios of expected and real damage and to hypothesise the possible evolution of the phenomena and the most critical areas for the concurrency of specific factors; to build homogenous and dynamic systems for the data comparison, for the monitoring of specific parameters and for the updating of the exposition values; to facilitate first aid operations, the

recognition of priority intervention orders, emergency management; to promote risk education and awareness.

2. *Geographers who use GIS in the geo-sanitary field*: to identify and geolocalize possible risk factors and polluting sources present on the territory; to assess causal links between risk factors or polluting sources and certain diseases; to evaluate the health offer and the possibility to meet the needs of the population also according to the age structure and the presence of foreign immigrants (who have different problems); to estimate the sanitary mobility and understand the reasons for eventual large movements; to analyse the temporal dynamics and the distribution of threatening phenomena or the main causes of death; to recognise the causes of inadequate first aid requests, which increase a number of related critical issues; to identify areas without certain services and envisage the optimal location of hospitals, Local Health Units and other healthcare services according to specific and multiple factors concerning residents and infrastructure and road network; to identify the minimum or optimal routes to reach hospitals and first aid centres in case of necessity.
3. *Geographers who use GIS in the touristic field and in the perspective of the exploitation of local resources*: to evaluate the actual offer and recognise critical situations for marked inadequacy (first of all the absence of hotels and complementary facilities which fuels the phenomenon of hit and run); to propose hypotheses for recovery and restructuring of structures according to an organic view, a system perspective and the local culture; to identify the optimal locations for specific services, sustainable attractions and light but attractive accommodation in harmony with nature; to recognise the territorial features and tourist vocations of the different areas; to define naturalistic and cultural–historical itineraries or mixed valence circuits; to conceive animations which through the flight of birds makes it possible to evocatively travel virtually in the most appealing places; to conduct georeferenced questionnaires which make it possible to collect many data useful for conceiving a shared and participatory offer or for collecting tourists' opinions and suggestions; to promote and spread a renewed image of places.
4. *Geographers who use GIS in the agricultural field*: to assess the number, size and type of farms which are present in the different territorial context; to consider parameters related to contour, hill shade, slope, aspect of the slopes; to identify the areas actually distinguished by specific crops and the ones with certain unvalued vocations; to recognise homogeneous plots, which have analogous characteristics and could be profitably destined to specific crops; to plan the sowing and harvesting activities according to many factors which can influence the final yield; to optimise the routes of the machines and to reduce costs and time; to identify zones subject to lower yield due to infestations, soil degradation or nutrient deficiency, or problems of water supply; to join the technological skills with the direct knowledge of older people who have worked on the land for many years; to identify new sales and commercial areas and to spread the image of more dynamic farms, ready to respond to market demands.

5. *Geographers who use GIS in the urban field and planning*: to represent all the land use categories or list specific classes; to rebuild the different phases of growth and the main development directions over time; to recognise the phases which have strongly compromised some balances, highly increasing space congestion; to make eventual development proposals based on the actual needs of the population and on the real estate market requirements (considering the age of the population and the position in the profession, the recent buy–sell transactions, the most required type of housing, the services that increase attractiveness, etc.); to plan the presence of equipped park and green areas, which can give huge benefits to psychophysical well-being and increase the soil value; to understand if there is the need for services which are undersized or inadequate; to evaluate the compatibility and the impact of the construction of certain activities in specific zones and to find alternative areas; to contribute to the planning of residential and commercial areas which respond to the criteria of functionality, harmony, sustainability and aesthetics.
6. *Geographers who use GIS in the field of crimes*: to identify areas which are particularly exposed to violent episodes and different types of crimes; to recognise positions which could represent privileged points for ambushes in case of particular events; to underline the presence of concentrations of structures that are easy targets, of strong appeal for crimes or and which possess particular economic value; to support the prediction of where certain crimes may occur next time on the basis of spatial and temporal hotspots for targeting and pattern distribution; to suggest the reinforcement of controls and surveillance in specific zones and positions, after the detection of ongoing trend, delicate contexts and fragile situations.
7. *Geographers who use GIS in the geomarketing field*: to support the reasoned opening of strategic sales points and to improve the performance of already active structures; to understand whether or not the different factors present on a territory can be attractive or not for specific economic activities, in order to provide the input for corrections and enhancements and to reduce reckless actions; to extend the potential commercial and trade areas, increasing the earnings and focusing on the real catchment area; to study in detail behaviour and purchasing dynamics, for single sales structures and territorial aggregates, reducing possible threats, increasing the profits, establishing collaborative relationships and raising the competitiveness of a supply chain system.
8. *Geographers who use GIS in the utilities field (water, electricity, gas and telecommunications networks)*: to ensure a continuous and effective monitoring of networks and to visualise weak elements for the efficient functioning before a temporary fault occurs; to geolocate and display failures and problems which need fast interventions or drastic upgrading solutions; to recognise areas that are less well-served and that require a performance boost with new nodes and network devices; to allow possible savings, maintaining the quality of services high, keeping supplies suitable and optimising procedures; to avoid impact phenomena and hazardous installations that do not comply with specific safety parameters.

9. *Geographers who use GIS in the re-demarcation and functional recovery of marginal lands*: for example, to recognise and border lands whose memory has been lost due to intensive migratory phenomena occurred in time and of which the current owners do not have direct knowledge; consequently to permit the concrete awareness and acquisition or the sale of abandoned lands which are sometimes reabsorbed by forest; to promote a profitable enhancement of forgotten lands which are subject to widespread processes of neglect and that in the past were hard worked for daily sustenance; and then to support shared and participatory planning actions aimed to socially and economically elevate marginal areas with original solutions.
10. *Geographers who use GIS with regard to big and open source data*: to select, analyse and validate georeferenced data and images which can be effectively used both in the research, education and planning activities and in the simulation of events, elaboration of scenarios, reconstruction of landscapes and evolutionary phases, etc.; to enhance the huge amount of available information, knowing how specifically and rigorously to search in the boundless sea of the web and in the different geoportals; to support the dynamic processes of technological innovation, socio-economic development, cultural progress, proactive sharing; to make up for more and more imminent and widespread needs which require the ability to connect, represent and quickly analyse multiple data in synthetic and relational forms.

Summary Considerations and Some Other Proposals

The ten application fields with high competences and abilities in the GIS environment here defined provide concrete examples of the possibilities in which Geographers can aspire to work putting into practice what they acquired during their university and post-university training. The ten application fields here shown tend to define specialist profiles but Geographers are characterised by a marked 'ductility', interdisciplinary approach and capacity to transfer abilities and *forma mentis* from one field to another on the basis of needs which represents a great added value. In all these cases, the Geographer can take advantage of the different views of GIS which can be synthetically translated into the creation of wide geodatabases with data coming from multiple sources, in the elaboration of digital maps and scenes in 2D, 3D and 4D, in the conduct of spatial and temporal analysis and in the production of simulations and modelling.

To fully carry out an assignment which comes into one of the different fields here underlined, the Geographer must be able to count both on a consolidated knowledge of the contents and a wide competence in GIS environments and applications. In addition to the customary functionalities and GIS tools, for example regarding 'ArcGIS Desktop', the Geographers will have to know specific products, which can permit them to propose alternative and original solutions, to overcome difficulties, to pack elaborations with great aesthetic effect and equally remarkable accuracy in the research and methodological application. The combined use of products, as for example ArcGIS Pro, ArcGIS Online, ArcPad and ArcGIS Mobile, Drone2Map, CityEngine, ArcGIS Indoors and ArcGIS for AutoCAD, can represent a solid foundation for building projects and operating according to the criteria of a modern geography which merges the theoretical contents with dedicated software solutions, giving concreteness and new life to the knowledge while always respecting the rigour of the approach, the importance of field survey and the passages from synthesis to details and from a small to large scale.

ArcGIS Pro offers very useful and impressive three-dimensional representations of the components both for physical and anthropic elements, facilitates the rapid and efficient comparison of 2D and 3D features and elaborations, in order to obtain additional information, and permits the creation of highly communicative 4D animation. ArcGIS Online gives the possibility to use many services, data and base maps that considerably enrich the elaboration possibilities also in the direction of the web applications, through which to disseminate and share contents and results, and quickly reach an enormous number of users who can consult the elaboration with a browser, desktop or mobile device. ArcPad and ArcGIS Mobile make it possible to record and geolocalize survey data since they are configured as mobile data collection, and field mapping programmes, useful also to share information in a dynamic way, to synchronise measurements and updates, to create an efficient horizontal and vertical network of operators. Drone2Map—by quickly capturing imagery—makes it possible to compose orthomosaics and to directly import data acquired by Unmanned Aerial Vehicles in the GIS platform for scenic visualisation and meticulous analysis, since imagery offers a considerable set of geographical and geometric information and measurements which are very useful for context interpretations and modelling. CityEngine offers—through realistic three-dimensional settings—multiple possibilities for planning new urban spaces and recovering degraded areas while respecting the criteria of sustainability and aesthetic harmony, using modular structures and components, varying and simulating different elements of decoration and equipment. ArcGIS Indoors provides a specific set of tools for planning and profitably organising the indoor spaces, through interactive detailed mapping and various elaboration solutions for increasing competitiveness, safety, aesthetics, level of information, etc. ArcGIS for AutoCAD makes it possible to combine the potentialities of the GIS and CAD platforms, enriching projects, enhancing drawings with satellite images and maps, integrating data and models of different sources to produce solid and dialoguing geotechnological infrastructures and to propose a rigorous and meticulous urban planning.

Another privileged educational and training opportunity, when well-structured, is represented by a Master in Geographical Information Systems and Science, which can profitably help: to acquire refined GIS and remote sensing knowledge and abilities which can be used in many fields; to obtain recognised certifications in special modules and software extensions; to develop critical theory and methodology; to promote dynamic and animated discussions on cross-cutting themes; to apply practical skills in specific projects (for example for the final thesis); to have an adequate training in some agencies and companies which make daily use of geotechnology application; to test an independent or coordinated group research under the supervision of experts.

In fact, Masters in Geographical Information Systems and Science have to provide adequate occasions to develop many competencies in a wide range of transferable and required skills, creating a strong motivation for career future (Solem et al. 2013: 92).

Moreover, important opportunities are to be found in the possibility of international mobility for students with high requirements, to universities and laboratories which—on the basis of collaboration agreements and cultural exchange—can offer assistance for the development and improvement of GIS skills. In this way, a stay abroad can become a privileged and enthusiastic chance enabling students to find themselves among the most deserving categories and which permits them to acquire baggage of highly requested knowledge with which to make their way in the world of work.

The fast development of geotechnologies, the importance of steady methodological approaches and the need for continuous renovation processes, in different applied fields and professions, make a profitable cooperation among educational and research institutions at international level necessary and it is particularly perceived for Geographers that wish to build their future employment also around the use of GIS applications (Lisec and Ruiz Fernández 2008: 185).

References

Artvinli E (2010) The contribution of geographic information systems (GIS) to geography education and secondary school students' attitudes related to GIS. Educ Sci Theory Prac 10(3):1277–1292

Bearman N, Jones N, André I, Cachinho HA, DeMers M (2016) The future role of GIS education in creating critical spatial thinkers. J Geogr High Educ 40(3):394–408

Bertazzon S (2013) Rethinking GIS teaching to bridge the gap between technical skills and geographic knowledge. J Res Didact Geogr (J-READING) 1(2):67–72

Cetraro F (2015) GIS per la cartografia e l'analisi territoriale. EPC, Rome

Chen XM (1998) Integrating GIS education with training: a project-oriented approach. J Geogr 97(6):261–268

Favier T, van der Schee J (2009) Learning geography by combining fieldwork with GIS. Int Res Geogr Environ Educ 18(4):261–274

Goodchild MF (2008) The fourth R? Rethinking GIS education. In ESRI (ed), GIS best practices. Essays on geography and GIS. Redlands

Hong JE (2016) Identifying skill requirements for GIS positions: a content analysis of job advertisements. J Geogr 115(4):147–158

Kemp KK, Goodchild MF, Dodson RF (1992) Teaching GIS in geography. Prof Geogr 44(2):181–191

Kerski JJ (2013) Understanding our changing world through webmapping based investigations. J Res Didact Geogr (J-READING) 2(2):11–26

Lisec A, Ruiz Fernández LÁ (2008) European project on higher education in the fields related to geomatics as support for mobility of students and teachers. In: The international archives of the photogrammetry, remote sensing and spatial information sciences, vol XXXVII, part B6a. Beijing

Longley PA, Goodchild MF, Maguire DJ, Rhind DW (2011) Geographic information systems & science. Wiley, Hoboken
Marta M, Osso P (2015) Story Maps at school: teaching and learning stories with maps. J Res Didact Geogr (J-READING) 2(4):61–68
Pesaresi C (2017) Applicazioni GIS. Principi metodologici e linee di ricerca. Esercitazioni ed esemplificazioni guida. UTET—De Agostini, Novara
Pesaresi C, Pavia D (2017) Progettualità mirate e corsi GIS: per un approccio geografico coinvolgente e professionalizzante. In: Pasquinelli d'Allegra D, Pavia D, Pesaresi C (eds) Geografia per l'inclusione. Partecipazione attiva contro le disuguaglianze. Franco Angeli, Milan
Ronza M (2012) Professione geografo: profili specialistici per committenza pubblico-privata. In: Morri R (ed), Insegnare il mare. Paesaggi costieri e vocazioni maritime. Proceedings of the 54° Convegno Nazionale AIIG, pp 97–105. Carocci, Rome
Sinton DS (2009) Roles for GIS within higher education. J Geogr High Educ 33(S1):S7–S16
Solem M, Cheung I, Schlemper MB (2008) Skills in professional geography: an assessment of workforce needs and expectations. Prof Geogr 60(3):356–373
Solem M, Kollasch A, Lee J (2013) Career goals, pathways and competencies of geography graduate students in the USA. J Geogr High Educ 37(1):92–116
Sui DZ (1995) A pedagogic framework to link GIS to the intellectual core of geography. J Geogr 94(6):578–591
Sui D (2015) Emerging GIS themes and the six senses of the new mind: is GIS becoming a liberation technology? Ann GIS 21(1):1–13
Sui D, Goodchild M (2011) The convergence of GIS and social media: challenges for GIScience. Int J Geogr Inf Sci 25(11):1737–1748
Tate NJ, Jarvis CH (2017) Changing the face of GIS education with communities of practice. J Geogr High Educ 41(3):327–340
Wiegand P (2001) Geographical information systems (GIS) in education. Int Res Geogr Environ Educ 10(1):68–71
Zamperlin P, Azzari M (2017) The smart city I would like. Maps and storytelling in teaching geography. J Res Didact Geogr (J-READING) 1(6):17–28

Index

A
Abstractions, 99, 101–103, 106, 109, 112, 118, 119, 195
Agents, 5, 13, 18, 22, 161
Analysis
component, 57, 67
structural, 59
Approaches
methodological, 51, 54, 263
qualitative, 51, 69
quantitative, 200, 255, 258
Areas, 10, 16, 17, 22, 27, 29, 32, 34, 38, 45, 47, 49, 53, 56–58, 62, 67, 68, 76, 78–80, 83–86, 90, 91, 93–95, 103, 105, 106, 126, 137, 139, 144, 145, 149, 150, 154, 166, 170, 178, 180, 183, 190, 193, 194, 198, 200, 211, 227, 246, 248, 249, 255, 258–262
Artifacts, 12, 13, 16, 17, 38, 152, 154, 155, 181

B
Bathymetry, 114, 116, 117
Behaviour, 19, 210, 241–244, 248
Beliefs, 139, 159–162, 165–173, 186–188, 196, 247, 248
Boundaries
bona fide, 145–147, 150, 151, 155, 156, 161, 162, 164, 167, 168, 186
crisp, 29, 141, 147, 148, 189, 197
fiat, 29, 94, 137–140, 155, 156, 161–168, 170, 171, 186, 224, 228–230
indeterminate, 29, 30, 139, 147, 148, 178, 195, 209, 211, 236
Jeffersonian, 145, 146
political, 59, 68, 94, 147, 148, 159, 186
property, 91, 159, 164, 187
spatial, 128, 129, 147, 191
state, 175

C
Cartography
digital, 51, 119
theoretical, 52, 56
Carto-linguistics, 51
Categories, 4, 13, 29, 58–61, 69, 85, 93, 95, 155, 180, 185, 186, 189, 190, 192–194, 199, 242, 246–249, 253, 260, 263
Classes, 9, 12, 13, 38, 58, 60, 61, 82, 105, 138, 180, 181, 186–188, 195, 198, 209–211, 223, 225, 237, 243, 260
Classifications, 7–9, 12, 15, 16, 76, 102, 185–187, 189, 190, 209, 211, 226, 228, 230–232, 234, 236
Cognition, 4, 5, 22, 51, 53–55, 69, 162, 172, 178, 183, 184, 187, 242–244, 249
Complexity, 3, 4, 89, 179
Conceptualizations
geographical, 60, 79, 182, 197
multiple, 18, 60, 177, 183, 201, 202, 255, 257
shared, 5, 21, 132
Construction, 3, 6, 13, 22, 38, 47, 58, 60, 68, 91, 94–96, 152, 201, 212, 221, 227, 260
Context, 3–8, 11, 17–20, 22, 29, 31, 32, 49, 57, 60, 62, 67–69, 75–78, 82, 83, 89, 96, 103, 106, 127, 129, 131, 148, 168, 177–179, 182, 188, 197, 199, 202, 209,

© Springer Nature Switzerland AG 2019

T. Tambassi (ed.), *The Philosophy of GIS*, Springer Geography,
https://doi.org/10.1007/978-3-030-16829-2

211, 212, 214, 219, 225–227, 230–232, 234, 236, 243, 259, 260, 262
Creation
direct, 163, 164, 166, 167, 169–171, 173
indirect, 163, 165, 166, 170, 173
Cultures, 5, 53, 102, 118, 144, 160, 161, 170, 171, 177, 179, 187, 188, 194, 197, 199–201, 259

D
2D, 27–29, 40, 42, 48, 261, 262
3D, 19, 27–29, 38, 40, 48, 256, 257, 261, 262
4D, 256, 257, 261, 262
Data, 4, 8, 12, 14, 16, 18, 21, 22, 27, 29, 33, 34, 36, 48, 49, 51–54, 69, 77–81, 91–93, 100–102, 105, 106, 109–113, 116–119, 170, 200, 254, 257–259, 261, 262
Databases, 4, 28, 33, 51, 102, 189, 246, 261
Definitions, 4, 5, 15, 19, 21, 28, 29, 62–66, 80, 126, 132, 138, 142, 179–182, 185–187, 199–201, 258
Delineations, 145, 146, 148, 150, 166, 209, 211, 230–232, 234, 236
Design, 5, 11, 13, 14, 20–22, 41, 48, 49, 57, 80
Diagrams, 80, 88, 100, 106, 109, 111–113, 215
Disorientation, 241–246, 248, 249
Domains, 4, 9–11, 21, 41, 46, 48, 53, 57, 63, 69, 76, 86, 88, 103, 142, 154, 177, 181, 182, 185, 188–190, 192, 194, 199–201, 223

E
Effects, 17, 22, 46, 58, 60, 63, 102, 128, 141, 143, 145, 146, 155, 199, 254, 257, 261
Entities
abstract, 13, 14, 132, 195
artificial, 13–15, 20–22, 170
complex, 102, 197–199
cultural, 187
geographical, 78, 96, 125–132, 177–187, 189, 190, 193, 194, 196–202, 210
historical, 196, 197
imaginary, 21, 131, 182
independent, 172–175
mind-dependent, 173, 175
natural, 13, 14, 17, 21, 22
non-existent, 195
physical, 4, 14, 21, 22, 170, 182
social, 142, 163, 173–175, 187
Epistemology, 53, 54, 109, 160, 167, 171
Exaggeration, 99, 100, 102, 106, 107, 112, 119
Existence, 4, 12, 25, 32, 46, 53, 96, 113, 129–132, 139, 147, 151, 160, 162–166, 168–175, 179, 181, 186, 196, 201, 248

F
Facts, 75, 78, 79, 81, 86, 87, 90, 145, 159–161, 163–173, 181, 186, 211, 213, 215, 217, 232, 235, 236, 254
Features, 16, 26, 27, 30, 32, 33, 42, 48, 56–66, 68, 76, 80, 83, 90, 95, 102, 103, 106, 148, 164, 166, 169, 170, 172, 178, 184, 189, 190, 225, 226, 243, 245, 259, 262
Fields, 5, 9, 12, 22, 23, 38–40, 51, 68, 69, 76–83, 85, 88, 91, 92, 94–96, 118, 119, 143, 154, 177, 194, 202, 241, 242, 244, 245, 248, 249, 253–255, 257–263
Frameworks
conceptual, 3, 5, 79, 81, 82, 88, 93, 119, 175
linguistic, 69, 83, 175
theoretical, 21, 209, 212
Functions, 8, 9, 11, 13–17, 19–21, 48, 49, 52, 54, 56, 58, 61, 63, 69, 80, 84, 86, 87, 91, 92, 147, 151, 170, 177, 186, 217–219, 237, 242, 255, 258

G
Geographical Information Science (GIS)
application, 77, 254–256, 263
environment, 77, 253, 258, 261
practice, 77, 101, 102, 118, 261
theory, 21, 55–57, 77
Geography
classical, 125, 127, 128, 131, 132, 184, 199, 201, 210
human, 78, 90, 159–161, 167, 169–174, 181, 190, 193, 199
non-classical, 125, 127, 131, 132, 184, 210
physical, 159, 181, 190, 198, 209, 225
political, 137, 145, 147–149, 184, 198
quantum, 209–211, 236
Geometry
computational, 51, 52, 69
Euclidean, 29, 66
spherical, 29
GIScience, 51–54, 61, 69
Granularity, 8, 9, 11, 42, 84, 86, 94, 95, 103, 139, 153, 183, 184, 226, 227, 229–234

H
Hierarchies, 198

I
Idealism, 173
Identity, 4, 58, 91–93, 127, 129, 130, 132, 139, 178, 181, 184, 193, 194, 197
Indeterminacy, 210, 211, 216–219, 221, 222, 225, 234, 236
Information
density, 230–233
geographical, 75–77, 238, 254, 257, 262, 263
geometric, 75, 262
quantum, 225, 234
spatial, 51, 56, 57, 63, 65, 67, 78
technologies, 3, 4, 25, 118, 256
temporal, 15, 77, 89
theory, 212, 221
Integration, 8, 20, 52, 77, 89, 106, 180
Intentionality, 7, 9, 11, 18, 22, 161, 162, 165, 166, 172–176, 186
Interoperability, 4, 40, 41, 255
Interpretations, 3–7, 18, 19, 22, 86, 94, 109, 112, 118, 132, 210–212, 220, 226–228, 233, 234, 236, 262

J
Job opportunities, 254

K
Kinds
geographical, 76, 190, 192, 193
natural, 169, 171, 174, 175
Knowledge
engineering, 26, 48, 52, 60, 69, 188
geographical, 25, 83
geospatial, 45, 46
Web, 4

L
Laboratories, 253, 256, 257, 263
Languages, 4, 7, 26, 51–55, 57–63, 66, 67, 77, 83, 86, 140–145, 177, 179, 185, 186, 189–191, 196, 200, 201, 221–223, 230, 258
Layers, 3, 7–11, 20–22, 79, 146, 151, 152, 155
Levels, 7–13, 16–19, 21, 22, 27, 53, 55, 57, 65, 66, 69, 76, 77, 79, 83, 84, 86, 88, 89, 92, 103, 105, 110, 116, 140, 153, 156, 162, 166, 169, 178, 181, 183, 184, 187, 198, 199, 210, 215, 219, 226–234, 244, 246, 249, 254, 256, 257, 262
Lines, 29, 34, 38, 68, 76, 91, 93, 95, 137, 138, 237
Linguistics, 6, 22, 51–55, 57, 69, 77, 83, 86, 96
Location, 8, 9, 14, 16, 26, 39, 46, 53, 68, 76, 78–82, 84, 85, 91–93, 95, 105, 107, 117, 127–132, 139, 142, 146, 147, 155, 159, 160, 163, 168, 171, 184, 185, 196, 197, 222–225, 229, 242, 243, 259

M
Mapping
digital, 51, 52, 62, 103, 109, 255, 261
Maps, 9, 14, 16, 27, 43, 52–54, 56, 57, 63, 69, 76, 77, 87, 90, 93–95, 99–107, 109, 111–119, 128–132, 137, 138, 144–146, 148, 149, 156, 166, 170, 178, 182–185, 197, 200, 237, 241–243, 246–249, 258, 262
Mathematics, 25, 29, 105, 110
Matters, 3, 4, 58, 62, 66, 77, 78, 83–86, 89, 90, 94–96, 120, 140, 141, 144, 145, 148, 149, 152–155, 160, 168, 171, 180, 183, 185, 192, 194
Meanings, 3–5, 13, 41, 46, 51, 53, 56–59, 61, 63, 167, 174, 182, 197, 198
Measurements, 3–7, 12, 17, 19, 22, 25, 28, 48, 49, 65, 99, 116, 148, 155, 210, 212, 214, 221–224, 226, 230–232, 262
Methodologies, 51, 52, 55, 69, 255, 257, 262, 263
Mind-dependence, 159, 160, 163, 166, 167, 170, 173–175
Minds, 4, 53, 54, 78, 118, 126, 159, 160, 163, 170, 172, 201, 253, 257
Modeling
geometric, 27, 48, 49
Models
conceptual, 3, 22, 51, 52, 55, 77, 79
spatial, 30, 77–79, 81, 89, 93, 102
temporal, 77, 89, 93
Morphology, 51, 54, 56, 61, 63

N
Neuroscience, 241, 242, 244, 249

O
Objects
bona fide, 29, 35, 139, 140, 142, 146, 147, 150, 152, 154, 155
composite, 9, 12, 15, 16, 84
connected, 35, 95
fiat, 35, 138, 139, 141, 143, 147, 148, 150, 154, 155, 161, 162, 171, 223–225, 228
geographical, 79, 150, 185, 188, 189, 193, 194
imaginary, 172

Objects (*cont.*)
mind-dependent, 174
natural, 92, 93, 174
scattered, 150
simple, 9, 12, 15, 16
Observations, 4, 15, 16, 20, 48, 54, 78, 209–212, 214, 221–224, 230–232, 243, 247, 248
Oceanography, 105
Ontology
applied, 137, 154
formal, 60, 137
micro, 6, 11, 13, 17, 21
of geography, 96, 125–127, 172, 202
of GIS, 96, 126
realist, 172
Operations
formal, 3, 5, 17

P
Paradigms, 51, 54, 55, 61, 69, 99, 167, 172, 177, 201, 202, 223–226, 228, 229, 231, 232
Parts, 4–7, 12, 18, 20, 28, 30, 58, 60, 68, 69, 77, 84, 85, 93, 105, 109, 125, 127, 138–141, 143, 144, 146, 148–151, 153–155, 159, 162–165, 168, 173, 175, 178, 184, 191, 192, 195–198, 200, 201, 211, 226, 228, 230, 244, 253, 255, 256
Perspectives, 3, 7, 9, 12, 14, 17, 19, 22, 23, 51, 52, 179, 200, 201, 222, 230, 234, 257
Phenomena, 7, 12, 21, 25, 38, 39, 56–58, 62, 77, 79, 81, 84, 86, 95, 109, 126, 141, 144, 185, 194, 200, 209–211, 222, 225, 230, 231, 234, 236, 237, 258–261
Philosophy
applied, 4
Phonetics, 51, 54–57, 60, 61, 63, 66, 67, 69
Phonology, 54, 55, 57
Places, 9, 11, 14, 21, 27, 29, 41, 46, 53, 82, 90, 114, 126, 139, 145, 147, 153, 159, 160, 178, 196–199, 210, 242–246, 256, 257, 259
Points, 8, 14–16, 19–21, 25–30, 33–38, 40–42, 44, 46–49, 51, 53, 56, 62, 64–68, 75, 79, 82, 83, 86, 90–92, 95, 107, 116, 118, 119, 126, 128–130, 138, 139, 141, 148–150, 155, 156, 162–165, 177, 179, 182, 183, 187, 189, 190, 194, 196–201, 230, 244, 246, 247, 253–256, 260
Polygons, 8, 29, 33–37, 43–45, 48, 57, 62, 66, 80, 91, 92, 107
Practices, 21, 77, 80–82, 99, 101–103, 118, 119, 146, 148, 159, 161–164, 166, 196, 253, 257, 261
Pragmatics, 6–8, 13, 17–19, 40, 54, 61, 62, 69, 82, 102
Predictions, 209, 211, 218, 235, 237, 246, 248, 260
Probability, 210, 211, 217–219, 221, 222, 225, 235–238, 241, 246–249
Professional profiles, 253, 257, 258
Properties, 9–14, 19, 21, 22, 29, 39, 53, 64, 66–68, 78, 85, 90, 103, 106, 128, 137, 138, 142, 147, 149, 159, 161, 162, 164, 165, 169, 184–187, 192, 193, 200, 210, 211, 217, 219, 221, 225, 243
Psychology, 52, 241, 242, 244, 249

Q
Quantum theory, 209–211, 225, 226, 234–237
Queries, 14, 16, 48

R
Raster, 38, 39, 75, 77–81, 86, 88, 89, 91, 96, 224, 226, 228, 231
Realism, 5, 160, 172, 173, 181
Reality, 4, 9, 16, 25, 27, 43, 53, 54, 56–58, 61, 68, 75, 81, 94, 95, 102, 105, 125, 137, 140–145, 147, 148, 153, 155, 156, 162, 167, 170, 179, 183, 185, 192, 194, 196, 200, 202, 244
Reasoning
geospatial, 46
spatial, 26, 46, 49
temporal, 77, 92
Regions, 6, 9, 14–16, 28, 82–84, 86, 91, 100, 126–131, 139, 141–144, 146–149, 154, 155, 159–161, 170, 178, 181, 184, 185, 189, 192, 193, 196–198, 209–211, 225, 226, 228, 230, 232, 236
Relations
direction, 62, 65
distance, 62
epistemic, 160, 168–171
metrical, 62
spatial, 12, 55, 59, 61–63, 65–69, 126, 127, 178, 193
topological, 41, 42, 44, 45, 48, 49, 62, 65, 66, 178, 190, 193
Relativism, 3, 5, 18
Representations
cartographic, 54, 105, 125, 128, 129, 132
continuos, 38, 76, 80, 194

discrete, 76, 80
geographic, 19, 75, 79, 125, 127, 194
multiple, 32, 33, 59
spatial, 78, 125–127, 131, 132, 243, 245
spatio-temporal, 8, 11, 93, 96

S

Scales, 8, 15, 27, 34, 38, 42, 58, 64, 66, 90, 94, 101–103, 105–107, 111, 112, 116, 119, 126, 147, 156, 181, 183, 185, 200, 209, 210, 226, 230, 262
Science
cognitive, 52, 69, 77, 185, 244
natural, 167, 172, 176
social, 52, 159–161, 173, 175
Selections, 3, 6–8, 14, 20, 99, 102, 103, 105, 119
Semantics
spatial, 57
Semiotics, 51, 52, 54, 56, 69
Sensors, 25, 49
Simplifications, 102, 106, 119
Skills, 253–255, 257–259, 262, 263
Software, 40, 41, 52, 107, 109, 255–257, 262
Space
absolute, 76, 80, 81, 185
relational, 76, 80, 81, 126, 178
Structures
algebraic, 214, 219
hierarchical, 60, 61, 185, 198
linguistic, 51, 55, 60, 66, 69
Syntax
spatial, 61
Systems, 9, 10, 25, 29, 51, 53–55, 57, 59–62, 66, 67, 69, 77, 82, 96, 101, 102, 105, 109, 114, 117, 126, 142, 150, 153, 183, 186, 192, 194, 210, 212, 214, 215, 217, 218, 220–222, 225, 226, 228, 232, 234, 235, 237, 243, 246, 254, 257–260, 262, 263

T

Taxonomies, 9, 185–187, 242
Technologies, 3, 4, 25, 49, 52, 106, 116, 118, 177, 200, 202, 256
Terms, 5, 7, 16, 23, 27, 38, 41, 60, 75–80, 83, 85, 87, 88, 91, 93, 100, 103, 109, 113, 118, 126, 127, 129, 140, 143–145, 147, 148, 152, 154, 155, 174–176, 179–182, 184, 185, 189, 191, 192, 196–199, 201, 245, 254, 255
Tessellations, 29, 35–37, 43–45, 78, 79, 91, 149, 247
Time, 4, 5, 7, 8, 10, 11, 14, 15, 17, 19, 21, 28, 48, 49, 53, 69, 75, 77, 81, 82, 84, 86–95, 101, 103, 113, 116, 120, 127, 129, 130, 146, 152, 154, 177, 179, 181, 186, 187, 193–198, 201, 210, 211, 224, 227, 231, 232, 246–248, 254, 256, 259–261
Tokens, 163–165, 167–173, 179
Tools
theoretical, 125, 127
Truthmakers, 143
Types
geometric, 38

U

Users, 3, 4, 6–8, 11–14, 17, 19–23, 52, 53, 77, 107, 109, 262

V

Vagueness, 12, 147–149, 191, 192
Vector, 11, 19, 38, 39, 75, 78–81, 86, 88, 89, 91, 96, 219–222, 235, 237, 242, 243
Views, 3, 5–7, 13, 17–19, 21, 22, 25, 26, 28, 35, 41, 42, 44, 46, 48, 76, 81, 93, 101, 116, 140, 142, 144, 145, 148, 149, 153, 167, 170, 172, 173, 177, 179, 183, 189, 195, 198, 201, 202, 242, 254, 255, 259, 261

W

Web
semantic, 4, 5, 18, 21, 23
Wilderness Search and Rescue (WiSAR), 241, 244, 246–249
World
artificial, 4
empirical, 5, 19, 22
geographical, 177, 179, 191, 200, 201, 210, 228
physical, 139, 173, 178, 184, 209
real, 4, 5, 21, 22, 62, 200
social, 3, 173

MIX
Papier aus verantwortungsvollen Quellen
Paper from responsible sources
FSC® C105338

If you have any concerns about our products,
you can contact us on
ProductSafety@springernature.com

In case Publisher is established outside the EU,
the EU authorized representative is:
Springer Nature Customer Service Center GmbH
Europaplatz 3, 69115 Heidelberg, Germany

Printed by Libri Plureos GmbH
in Hamburg, Germany